GEOTHERMAL ENERGY

SECOND EDITION

Renewable Energy and the Environment

ENERGY AND THE ENVIRONMENT

SERIES EDITOR
Abbas Ghassemi
New Mexico State University

PUBLISHED TITLES

Geothermal Energy: Renewable Energy and the Environment, Second Edition
William E. Glassley

Energy Resources: Availability, Management, and Environmental Impacts
Kenneth J. Skipka and Louis Theodore

Finance Policy for Renewable Energy and a Sustainable Environment
Michael Curley

Wind Energy: Renewable Energy and the Environment, Second Edition
Vaughn Nelson

Solar Radiation: Practical Modeling for Renewable Energy Applications
Daryl R. Myers

Solar and Infrared Radiation Measurements
Frank Vignola, Joseph Michalsky, and Thomas Stoffel

Forest-Based Biomass Energy: Concepts and Applications
Frank Spellman

Introduction to Renewable Energy
Vaughn Nelson

Solar Energy: Renewable Energy and the Environment
Robert Foster, Majid Ghassemi, Alma Cota,
Jeanette Moore, and Vaughn Nelson

GEOTHERMAL ENERGY

SECOND EDITION

Renewable Energy and the Environment

William E. Glassley

CRC Press
Taylor & Francis Group
Boca Raton London New York

CRC Press is an imprint of the
Taylor & Francis Group, an **informa** business

CRC Press
Taylor & Francis Group
6000 Broken Sound Parkway NW, Suite 300
Boca Raton, FL 33487-2742

© 2015 by Taylor & Francis Group, LLC
CRC Press is an imprint of Taylor & Francis Group, an Informa business

Printed on acid-free paper
Version Date: 20140502

International Standard Book Number-13: 978-1-4822-2174-9 (Hardback)

Library of Congress Cataloging-in-Publication Data

Glassley, William E.
    Geothermal energy : renewable energy and the environment / William E. Glassley. -- Second edition.
        pages cm. -- (Energy and the environment)
    "A CRC title."
    Includes bibliographical references and index.
    ISBN 978-1-4822-2174-9 (alk. paper)
    1. Geothermal engineering. 2. Geothermal resources. 3. Geothermal resources--Environmental aspects.
4. Renewable energy sources. I. Title.

TJ280.7.G55 2014
333.8'8--dc23   2014015961

Visit the Taylor & Francis Web site at
http://www.taylorandfrancis.com

and the CRC Press Web site at
http://www.crcpress.com

Contents

Series Preface

By 2050, the demand for energy could double or even triple as the global population rises and developing countries expand their economies. According to the data from the United Nations, it is projected that the world population will increase from 7.2 billion to more than 9 billion in 2050. This increase coupled with continued demand for the same, limited natural resources will cause significant increase in consumption of energy. All life on Earth depends on energy and the cycling of carbon. Affordable energy resources are essential for economic and social development as well as food production, water supply availability, and sustainable healthy living. In order to avoid long-term adverse and potentially irreversible impact of harvesting energy resources, we must explore all aspects of energy production and consumption including energy efficiency, clean energy, global carbon cycle, carbon sources and sinks, and biomass as well as their relationship to climate and natural resource issues. Knowledge of energy has allowed humans to flourish in numbers unimaginable to our ancestors. The world's dependence on fossil fuels began approximately 200 years ago. Are we running out of oil? No, but we are certainly running out of the affordable oil that has powered the world economy since the 1950s. We know how to recover fossil fuels and harvest their energy for operating power plants, planes, trains, and automobiles which results in modifying the carbon cycle and additional greenhouse gas emissions. This has resulted in the debate on availability of fossil energy resources, peak oil era, and timing for anticipated end of fossil fuel era and price and environmental impact versus various renewable resources and use, carbon footprint, emission and control including cap and trade, and emergence of "green power."

Our current consumption has largely relied on oil for mobile applications and coal, natural gas, nuclear, or water power for stationary applications. In order to address the energy issues in a comprehensive manner, it is vital to consider the complexity of energy. Any energy resource including oil, gas, coal, wind, and biomass is an element of a complex supply chain and must be considered in the entirety as a system from production through consumption. All of the elements of the system are interrelated and interdependent. Oil, for example, requires consideration for interlinking all of the elements including exploration, drilling, production, transportation, water usage and production, refining, refinery products and by-products, waste, environmental impact, distribution, consumption/application, and finally emissions. Inefficiency in any part of the system has impact on the overall system, and disruption in one of these elements causes major interruption and a significant cost impact. As we have experienced in the past, interrupted exploration will result in disruption in production, restricted refining and distribution, and consumption shortages; therefore, any proposed energy solution requires careful evaluation and as such, may be one of the key barriers to implement the proposed use of hydrogen as a mobile fuel.

Even though an admirable level of effort has gone into improving the efficiency of fuel sources for delivery and use of energy, we are faced with severe challenges on many fronts. They include population growth, emerging economies, new and expanded usage, and limited natural resources. All energy solutions include some level of risk such as technology snafus, changes in market demand, economic drivers, and others. This is particularly true when proposing energy solutions involving implementation of untested alternative energy technologies.

There are concerns that emissions from fossil fuels lead to a changing climate with possibly disastrous consequences. Over the past five decades, the world's collective greenhouse gas emissions have increased significantly even as efficiency has increased resulting in extending energy benefits to more of the population. Many propose that we improve the efficiency of energy use and conserve resources to lessen greenhouse gas emissions and avoid a climate catastrophe. Using fossil fuels more efficiently has not reduced overall greenhouse gas emissions due to various reasons, and it is unlikely that such initiatives will have a perceptible effect on atmospheric greenhouse gas content. Although there is a debatable correlation between energy use and greenhouse gas emissions, there are effective means to produce energy, even from fossil fuels, while controlling emissions. There are also emerging technologies and engineered alternatives that will actually manage the makeup of the atmosphere, but will require significant understanding and careful use of energy.

We need to step back and reconsider our role and knowledge of energy use. The traditional approach of micromanagement of greenhouse gas emissions is not feasible or functional over a long period of time. More assertive methods to influence the carbon cycle are needed and will be emerging in the coming years. Modifications to the carbon cycle mean we must look at all options in managing atmospheric greenhouse gases including various ways to produce, consume, and deal with energy. We need to be willing to face reality and search in earnest for alternative energy solutions. There appear to be technologies that could assist; however, they may all not be viable. The proposed solutions must not be in terms of a "quick approach," but a more comprehensive, long-term (10, 25, and 50 plus years) approach that is science based and utilizes aggressive research and development. The proposed solutions must be capable of being retrofitted into our existing energy chain. In the meantime, we must continually seek to increase the efficiency of converting energy into heat and power.

One of the best ways to define sustainable development is through long-term, affordable availability of limited resources including energy. There are many potential constraints to sustainable development. Foremost of these is the competition for water use in energy production, manufacturing, farming, and others versus a shortage of freshwater for consumption and development. Sustainable development is also dependent on the Earth's limited amount of productive soil. In the not too distant future, it is anticipated that we will have to restore and build soil as a part of sustainable development. We need to focus our discussions on the motives, economics, and benefits of natural resource conservation, as well as the limitation of technology improvement in impacting sustainability, that is, we are limited on catching fish from the ocean due to the number of fish that is available not by a bigger boat or better net. Hence, possible sustainable solutions must not be based solely on technology enhancement and improvement, specifically in obtaining the fossil resources, but be comprehensive and based on integrating our energy use with nature's management of carbon, water, and life on Earth as represented by the carbon and hydrogeological cycles. The challenges presented by the need to control atmospheric greenhouse gases are enormous and require "out of the box" thinking, innovative approaches, imagination, and bold engineering initiatives in order to achieve sustainable development. We will need to ingeniously exploit even more energy and integrate its use with control of atmospheric greenhouse gases.

The continued development and application of energy are essential to the sustainable advancement of society. Therefore, we must consider all aspects of the energy options including performance against known criteria, basic economics and benefits, efficiency, processing and utilization requirements, infrastructure requirements, subsidies and credits, waste and ecosystem, as well as unintended consequences such as impacts to natural resources and the environment. Additionally, we must include the overall changes and the emerging energy picture based on current and future efforts in renewable alternatives and modified and enhanced fossil fuels and evaluate the energy return for the investment of funds and other natural resources such as water. Water is a precious commodity in the west in general and the southwest in particular and has a significant impact on energy production, including alternative sources due to the nexus between energy and water and the major correlation with the environment and sustainability-related issues.

A significant driver in creating this book series that is focused on alternative energy and the environment was provoked as a consequence of lecturing around the country and in the classroom on the subject of energy, environment, and natural resources such as water. Although the correlation between these elements, how they relate to each other and the impact of one on the other, is understood, it is not significantly debated when it comes to integration and utilization of alternative energy resources into the energy matrix. Additionally, as renewable technology implementation grows by various states, nationally and internationally, the need for informed and trained human resources continues to be a significant driver in future employment, resulting in universities, community colleges, and trade schools offering minors, certificate programs, and even in some cases majors in renewable energy and sustainability. As the field grows, the demand for trained operators, engineers, designers, and architects who would be able to incorporate these technologies into their daily activity is increasing. We receive daily deluge of flyers, emails, and texts on various short courses available for interested parties in solar, wind, geothermal, biomass, and so on under the umbrella of retooling an individual's career and providing trained resources needed to interact with financial, governmental, and industrial organizations.

In all my interactions throughout the years in this field, I have conducted significant searches in locating integrated textbooks that explain alternative energy resources in a suitable manner and would complement a syllabus for a potential course to be taught at the university while providing good reference material for interested parties getting involved in this field. I have been able to locate a number of books on the subject matter related to energy, energy systems, resources such as fossil and nuclear, renewable and energy conversion, as well as specific books on the subjects of natural resource availability, use, and impact as related to energy and environment. However, specific books that are correlated and present the various subjects in detail are few and far between. We have therefore started a series of texts each addressing specific technology fields in the renewable energy arena. As a part of this series, there are textbooks on wind, solar, geothermal, biomass, hydro, and others yet to be developed. Our texts are intended for upper-level undergraduate and graduate students and for informed readers who have a solid fundamental understanding of science and mathematics as well as individuals/organizations involved with design development of the renewable energy field entities who are interested in having reference material available to their scientists and engineers, consulting organizations, and reference libraries. Each book presents fundamentals as well as a series of numerical and conceptual problems designed to stimulate creative thinking and problem solving.

The series editor wishes to express his deep gratitude to his wife Maryam who has served as a motivator and intellectual companion and too often was victim of this effort. Her support, encouragement, patience, and involvement have been essential to the completion of this series.

Abbas Ghassemi
Las Cruces, New Mexico

Preface to 2nd Edition

Since the first edition of this book was written, changes in the national and international energy markets have been substantial. Likewise, technological achievements within the geothermal industry have been vast. Particularly important has been the advent of new technologies that influence how exploration is done and the efficiency of the process, rapid expansion and success with enhanced geothermal systems, and new drilling techniques that have changed the ability to access resources. In addition, as more renewable energy resources are brought into the energy grid, particularly the rapid deployment of wind and solar resources, new concepts for distributed generation, hybrid technologies, and the use of smart grid concepts have affected how energy is conceived and managed. Advances in understanding the economics and valuation of resources, particularly the expanded attention to life-cycle analysis on "energy returned on energy invested" approaches, have made it important to view energy generation and use more holistically. As a result, it was decided that an updated version of the textbook was needed. It is for this reason that the current work is provided.

Preface to 1st Edition

The rapidly growing influence of human activity on the environment has changed the way human beings view the world and their relationship with it. Until the middle of the twentieth century, the world was seen as an essentially stable, unchanging landscape. What changes occurred were either of small global impact or constrained to play out on timescales more familiar to geologists than the average worker, politician, or student. However, over the past 50 years, the cumulative effects of industrial activity, coupled in complex ways with population growth and economic development, have become more apparent. We are now capable of monitoring every aspect of the planet's environment and have come to realize that the world and the biology it supports have long been evolving in response to our actions.

Underlying every aspect of the human juggernaut has been the ability to access and utilize what seemed to be boundless and benign fossil energy resources. With the realization that those energy resources are, in fact, exhaustible and that their use is affecting the global hydrosphere, biosphere, and atmosphere, there has developed an interest in finding and developing energy resources that have minimal environmental impact and are sustainable, and geothermal energy is one such resource.

Geothermal energy is ubiquitous, abundant, and inexhaustible. It powers the movement of the continents across the face of the planet, it melts rock that erupts as volcanoes, and it supplies the energy that supports life in the ocean depths. It has been present for 4500 million years and will be present for billions of years into the future. It flows through the earth constantly, 24 hours a day, seven days a week, rain or shine, eon upon eon. It has the potential to provide power to every nation in the world—in the United States alone, it has been noted that the amount of geothermal energy available for power generation exceeds by several times the total electrical power consumption of the country. All of this is possible and with minimal environmental consequence.

This book is about where that energy comes from and how to find it, how it can be accessed, the kinds of applications that have been successfully developed in the past, and what could be done to improve its use in the future. This book also considers the constraints that affect use of geothermal energy—how water must be managed, what emissions must be controlled, and when utilization may not be appropriate. Finally, this book also discusses the economic and social issues that must be addressed for wise and orderly development of this robust and bountiful resource.

The audience for this book is anyone seeking an in-depth introduction to geothermal energy and its applications. It is intended for course work at the undergraduate level; as a reference book for designers, planners, engineers, and architects; and as a source for background material for policy makers, investors, and regulators.

Geothermal energy, wisely used, can contribute in many important ways to resolving one of the fundamental challenges faced by the global community—how to acquire energy to assure the health, prosperity, and security of the global community. It is hoped this book will contribute to achieving that goal.

Acknowledgments

The author thanks the following individuals for use of materials they provided for this book: Tonya Boyd of the Geo-Heat Center at the Oregon Institute of Technology; Mark Coolbaugh of the SpecTIR Corporation, Reno, Nevada; Mariana Eneva of Imageair, Inc., San Diego, California; Christopher Kratt of the Great Basin Center for Geothermal Energy at the University of Nevada, Reno, Nevada; Dale Merrick of I'SOT, Canby, California; and Colin Williams of US Geological Survey, Menlo Park, California. Discussions with Bill Bourcier, Elise Brown, Carolyn Cantwell, Judy Fischette, Andrew Fowler, Karl Gawell, Samuel Hawkes, James McClain, Dale Merrick, Curt Robinson, Peter Schiffman, Charlene Wardlow, Jill Watz, and Maya Wildgoose are gratefully recognized. Research assistance provided by Adam Asquith, Tucker Lance, and Gabriel Perez is gratefully acknowledged. Discussions with Trenton Cladohous, Yini Nordin, and Susan Petty materially improved discussions regarding enhanced geothermal systems applications. Review of the manuscript by Carolyn Feakes, Marcus Fuchs, Abbas Ghassemi, Joe Iovenitti, and Robert Zierneberg greatly improved the content and presentation and their comments are appreciated. Drafting of many of the figures was expertly accomplished by Ingrid Dittmar.

Series Editor

Dr. Abbas Ghassemi is the director of the Institute for Energy and Environment (IEE) and professor of chemical engineering at New Mexico State University. He earned an MS and a PhD in chemical engineering, with minors in statistics and mathematics, from New Mexico State University, and a BS in chemical engineering, with a minor in mathematics, from the University of Oklahoma. As the director of IEE, he is the chief operating officer for programs in education and research, and outreach in energy resources including renewable energy, water quality and quantity, and environmental issues. He is responsible for the budget and operation of the program.

Dr. Ghassemi has authored and edited several textbooks and has many publications and papers in the areas of energy, water, carbon cycle including carbon generation and management, process control, thermodynamics, transport phenomena, education management, and innovative teaching methods. His research areas of interest include risk-based decision making, renewable energy and water, carbon management and sequestration, energy efficiency and pollution prevention, multiphase flow, and process control.

Dr. Ghassemi serves on a number of public and private boards, editorial boards, and peer review panels.

Author

William E. Glassley has more than 40 years of experience in the analysis, modeling, and evaluation of geological processes that drive geothermal systems and the evolution of continents. He earned his BA at the University of California, San Diego, and his MSc and PhD from the University of Washington. He has authored and coauthored over a 100 scholarly and technical reports and publications. He is a senior researcher in the Department of Earth and Planetary Sciences at the University of California, Davis.

Dr. Glassley is also the executive director of the California Geothermal Energy Collaborative, which is part of the Energy Institute, the University of California, Davis. He holds an emeritus researcher position at the University of Aarhus, Denmark, and has held research, teaching, and management positions at the University of Washington, Middlebury College, and Lawrence Livermore National Laboratory. He has been a member of the scientific review panels for the National Science Foundation, the European Commission, the International Atomic Energy Agency, and research councils for several nations.

Dr. Glassley has been a reviewer for international scientific journals, and his research has been featured in several popular scientific magazines. He was awarded a G. Unger Vetlesen Foundation Fellowship for his postdoctoral research at the University of Oslo.

1 Introduction

It is a well-known fact that the interior portions of the globe are very hot, the temperature rising, as observations show, with the approach to the center at the rate of approximately 1 degree C. for every hundred feet of depth. The difficulties of sinking shafts and placing boilers at depths of, say, twelve thousand feet, corresponding to an increase in temperature of about 120 degrees C., are not insuperable, and we could certainly avail ourselves in this way of the internal heat of the globe. In fact, it would not be necessary to go to any depth at all in order to derive energy from the stored terrestrial heat. The superficial layers of the earth … are at a temperature sufficiently high to evaporate some extremely volatile substances, which we might use in our boilers instead of water.

Tesla (1900)

As the above quote demonstrates, the vision of utilizing the earth's internal heat to benefit the world is not new. It is obvious to even the most casual observer that energy, in the form of heat, is present below the earth's surface—volcanoes spew scorching lava from their summits and water in hot springs bubbles up from below—attesting to the fact that the interior of the earth is hot. But the nature of that heat remained, for many years, a source of mystery. Why volcanoes are unevenly distributed over the continents and why hot springs are abundant in some places and nonexistent in others seemed inexplicable. But, over the last 200 years, intensive research by cadres of earth scientists has provided a clearer understanding of the source, distribution, and properties of this massive thermal reservoir.

It is now well understood that heat constantly radiates from the earth's surface into space. Some of that energy is solar energy that has been absorbed by soil and rock and re-radiated as infrared radiation. But, using a variety of measurement techniques, earth scientists have been able to establish that on average about 1% of the total energy that radiates into space is heat that comes from the interior of the earth itself. However, it may seem insignificant that 1% of the radiated heat represents a miniscule fraction of the amount of heat energy the earth contains. In fact, the amount of energy, in the form of heat, that is present a few thousands of feet below the surface is more than enough to satisfy the energy needs of every nation of the world many times over.

That heat energy is *geothermal energy*. It is remnant heat derived from the formation of the planet 4.5 billion years ago, as well as heat from the radioactive decay of naturally occurring radioactive isotopes. That heat is sufficient to power plate tectonics, which is the slow movement of the continents and the ocean floor that make up earth's crust, and the upper mantle. It provides the energy to drive mountain building processes that occur when continents and oceans collide. It is also sufficient to melt rocks, generate volcanoes, heat water to form hot springs, and keep basements of buildings at a constant temperature. It is a perpetual, renewable, and inexhaustible energy resource.

With a few important exceptions, geothermal energy did not play a significant role in the energy mix associated with electrical power generation or other applications until the latter half of the twentieth century. At that time, growing interest in the environmental, economic, and social aspects of energy generation and use encouraged exploration of energy sources that would diminish reliance on *fossil fuels*. This chapter will discuss the context of those changes and their implication for the development of geothermal energy. The remainder of this book will consider specific topics that, if taken together, provide a comprehensive body of knowledge for informed consideration of geothermal energy use.

GLOBAL ENERGY LANDSCAPE

HISTORICAL ROLE OF FUEL

One of the hallmarks of the human species is the creative use of energy. Over many centuries, humanity learned through experience, insight, and experimentation that fire could be controlled and used to our mutual benefit. And, with that ability and skill, the quality of life has rapidly improved for an ever-growing proportion of the planets' people.

The use of fire to support life and industry is dependent upon a fuel source. It is generally assumed that biomass in the form of wood, grasses, and animal solid wastes was the first fuel to be systematically utilized by humans. Such fuels, although easy to acquire in most places on the planet, provide a relatively low-quality energy source, meaning that the amount of energy they provide, for a given weight or *mass* of material burned, is low. The discovery of buried fuels, such as coal, made use of that new commodity attractive because the amount of energy it provides, per unit of mass, is much greater than most untreated biomass resources. In addition, coal is compact and easily transportable, which made it the fuel of choice for most instances where heat was needed. Later discoveries of oil and natural gas expanded the choices available for applications needing a high-quality energy source. Oil, in particular, grew dramatically in use because of its very high *energy density* (amount of energy per mass of fuel). Figure 1.1 graphically illustrates how the use of these fuels changed over time in the United States.

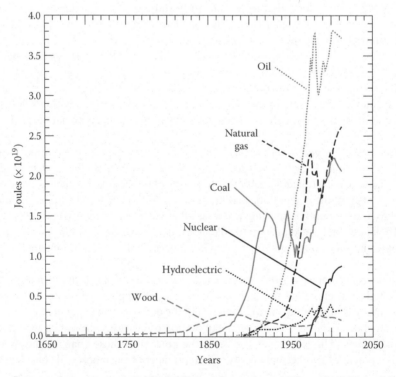

FIGURE 1.1   Sources of energy consumed in the United States, from 1650 to 2011. (United States Energy Information Administration, http://www.eia.doe.gov/emeu/aer/eh/intro.html.)

The ability to access, control, and maintain a source of fuel has become a prerequisite for supporting industrial activity and economic growth. As a result, discussions about energy have become inextricably linked with the necessity that a fuel source exists to support energy production. When fuel is readily available and the population competing for it is relatively small, growth and development are not constrained by the fuel source. Population growth and technological evolution, however, have changed this condition.

IMPACT OF POPULATION GROWTH AND PER CAPITA ENERGY USE

Plotted in Figure 1.2 are two trends that demonstrate the nature of the challenge posed by population growth and energy demand. One important fact is that the population of the planet has been growing exponentially. Between 1850 and 2010, the population increased from about 1.3 billion to about 6.9 billion, an increase of just over five times. Current estimates are that, between the years 2000 and 2050, the population will grow from about 6.1 billion to about 9.6 billion, an increase of 57% in 50 years (United Nations 2012). Although this projected growth rate is slower than it has been historically, it does demonstrate that the population of the planet will continue to grow at a fast pace.

The other part of the figure shows the average use of energy per person on the planet per year. Between the years 1850 and 2010, the average per person energy consumption increased from about 4.85×10^9 to over 77.3×10^9 J/person/yr, an increase of more than 15 times. In other words, not only

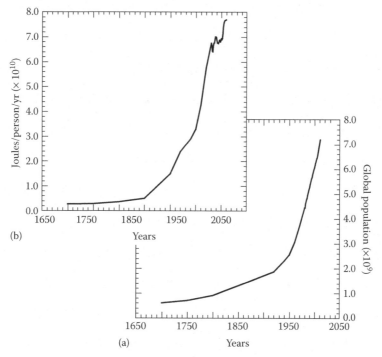

FIGURE 1.2 (a) Population of the planet since 1700. Population estimates for 1950 through 2006 are from the US Census Bureau website. Estimates for 2010 are from the United Nations (2012) (http://www.census.gov/ipc/www/idb/informationGateway.php). Earlier population estimates were derived from data in Grübler, 1998. (b) Energy consumption per person since 1700. Energy consumption through 1979 is based on data from Hafele and Sassin (1977). Data from Grübler, A., *Technology and Global Change*, Cambridge University Press, Cambridge, 1998, 464; data for 1980 through 2006 are from the United States Energy Information Administration, International Energy Annual 2006. Data for 2010 are from International Energy Agency, Key World Energy Statistics, 2012.

is the population of the planet growing rapidly, the average person is using more energy each year than they used previously.

In this situation, access to fuel becomes an important factor affecting economic, political, industrial, and social activities. Situations that disturb the free flow of fuel can have important global impacts. Upheavals in the supply of oil and the resulting chaos in oil markets, such as that occurred in the early 1970s and in 2007, underscore this point. It therefore becomes important to seek reliable energy sources that can meet societal needs at reasonable prices and in a sustainable and secure manner.

Fuel Emissions and Environmental Considerations

An additional issue that has taken on importance concerns the environmental impact of fuel extraction and energy use. There now is little scientific debate that the use of carbon-based fuels for energy generation has affected the atmosphere, and with it the global climate (Solomon et al. 2007; Rohde et al. 2013). Combustion of carbon-based fuels and human activity produce gases such as carbon dioxide (CO_2), oxides of nitrogen, and methane (CH_4), among others, all of which affect the ability of the atmosphere to absorb or transmit radiation. As the abundance of these *greenhouse gases* increases in the atmosphere (Figure 1.3), the transmissivity of the atmosphere to thermal energy drops. The result of this change in the composition of the atmosphere is that the atmosphere traps an increasing proportion of thermal energy that otherwise would have been radiated back into space, leading to an increase in the average surface temperature of the planet (Figure 1.4). It is precisely this process that has kept the surface of Venus at a nearly uniform temperature of 462°C (approximately 736 K or 864°F). The absence of greenhouse gases is also one of the reasons Mars never gets above 0°C.

Figure 1.5 shows the history of the annual total global CO_2 emissions since 1750. Between 1850 and 2011, total emissions of CO_2 to the atmosphere from burning of fossil fuels by humans increased from 1.98×10^{11} to 32.6×10^{12} kg/yr. This change represents an increase of more than 163 times. Also shown in the figure is the change in per capita annual emissions of CO_2. For the same time period, per capita emissions increased by almost 30 times. This clearly shows that there has been an historical trend in which each human being on the planet, on average, is putting more CO_2 into the atmosphere each year.

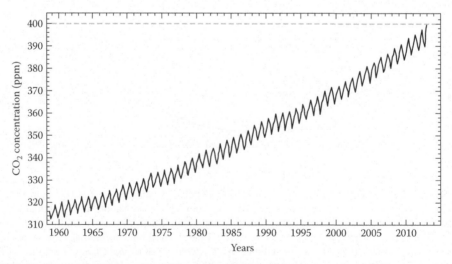

FIGURE 1.3 Measured concentration of atmospheric CO_2 at Mauna Loa, from 1958 to July 2013. The annual seasonal variation is shown by the saw-toothed curve. (From the National Oceanic and Atmospheric Administration website, http://www.esrl.noaa.gov/gmd/obop/mlo/.)

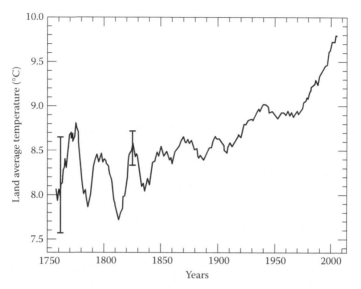

FIGURE 1.4 Change in the global average land surface temperature, from 1753 to 2011. (Modified from Rohde, R. et al., *Geoinformatics & Geostatistics: An Overview*, 1, 1–7, 2013.)

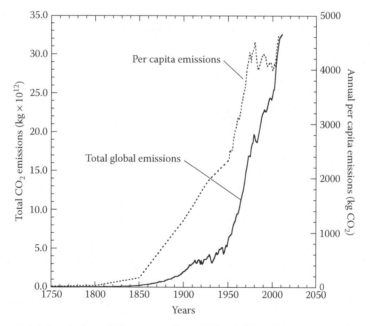

FIGURE 1.5 Total global emission of CO_2 per year from 1750 to 2011 and the per capita CO_2 emission for the same period. (Data from Boden, T.A. et al., Global, regional, and national fossil-fuel CO_2 emissions. Carbon Dioxide Information Analysis Center, Oak Ridge National Laboratory, US Department of Energy, Oak Ridge, TN, 2010; US Energy Information Agency, 2013, http://www.eia.gov/cfapps/ipdbproject/iedindex3.cfm?tid = 90&pid = 44&aid = 8&cid = ww,&syid = 2007&eyid = 2011&unit = MMTCD.)

Whether viewed from the perspective of economics, environmental stewardship, social stability, or national interest, these patterns of energy consumption and greenhouse gas emissions have commanded global attention. It is precisely for these reasons that the United Nations Environment Programme (UNEP) and the World Meteorological Organization (WMO) established the Intergovernmental Panel on Climate Change (IPCC) "to provide the world with a clear scientific view on the current state of climate change and its potential environmental and socio-economic consequences," as stated on the IPCC website (http://www.ipcc.ch). It is within this context that renewable energy sources such as geothermal energy take on significance.

GEOTHERMAL ENERGY AS A RENEWABLE ENERGY SOURCE

NO FUEL, FEW EMISSIONS, AND REDUCED VOLATILITY

As noted in Section "Historical Role of Fuel", the historical use of energy has relied on fuel to make heat. The sources of fuel have traditionally been wood, coal, oil, and natural gas. The changes in use of these fuels over time within the United States, as well as the contribution of hydroelectric and nuclear generation, are shown in Figure 1.1. Coal, natural gas, and oil account for more than 85% of the energy used in the United States. Although these fuels may, in combination, provide decades to several hundred years of supply, the combination of greenhouse gas emissions and competition for the petroleum-based fuels that are implicit in Figure 1.2 have made problematic reliance on these energy sources. In addition, these fuels are derived from a resource base that is not being replenished. Instead, these resources are being extracted from geologically derived materials that took millions of years to form. As a commodity, these fossil fuels are not renewable and will become more and more difficult, and hence more costly, to extract.

Resource depletion is a characteristic of nonrenewable energy resources. However, how depletion is measured is a matter of considerable debate. This issue is discussed in detail in Chapter 8, but several facets of this topic are relevant for consideration here. One aspect of this issue concerns the accessibility of fossil fuel energy resources, such as oil or natural gas or coal. Oil, for example, occurs in underground reservoirs (oil "fields" or "pools"). Some of these reservoirs are within a few hundred meters of the ground surface and occur in very porous rocks through which the oil can flow relatively easily. Other oil reservoirs are present at much deeper levels (thousands of meters below the surface) in rocks that have very little interconnected porosity, thus making it difficult for the oil to flow. The shallow reservoirs are easy to access through standard drilling techniques and the oil is easily and readily pumped out of them. The cost of producing such oil is relatively low. The converse is true for the deeper reservoir—drilling costs will be high and production from a given well will be low. Thus, as long as the price per barrel of oil that a customer is willing to pay is low, the shallow reservoir will be developed and considered an oil "reserve," whereas the deeper reservoir may be a resource but it will not be possible to develop it and consider it a reserve because the cost of doing so would be greater than a potential customer would be willing to pay.

However, once the shallow reservoir is depleted (it contains a finite amount of oil), the deeper reservoir will become economic to develop if there is a shortage of oil and the price a customer is willing to pay increases sufficiently. At that point, a new reserve is now possible to develop and oil availability increases. If the size of that deeper reservoir is larger than the shallow reservoir, the total available oil inventory will then increase, even though the original reservoir is totally depleted, provided customers remain willing to pay the higher cost for the fuel. In other words, the size of a resource depends, in part, on economic considerations. Generally, as markets increase the value of a commodity, as that commodity becomes more difficult to access, the price of the commodity increases.

Recently, it has become important to also consider the amount of energy that is required to produce an energy resource (e.g., Murphy and Hall 2010). This measure, called energy returned on energy invested (EROEI or EROI), is the ratio between the amount of energy that is expended in developing, producing, and processing an energy source, such as oil, and the amount of energy that is obtained in using that fuel. EROEI is difficult to establish in an absolute sense, because it is dependent on which factors one chooses to include in the analysis. For example, with oil production, the energy invested includes the energy expended in drilling the well, pumping the oil out of the ground, transporting the oil to a refinery, processing the oil in the refinery, etc. What *et cetera* might include would be the energy cost of delivering the product, the energy expended for environmental maintenance and cleanup once production ceases, energy expended in marketing, and energy expended during initial exploration for the resource. Obviously, each analysis that is conducted will provide different results, depending upon how the analysis is done. One example of such an analysis that was recently done (Figure 1.6) shows how the EROEI for some fossil fuels has dropped significantly as easily accessed fossil fuel resources are depleted and more challenging resources become economically attractive. These results show that, over time, less net energy is available from the extracted resource, as we expend more energy to obtain it. This analytical approach to the economics of energy extraction and use is discussed in more detail in Chapter 14.

It is for these reasons that there is growing interest in finding sources of energy that reduce reliance on fossil fuels to generate heat to produce work and that can reduce or eliminate production of greenhouse gases. In addition, more attention is being given to innovative direct uses of heat in applications where a process can be accomplished without burning a fuel to obtain heat. For example, fruit drying can be accomplished either by generating electricity or burning fuel to heat a drying oven, or by using hot waters obtained directly from natural hot springs to heat a drying oven.

Criteria that are generally used to establish the viability of an energy source that would supplant or displace reliance on fossil fuels are the following:

- It is sufficiently abundant to meet a significant percentage of the market demand.
- It can be obtained at a cost competitive with existing energy sources.
- Its use will reduce or eliminate greenhouse gas emissions.
- It is self-replenishing (i.e., renewable).

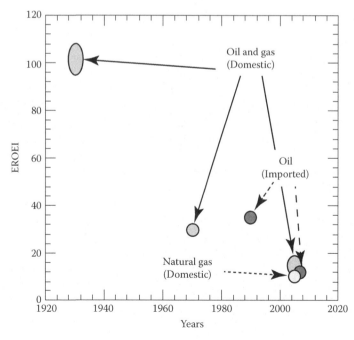

FIGURE 1.6 EROI for oil and natural gas from the indicated sources, from 1930 to 2008. (Data from Cleveland, C.J., *Energy*, 30, 769–782, 2005; Hall, C.A.S., *The Oil Drum*, April 8, 2008. http://theoildrum.com/node/3810.)

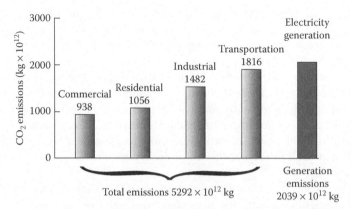

FIGURE 1.7   The amount of CO_2 emissions from commercial, residential, industrial, and transportation sectors of the US economy in 2012. Those four sectors generated a total of 5292×10^{12} kg of CO_2 in that year. Also shown is the total amount of CO_2 emitted through electricity generation, which contributed to all of those sectors. (Data from the United States Energy Information Administration, Preliminary Data for 2012, http://www.eia.gov/tools/faqs/faq.cfm?id = 75&t = 11.)

These criteria can be best employed when selecting alternative energy sources by first understanding how fossil-fueled energy is used and how it contributes to greenhouse gas emissions.

Figure 1.7 shows the energy-consuming sectors of the US economy and their respective greenhouse gas emissions in 2012. Currently, the transportation sector is the largest contributor to greenhouse gas emissions. Transportation relies, for the most part, on liquid fuels. This fact places important limitations on energy sources being considered for this sector. The other three sectors together contribute about two-thirds of the total emission of greenhouse gases. Significantly, more than half of that contribution comes from the generation of electricity that supports those sectors. Electricity generation alone accounts for nearly half of all greenhouse gas emissions. This situation reflects the fact that power generation is accomplished primarily by burning coal, natural gas, and oil.

In combination, these facts and observations suggest that the greatest reduction in the use of fossil fuels can be accomplished by the following:

• Reducing the demand for electricity
• Replacing fossil fuel-based electrical generation capacity with renewable energy sources
• Replacing fossil fuel-based liquid fuels with other forms of portable energy sources

Geothermal energy can significantly contribute to all of these needs. For example, geothermal heat can be used directly and indirectly to heat and cool buildings (discussed in Chapters 11 and 12), thus reducing electrical demand for heating, ventilation, air conditioning, and cooling (HVAC). Geothermal energy can also be used to generate electricity, as detailed in Chapter 10. Geothermal resources can also be used to provide heat necessary for processing biofuels, as discussed in Chapters 12 and 16. In fact, when deployed in combination with other renewable energy sources and astute conservation efforts, most of the energy needs for the world can be satisfied using geothermal resources along with wind, solar, biomass, and hydrodynamic resources.

Geothermal energy has several attributes that allow it to satisfy the criteria listed above. One attribute is that geothermal energy requires no fuel. Because it relies on the persistent flow of heat from the earth's interior, it can be tapped without recourse to a fuel supply infrastructure. For every kilowatt of electrical energy displaced by geothermal energy use, the greenhouse gas emissions that would have been produced from a fossil-fueled power plant are reduced by a minimum of 90%, and in many cases, they are eliminated completely.

GEOTHERMAL ENERGY IS A FLEXIBLE ENERGY RESOURCE

Another significant attribute is that geothermal heat occurs in diverse ways, making it possible to use it for different purposes. One such application is for heating and cooling buildings, as mentioned in Section "No Fuel, Few Emissions, and Reduced Volatility". Every square meter of land surface has heat flowing through it. Although the temperature of the upper few meters of soil and rock fluctuates with the effects of local weather patterns and solar insolation, the temperature at depths of three to ten meters is usually constant because of the flow of heat from the interior of the earth. That heat can be used as a source of energy for HVAC purposes in buildings using geothermal heat pumps (GHP; Chapter 11). Because of their high efficiency, GHP use diminishes electrical demand. When combined with programs that improve building efficiency, such applications of geothermal energy can displace a significant percentage of electricity generation and natural gas usage required to meet the HVAC load.

In many areas, modest heat flow is available to allow the use of geothermal energy for industrial applications that currently rely on fossil fuels. These *direct-use* applications include food processing, drying materials, agricultural activities and greenhouses, aquaculture, and paper manufacturing. Although such applications have been developed and successfully used throughout the world, they remain relatively unknown and vastly underutilized. Chapter 12 discusses many of these applications.

In regions where geothermal energy occurs at higher density, as discussed in Chapters 2 and 10, temperatures are high enough to allow the generation of electrical power. Figure 1.8 shows a comparison of CO_2 emissions for fossil-fueled power generation compared to geothermal plants. Geothermal power generation in appropriate settings can effect a very significant reduction in greenhouse gas emissions. As discussed in Chapter 2, current geothermal power generation technology has application in specific geological settings and thus is not currently able to provide electrical power in all settings. For this reason, its application is restricted to about 30% of the geographical area of the United States. However, as discussed in Chapter 13, the development of new technology that allows the drilling and use of deeper wells and geothermal reservoirs will expand deployment of geothermal generation capabilities to most regions of every continental land mass. It is currently

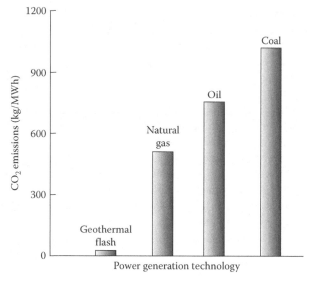

FIGURE 1.8 CO_2 emissions (in kg of CO_2/MWh) for different power generating technologies. The values for the generating systems that use fossil fuels are from the United States Environmental Protection Agency eGrid 2000 database. Binary geothermal power generation produces no emissions. (Data from Slack, K., Geothermal resources and climate emissions. Draft Report for Public Review. Geothermal Energy Association, Washington, DC, 2009.)

anticipated that this new technology could supply a large fraction of US electrical power by 2050 (Tester et al. 2006).

Deployment of geothermal energy technologies will be most effective where the attributes of the technology are matched to the characteristics and needs of the energy sector they are designed to serve. In the following section, the use of geothermal power in the electricity generation sector is discussed, as an example of how resource attributes and sector needs can be matched for greatest efficiency.

ELECTRICAL DEMAND AND THE CHARACTERISTICS OF GEOTHERMAL ENERGY

GENERATING ELECTRICAL POWER FOR THE GRID

The modern electrical grid has evolved over the last century to reliably supply power to an increasingly complex market. The grid is a network that links together power generators and power users through a system of transmission and distribution lines. In principle, a sophisticated electrical grid can allow a customer on one side of a country to purchase power from a generator on the other side of a country, or even another country, and reliably have that power provided on demand. For historical and economic reasons, however, power grids are commonly segmented into regions supplied and administered by operators and regulators. Such a system allows responsiveness to local and regional needs, while reducing transmission losses, and provides a measure of security from failures in the grid that could be catastrophically transmitted through the network if the grid were not segmented.

Such a system has several important characteristics and limitations. One characteristic is that the demand or *load* on the grid will vary during the day. In a given region, the demand will be lowest in the earliest morning hours and greatest at some point during the day or early evening. Seasonal variability will influence the timing of the load, and unusual weather can cause spikes in demand at nontraditional times. Because of this variability, the concepts of *baseload*, *peak load*, and *load following* have evolved and play an important role in designing grid components.

Baseload is the minimum amount of power a supplier must make available to its customers. The amount of baseload power can vary from hour to hour, depending upon region and the power demands that exist there. Usually a utility or power administrator will have an historical record and contractual obligations on which to establish what the baseload demand will be.

The peak load is that load placed on the grid by the immediate conditions that are being experienced at that moment that exceed the baseload. Peak load can vary from day to day and month to month. It can be strongly affected by such things as extremes in weather or local emergencies. The capacity to meet this temporary increase in load that exceeds the baseload is called *peaking capacity*. A power provider uses historical records to estimate what the likely peak demand will be. It is on this basis that the maximum required generating capacity in a region is determined. Generally, a utility or other provider will design the capacity of the local or regional system to exceed by a few percent the estimated maximum possible peak load.

Load following is the ability to respond to changes in demand for power. Load following requires the ability to increase power output on a timescale of minutes to tens of minutes. This can be accomplished by either having generating plants capable of relatively rapid changes in power output or having in place contractual agreements to buy power on short notice from suppliers who can quickly respond to demand changes.

How demand changes over a 24-hour cycle varies with location, season, and local weather conditions on any particular day. Currently, renewable energy resources are not able to meet the full energy demand anywhere in the United States, except for very local settings. Even so, they currently are capable of contributing significantly to the power supply in many regions. Figure 1.9 shows the power generation achieved by California from renewable resources during a specific 24-hour period. This figure shows the characteristic attributes of the current renewable energy resources,

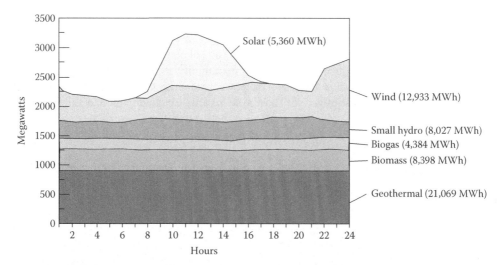

FIGURE 1.9 Variation in electrical generation for renewable energy resources for December 21, 2012, in California. Plotted for each renewable resource is the amount of power it generated for the 24 hours of December 21. Also shown (in parentheses) is the total MWh of produced power for each renewable resource.

such as the variability of solar and wind with time of day, the baseload nature of geothermal energy, and the intermediate (i.e., nearly baseload) characteristics of biofuels.

Managing transmission and distribution for a power grid have historically been accomplished manually through centralized facilities that can locally or remotely call up generation capacity, or reduce it, as needed. However, recent technological developments have made it conceivable that an electrical grid can be monitored and managed electronically using sophisticated computer systems that quickly compute demand, load variation, and changes in wind and solar generation resulting from local climate conditions. This has given rise to the concept of a *smart grid* that, essentially, manages itself. Although some years away from full implementation, the concept and framework for putting in place such a capability is moving forward in various parts of the world. Such a capability will reduce transmission losses, manage generation facilities more efficiently, and allow dispatchability of the most cost-effective generation methods, and do so in a way that takes into account local and regional needs.

Geothermal power plants have historically been viewed and operated as baseload power plants. In this sense, they are different from solar and wind power generating facilities, which are *intermittent*. Although all of these facilities have the advantage that they do not require a fuel supply infrastructure to provide energy, power generation from solar or wind facilities is interrupted (i.e., is intermittent) on a diurnal and/or seasonal cycle and thus cannot provide baseload capability. In contrast, geothermal energy never stops being supplied from the earth's interior and can thus be part of a baseload supply system. However, there is no theoretical reason why a geothermal reservoir could not be managed in a way that allowed it to be load following, that is, dispatchable. Research and development efforts are underway to develop the means to manage geothermal power facilities to allow a load following and peaking capability, as discussed in Chapters 10 and 16.

GENERATING ELECTRICAL POWER FOR LOCAL USE

Electrical power generation as described in Section "Generating Electrical Power for the Grid" is the primary means, whereby residential, commercial, and industrial customers obtain electrical power. However, renewable energy resources readily lend themselves to situations where power generation can be accomplished with modest generators that supply power to a limited number of customers in a restricted area. Such *distributed generation* capabilities are growing in number. The compact nature of new generation binary geothermal generators, for example, has the potential to supply a few hundred to a few hundred thousand kilowatts of power from a single well. Such a facility can supply power to an industrial site, a small community, or any other type of operation that requires power at that scale without accessing the power grid. Such facilities require minimal operational oversight and can be cost-effective. These systems are discussed in more detail in Chapter 10.

HOW THIS BOOK IS ORGANIZED

Geothermal energy is a complex energy source with diverse applications. For that reason, this book develops the fundamental scientific principles that apply to geothermal energy resources and their use in Chapters 2 (earth sciences), 3 (thermodynamics), 4 (hydrology), and 5 (geochemistry). Using the principles developed in those chapters, the topics of exploring for resources (Chapters 6 and 7), assessing their properties and magnitude (Chapter 8), and drilling for the resource (Chapter 9) are then discussed. The properties, characteristics, and operational fundamentals are considered for each of the applications that employ geothermal energy in Chapters 10 (power generation), 11 (geothermal heat pumps), and 12 (direct-use applications). Chapter 13 discusses the advancements that have recently been made in enhanced geothermal systems (EGS). Chapters 14 and 15 present the economic and environmental issues, respectively, associated with geothermal applications. Chapter 16 discusses possible future developments that can significantly affect the role of geothermal energy in the overall energy landscape.

Within each chapter, certain conventions are followed in the presentation of material. The first time a word or phrase is used that contains important conceptual material and will appear repeatedly in the future, it is italicized. To assure that the material is presented with sufficient rigor that the basic concepts and principles are precisely accurate, mathematical descriptions are used when appropriate. It is assumed that the reader has a basic command of algebra, trigonometry, and calculus. Finally, resources that contain important information are referenced in the body of the material in each chapter. When appropriate, websites are included. There is also, at the end of each chapter, an annotated list of a few key information sources that can provide useful information beyond that contained in the body of the chapter.

SYNOPSIS

Growth in the human population and energy use, as well as the consequent environmental impacts, have led to interest in finding new energy resources that are renewable and have reduced greenhouse gas emissions. Geothermal energy is a versatile resource that can be used in many situations to meet these goals. It requires no fuel supply and related infrastructure and can be deployed in a variety of settings. Geothermal energy can be used to provide heat for HVAC or other purposes, as well as for power generation. It has the potential to play an important role in the transition from fossil fuels to energy sources that have minor environmental impacts. Successful deployment of renewable energy resources, however, requires that the resource be carefully matched to the application being developed.

PROBLEMS

1.1. What is baseload energy?
1.2. Using the data in Figure 1.2, plot the doubling time for per capita power use at 25-year intervals. The doubling time is the time required for a value to double. Discuss the implications of this plot.
1.3. From an environmental perspective, how does geothermal energy use differ from that of fossil-fueled systems? What are the benefits? What are the disadvantages?

1.4. If one were to use geothermal, solar, wind, and biomass technologies together, what would be the considerations that would have to be addressed in order to satisfy the daily and seasonal load?

1.5. Assume that protocols are put in place to reduce global CO_2 emissions to the levels in 1980. Using the data in Figures 1.5, 1.7, and 1.8, suggest, quantitatively, how this might be achieved? What assumptions must you make to do these calculations?

1.6. Using the data in Figures 1.4 and 1.5, plot the correlation between global emissions and global temperature. What are the implications of this figure?

REFERENCES

Boden, T.A., Marland, G., and Andres, R.J., 2010. Global, regional, and national fossil-fuel CO_2 emissions. Carbon Dioxide Information Analysis Center, Oak Ridge National Laboratory, US Department of Energy, Oak Ridge, TN. doi:10.3334/CDIAC/00001_V2010.

Cleveland, C.J., 2005. Net energy from the extraction of oil and gas in the United States. *Energy*, 30, 769–782.

Grübler, A., 1998. *Technology and Global Change*. Cambridge: Cambridge University Press. 464 pp.

Hafele, W. and Sassin, W., 1977. The global energy system. *Annual Review of Energy*, 2, 1–30.

Hall, C.A.S., 2008. Provisional results from EROI assessments. *The Oil Drum*. April 8. http://theoildrum.com/node/3810.

International Energy Agency, 2012. *Key World Energy Statistics*. Organisation for Economic Co-operation and Development. 80 pp. http://www.iea.org/publications/freepublications/publication/kwes.pdf.

Murphy, D.J. and Hall, C.A.S., 2010. Year in review—EROI or energy return on (energy) invested. *Annals of the New York Academy of Sciences*, 1185, 102–118.

Rohde, R., Muller, R.A., Jacobsen, R., Muller, E., Perlmutter, S., Rosenfeld, A., Wurtele, J., Groom, D., and Wickham, C., 2013. A new estimate of the average earth surface land temperature spanning 1753 to 2011. *Geoinformatics & Geostatistics: An Overview*, 1(1), 1–7.

Slack, K., 2009. Geothermal resources and climate emissions. Draft Report for Public Review. Geothermal Energy Association, Washington, DC, 39 pp.

Solomon, S., Qin, D., Manning, M., Chen, Z., Marquis, M., Averyt, K.B., Tignor, M., and Miller, H.L. (eds.), 2007. *Climate Change 2007: The Physical Science Basis*. Contribution of Working Group I to the Fourth Assessment Report of the Intergovernmental Panel on Climate Change. Cambridge University Press, Cambridge; New York, NY, 996 pp.

Tesla, N., 1900. The problem of increasing human energy: With special references to the harnessing of the sun's energy. *Century Illustrated Magazine*, June.

Tester, J.W., Anderson, B.J., Batchelor, A.S., Blackwell, D.D., DiPippio, R., Drake, E.M., Garnish, J. et al., 2006. *The Future of Geothermal Energy*. Cambridge, MA: MIT Press. 372 pp.

United Nations, 2012. World population prospects: The 2012 revision. United Nations Department of Economic and Social Affairs, Population Division. http://esa.un.org/wpp/.

US Energy Information Agency, 2013 Washington, DC. http://www.eia.gov/cfapps/ipdbproject/iedindex3.cfm?tid = 90&pid = 44&aid = 8&cid = ww,&syid = 2007&eyid = 2011&unit = MMTCD.

FURTHER INFORMATION SOURCES

The European Commission, Joint Research Centre (JRC; http://ec.europa.eu/dgs/jrc/index.cfm).

The JRC for the European Commission provides high-quality analyses of a broad range of issues that relate to energy, environment, and technology, within the European context. It can be an important resource for obtaining data and information relating to present-day energy-related challenges and actions within the European Union.

The Geothermal Energy Association (GEA; http://www.geo-energy.org/).

The GEA is a trade organization dedicated to providing timely and accurate information about many aspects of geothermal energy. They are an excellent data and information resource. They are particularly useful for obtaining timely information regarding topics that have been in the media recently.

The Geothermal Resources Council (GRC; http://www.geothermal.org/).

The GRC is a membership organization dedicated to supporting geothermal energy activities of all kinds. It has the world's most complete library on geothermal topics and is an excellent resource for data, links to information sources and industry members.

The International Geothermal Association (IGA; http://www.geothermal-energy.org/index.php).

The IGA performs a function complementary to that of the GRC, but with a broader international focus. It is particularly useful for information concerning geothermal activities outside the United States.

United States Energy Information Agency (USEIA; http://www.eia.gov/).

The USEIA provides basic information about every energy resource in use in the United States. Its website has a variety of tools for tracking current levels of use and production, as well as costs. It also has a host of resources for exploring the technology behind energy resource extraction.

United States Environmental Protection Agency (USEPA; http://www.epa.gov).

The USEPA maintains a website at which it is possible to obtain information regarding a broad range of energy-related topics and their environmental attributes. The website for "Clean Energy" (http://www.epa.gov/RDEE/energy-and-you/affect/air-emissions.html) is a good portal to begin exploration of that topic.

2 Sources of Geothermal Heat
The Earth as a Heat Engine

The earth gives the impression that it is dependably constant. Over the timescale of a human lifetime, little seems to change, as John Burroughs noted when he wrote of, "the unshaken permanence of the hills" of Ireland (John Burroughs 1876, Winter Sunshine, vol. II). The reality, however, is quite contrary to that experience. Every cubic centimeter of the earth is in motion and has been since the earth was formed 4.5 billion years ago. Indeed, the earth is a profoundly dynamic entity. On the timescale of seconds, earthquakes jar the earth; over the time span of a few years, volcanoes appear and grow; over millennia, landscapes slowly evolve; and over millions of years, the continents rearrange themselves on the planet's surface.

The energy source to drive these processes is heat. Although the extrusion of molten rock at volcanoes is perhaps the most dramatic evidence that heat energy exists in the earth's interior, there is, in fact, a constant flux of heat from every square meter of the earth's surface. The average heat flux for the earth is 87 mW/m^2 (Stein 1995; see Sidebar 2.1 for a discussion of the units). For a total global surface area of 5.1×10^8 km^2, this heat flux is equivalent to a total heat output of more than 4.4×10^{13} W. For comparison, it is estimated that the total power consumed by all human activity in 2006 was approximately 1.57×10^{13} W (US Energy Information Agency 2008). Clearly, heat in the earth has the potential to significantly contribute to satisfying human energy needs. This heat is the source of geothermal energy. The remainder of this chapter will consider the origin, distribution, and properties of that heat.

ORIGIN OF THE EARTH'S HEAT

In order to intelligently use the heat available in the earth, it is important that the sources of that heat be understood. Geothermal resources can heat homes, air condition greenhouses, dry spices and vegetables, and generate electricity. Some of these applications can be pursued anywhere on the planet, others require special circumstances. Using this resource in a way that is both economical and environmentally sound requires that the characteristics of the resource be understood. This chapter describes the origins of earth's heat, the processes that determine how it is distributed over the surface of the planet, and what determines its intensity.

HEAT FROM FORMATION OF THE CORE

Studies of meteorites and models of astronomical processes have posited that the earth formed 4.56 billion years ago (Göpel et al. 1994; Allégre et al. 1995) by accretion of material from the early solar nebula. Dust, sand-sized particles, and other objects collided and aggregated, forming terrestrial-sized planets within a time frame of a few tens of millions of years (Wetherill 1990; Canup and Agnor 2001; Chambers 2001; Kortenkamp et al. 2001; Kleine et al. 2002; Yin et al. 2002). The aggregating materials were composed of a variety of minerals, primarily silicates similar to those that make up the rocks of the earth, as well as native metals (primarily iron) and frozen volatiles such as water and simple hydrocarbons.

As the planet accreted material, the kinetic energy of incoming bodies was transformed, in part, to heat when the bodies impacted the planetary surface. That process leads to an increase in temperature of the growing planet. In addition, as the planet grew in size, pressures in the interior increased, compressing the silicate minerals and other materials, ultimately contributing to elevating the internal temperature of the planet.

The early solar nebula also had a significant abundance of radioactive elements with short half-lives (e.g., ^{26}Al, $t_{1/2}$ = 740,000 years; ^{182}Hf, $t_{1/2}$ = 9 million years; ^{53}Mn, $t_{1/2}$ = 3.7 million years, where $t_{1/2}$ is the half-life, which is the time required for one half of a given amount of a radioactive element to decay). Radioactive decay occurs via one of several mechanisms: α-decay, which is the emission of a helium nucleus; β-decay, which occurs when a neutron becomes a proton with the emission of an anti-neutrino and an electron; β'-decay, which occurs when a proton becomes a neutron with the emission of a neutrino and a positron; γ-decay, which occurs when a gamma ray is emitted; and electron capture, when an inner electron is captured by a nucleus. Although neutrinos do not significantly interact with matter, the other products of radioactive decay, as well as the recoil of the decaying nucleus, heat the immediately surrounding environment when the particles collide with the atoms surrounding the decaying isotopes. The kinetic energy of the particles is transformed to heat, which results in raising the local temperature. Table 2.1 summarizes the heat production for the primary radioactive elements that currently provide heat to the earth. Although ^{26}Al is no longer present on the planet, it was abundant during the early formation of the earth and contributed substantially to raising its internal temperature.

In combination, these processes resulted in the temperature of the earth's interior exceeding that of the melting point of iron. Because of its high density relative to that of the silicates with which it was in contact, and because of its mobility in the liquid state, the liquid iron migrated to the center of the earth, forming a liquid core. That migration process also contributed to heating the earth, because movement of the iron to a position of lower gravitational potential resulted in the release of gravitational potential energy.

In combination, these processes led to the formation of a differentiated planet with a hot, liquid metal core within less than 30 million years of the planet's formation (Kleine et al. 2002; Yin et al. 2002). Since that time, the core has been slowly cooling, resulting in the growth of a solid inner core and diminution in the size of the liquid outer core. Today, the solid inner core has a radius of approximately 1221 km (Figure 2.1). The liquid outer core extends to about 3480 km from the center of the earth, making it about 2200 km thick. Although difficult to establish, and subject to considerable debate, the temperature at the boundary between the inner and outer core is probably between 5400 and 5700 K. The temperature at the outer edge of the liquid core is about 4000 K (Alfé et al. 2007).

Some of the heat (approximately 40%; Stein 1995) used in geothermal applications ultimately derives from this remnant heat from the early formation of the earth's core. The remaining 60% is derived from the decay of long-lived radioactive isotopes.

TABLE 2.1
Heat Generation of the Primary Heat Producing Elements

Material	K	U	Th
Heat production (W/kg of element)	3.5×10^{-9}	96.7×10^{-6}	26.3×10^{-6}

Source: Beardsmore, G.R. and Cull, J.P., *Crustal Heat Flow: A Guide to Measurement and Modeling*, Cambridge University Press, Cambridge, 2001.

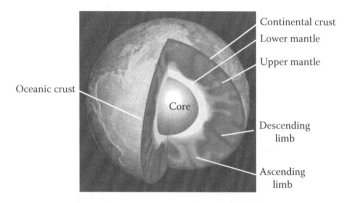

FIGURE 2.1 **(See color insert.)** Interior of the earth, shown in a cutaway that depicts the outer edge of the liquid core (reflecting orange sphere), the lower mantle (yellow), the upper mantle (pink and purple), and the crust. Ascending limbs of convection cells are shown as the orange-tinted plumes extending from the lower mantle through the upper mantle to the base of the crust. Descending limbs of convection cells are shown as the darker purple features extending into the mantle from the base of the crust. (From United States Geological Survey, http://geomag.usgs.gov/about.php.)

HEAT FROM THE RADIOACTIVE DECAY OF LONG-LIVED ISOTOPES

The formation of the core was a fundamental event in the history of the earth. The redistribution of metal was accompanied by density stratification within the remaining planetary body. Although the details and timing of the process remain a matter of substantial scientific uncertainty and interest, the end result was that the earth segregated into layers with distinct chemical compositions (Figure 2.1).

The mantle, which surrounds the core, has a thickness of about 2890 km. It is composed of minerals that are relatively low in silica and that have relatively high density. The density of the minerals is a reflection of the fact that the high pressures in the interior of the earth favor mineral structures that have relatively small volumes per formula unit (i.e., per mole). Such structures do not readily accommodate large atoms, such as potassium (K), rubidium (Rb), thorium (Th), or uranium (U), as well as a host of other elements. As a result, the mantle tends to be composed of silicate minerals; oxides; and other high-density minerals with high contents of atoms with relatively small atomic radii such as magnesium (Mg), titanium (Ti), calcium (Ca), and some aluminum (Al); and low abundances of larger atoms. The exclusion of atoms with relatively large atomic radii has the consequence of depleting the mantle in elements that have relatively high proportions of radioactive isotopes, for example, K, Rb, Th, and U.

The crust, which floats on the mantle, is of two types. Oceanic crust underlies the global oceans and has a thickness that varies between 6 and 10 km. Continental crust, of which all the major landmasses of the earth are composed, varies in thickness from 30 to 60 km. Oceanic crust is formed at regions where the upwelling portions of convection cells in the mantle reach the surface (see Section "Plate Tectonics and the Distribution of Geothermal Resources"). Magma that forms during the upwelling process is extruded at ocean ridges and solidifies as oceanic crust. Because this crust is formed directly from mantle, which is low in radioactive elements, the oceanic crust also has a low abundance of radioactive elements.

The continental crust (Figure 2.2), however, is largely composed of the material that was incompatible with the high-density minerals in the mantle. As a result, the continental crust is richer in lower density minerals that are made up of relatively large atoms. Included in this suite of minerals are those that can readily accommodate K, Rb, Th, and U. Consequently, continental crust holds the largest global reservoir of radioactive elements (Table 2.2). Approximately 60% of the heat that exists in continents is derived from the radioactive decay of these four elements.

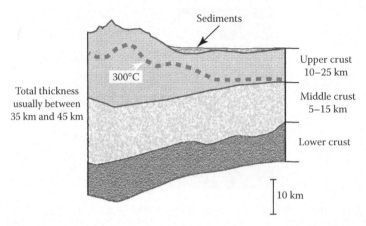

FIGURE 2.2 A schematic cross section through continental crust. The known ranges of thickness of the upper, middle, and lower crust segments are shown, along with a hypothetical 300°C isotherm. Note that the isotherm may be suppressed in mountain ranges, rather than elevated as shown here, depending upon the type of mountain system and its geological history.

TABLE 2.2
Heat Production from Radioactivity (J/kg-s)

Material	K	U	Th	Total
Upper continental crust	9.29×10^{-11}	2.45×10^{-10}	2.77×10^{-10}	6.16×10^{-10}
Average continental crust	4.38×10^{-11}	9.82×10^{-11}	6.63×10^{-11}	2.07×10^{-10}
Oceanic crust	1.46×10^{-11}	4.91×10^{-11}	2.39×10^{-11}	8.76×10^{-11}
Mantle	3.98×10^{-14}	4.91×10^{-13}	2.65×10^{-13}	7.96×10^{-13}
Bulk earth	6.90×10^{-13}	1.96×10^{-12}	1.95×10^{-12}	4.60×10^{-12}

Source: Van Schmus, W.R., Natural radioactivity of the crust and mantle. In *Global Earth Physics*, ed. T.J. Ahrens, American Geophysical Union, Washington, DC, 1995.

The structure of continents is, superficially, quite simple if one considers just the density structure. Seismological observations have led to the concept that the continental crust has three basic components—lower, middle, and upper crust (Figure 2.2). These components vary in thickness from place to place, as indicated in the figure and are sometimes difficult to actually distinguish. The lower crust tends to be composed of igneous and metamorphic rocks that have relatively low concentrations of radioactive elements. The middle and upper crusts are, respectively, more enriched in radioactive elements. This fact, plus the significant variation in thickness due to tectonic activity that causes mountain belts and sedimentary basins to form, results in substantial variation in heat output from the heterogeneous distribution of radioactive elements. The consequence of this variability is that the distribution of heat in the shallower levels of the crust can be quite uneven, as indicated schematically by the 300°C isotherm drawn in the diagram.

The contribution of radioactive decay of these elements to the surface heat flow is described by the following relationship:

$$q = q_0 + D \times A \tag{2.1}$$

where:
q is the heat flow at the surface (mW/m^2)

q_0 is an initial value for heat flow unrelated to the specific decay of radioactive elements at the
  current time (mW/m^2)
D is the thickness of crust over which the distribution of radioactive elements is more or less
  homogeneous (for a general model of the crust this value is usually taken as 10–20 km,
  more or less corresponding to the upper crust) (m)
A is the heat production rate per volume of rock (W/m^3)

The variable A can be recast as

$$A = h_0 e^{-\lambda t} \qquad (2.2)$$

where:
  h_0 is the initial heat production (W/m^3)
  λ is the decay constant (which is equal to $0.693/t_{1/2}$) for the isotope of interest
  t is the time (s)

Table 2.2 provides a summary of heat production in various parts of the crust and mantle. Using the
values for the upper crust, and assuming that the background heat flow is approximately 40 mW/m^2
(based on the heat from the cooling core), the average heat flow at the surface of continental crust
would be approximately

$$q = 0.04\,\mathrm{W/m}^2 + 10{,}000\,\mathrm{m} \times 1.35e - 6\,\mathrm{W/m}^3 = 53.5\,\mathrm{mW/m}^2$$

assuming an average rock density of 2200 kg/m^3 and a background heat flux of 40 mW/m^2. If the
average heat flow for the planet is 87 mW/m^2, why is this result so low? The answer lies in how heat
is transferred in the earth.

TRANSFER OF HEAT IN THE EARTH

It is common experience that heat does not stay where it is generated—heat obviously and always
moves from a warm body or region to a cooler body or region. That simple observation gave rise to
one of sciences most fundamental breakthroughs, the principles of thermodynamics, which will be
discussed quantitatively in Chapter 3. At this point, what we will consider are the primary mechanisms (radiation, conduction, and convection), whereby heat transfer takes place, and how those
mechanisms influence the availability of geothermal energy. Heat transfer mechanisms will be
revisited in Chapter 12, where its role in *direct use* geothermal applications is considered in detail.

RADIATION

Heat can be transferred radiatively by the emission and absorption of thermal photons. Thermal
photons are similar to light photons (wavelengths of 380–750 nm), but fall at the longer wavelength
(approximately 700 to beyond 15,000 nm) infrared portion of the electromagnetic spectrum. Most
earth materials are relatively opaque to infrared radiation, so the primary role radiation plays in heat
transfer is through infrared emission at the earth's surface. It is for that reason that radiation contributes relatively little in the way of energy or mass transfer within the earth. That is not the case,
however, for those instances in which it is important to sense or measure heat flow at the earth's
surface. Remote sensing techniques rely on heat radiation to characterize surface properties.
  Within the last decade, infrared sensing aircraft and satellites have been used to map the intensity of thermal or infrared emissions at the surface as a means of identifying thermal anomalies.
Although complex and still in the formative stages of routine deployment, such efforts have the

potential to significantly impact exploration for and assessment of geothermal resources. We will leave a detailed examination of thermal radiation for Chapter 7, where the topic of exploration for geothermal resources is considered in detail.

CONDUCTION

A slab of rock, such as a granite countertop in a kitchen, will achieve a uniform temperature if left undisturbed for a few hours. By definition, when it reaches that state of uniform temperature, it will be in a state of thermal equilibrium. If a pot of boiling water is placed on the countertop, that state of thermal equilibrium will be perturbed. Careful observation and measurement will document that the granite in contact with the pot will quickly rise in temperature, reaching a maximum temperature somewhat less than that of the boiling water. Over time, the temperature of the granite slab will increase at progressively greater distances from the pot, whereas at the same time the temperature of the pot, and the granite immediately in contact with it, will drop. Eventually, in an ideal case, the granite and the pot will reach a state of thermal equilibrium, at a temperature just slightly higher than that which the countertop had achieved before the pot was placed on it. Much of this progressive change in temperature is due to *conduction*, which is the transfer of heat through direct physical contact. It is a diffusive process, involving no transfer of mass.

Conduction occurs via transfer of energy between atoms (and electrons) of a material. This process is often conceptualized as a change in the vibrational frequency of atoms in the material being thermally perturbed. In the case of the granite countertop and pot of boiling water described in this section, the frequency of the vibrating atoms in the minerals making up the granite of the countertop would increase when they come in contact with the rapidly vibrating atoms of the pot. Because the atoms in the minerals of the countertop are in physical contact with each other, the increased vibrational frequency will propagate throughout the granite slab, eventually achieving a state where the temperature is the same everywhere.

The rate at which thermal equilibrium is achieved will depend, primarily, on the thermal conductivity and diffusivity of the material. Thermal conductivity is a measure of the ability of a material to conduct heat. Thermal conductivity (k_{th}) has units of W/m-K and must be measured for each material of interest, because it depends upon the microscopic (e.g., atomic structure, bond strength, and chemical composition) and macroscopic (e.g., porosity and phase state) properties of a material. Because the microscopic and macroscopic properties of a material change with temperature, k_{th} will also be a function of temperature, thus requiring that it also be measured at the temperatures of interest. Table 2.3 provides a list of thermal conductivities for some common materials, as a function of temperature.

The flow of heat (q_{th}) through a material depends directly upon k_{th}, as well as on the temperature gradient (∇T) over some specified distance (x):

$$q_{th} = k_{th} \times \frac{\nabla T}{x} \tag{2.3}$$

The resulting heat flow rate is specified for an area (A) by

$$\frac{dq_{th}}{dt} = k_{th} \times A \times \left(\frac{d\nabla T}{dx} \right) \tag{2.4}$$

The units for heat flow are J/m²-s, which is equivalent to W/m².

The relationship represented by Equation 2.4 applies directly to instances where temperature measurements are made in a well or borehole, allowing projection to depth of the temperature of a potential geothermal resource. In essence, the geometry represented by this problem consists of a

TABLE 2.3
Thermal Conductivity of Some Common Materials (W/m-K)

Material	25°C	100°C	150°C	200°C
Quartz[a]	6.5	5.01	4.38	4.01
Alkali feldspar[b]	2.34	–	–	–
Dry sand[a]	1.4	–	–	–
Limestone[a]	2.99	2.51	2.28	2.08
Basalt[a]	2.44	2.23	2.13	2.04
Granite[a]	2.79	2.43	2.25	2.11
Water[c]	0.61	0.68	0.68	0.66

Sources: [a] Clauser, C. and Huenges, E., Thermal conductivity of rocks and minerals. In *Rock Physics and Phase Relations*, ed. T.J. Ahrens, American Geophysical Union, Washington, DC, 1995
[b] Sass, J.H., *Journal of Geophysical Research*, 70, 4064–4065, 1965
[c] Weast, R.C., *CRC Handbook of Chemistry and Physics*, CRC Press, Boca Raton, FL, 1985.

slab of material across which a temperature difference exists. By measuring the heat capacity of the geological material and the temperature difference between two different depths in a well, it is possible to deduce how many J/s (i.e., watts) are flowing through an area, and thus project to depth what the temperature may be. Suppose, for example, that, at a potential geothermal site located in basalt, an exploration well was drilled to a depth of 2000 m and the temperature measured at the bottom of the well was 200°C. If we average the thermal conductivity of basalt between 25°C (the temperature at the ground surface) and 200°C (Table 2.3), which is reasonable given the nearly linear change in k_{th} with temperature over this temperature interval, and assume that our measurements are representative of each square meter of surface area at the site, the rate of heat flow at this site is

$$2.2\,\text{W/m-K} \times 1\,\text{m}^2 \times \left(\frac{473 - 298\,\text{K}}{2000\,\text{m}} \right) = 0.193\,\text{W}$$

Given that the global average heat flow is 87 mW/m², the value at our hypothetical site is suggestive of a significant heat source at depth that could warrant further investigation.

However, this approach is not suitable for calculating how much heat a geological body, such as a magma chamber, will transfer to its surroundings over time because the geometry of the magma system is not planar. If we assume the heat source can be conceptualized as a cylindrical body, then the form of the equation becomes

$$\left(\frac{dq_{th}}{dt} \right) = k_{th} \times 2\pi r \times l \times \left(\frac{dT}{dr} \right) \tag{2.5}$$

where:
r is the radius of the cylinder
l is the length

By integrating this equation, heat conduction as a function of radial distance from the body can be determined from

$$\frac{dq_{th}}{dt} = k_{th} \times 2\pi l \times \left[\frac{T_1 - T_2}{\ln(r_2/r_1)} \right] \tag{2.6}$$

where subscripts 1 and 2 refer to the inner and outer locations relative to the center of our heat source, respectively.

Careful examination of Table 2.3 reveals the magnitude of several of the important dependencies for k_{th}, all of which have implications for the performance of geothermal energy systems. Plotted in Figure 2.3 is the temperature dependence of k_{th} for quartz (a common mineral in rocks, sand, and soil) and several rocks (limestone, basalt, and granite). Also plotted in Figure 2.3 is the temperature dependence of k_{th} for water. Note that the effect of temperature is different for each material and is not linear. From room temperature to 200°C (473 K), k_{th} changes by 61% for quartz, but nearly 70% for basalt. Over the same temperature interval, k_{th} for water slightly increases then decreases.

The feldspar minerals, for example, alkali feldspar, and quartz are the most abundant minerals that make up the bulk of the rocks composing the continental crust. Note in Figure 2.3 the nearly threefold difference in k_{th} between these minerals. Obviously, the thermal conductivity of rocks will be strongly influenced by the relative proportion of these two minerals in any particular rock sample.

Thermal diffusivity (κ) is a measure of the rate at which heat transfer occurs. Thermal diffusivity has the units of m^2/s. It is defined as the ratio of the thermal conductivity, k_{th} (in W/m-K), to the heat capacity (by volume) of a material (C_v, in J/m^3-K):

$$\kappa = \left(\frac{k_{th}}{C_v} \right) \tag{2.7}$$

Heat capacity is the amount of heat required to raise the temperature of a unit volume of a material by 1 K. Note that the heat capacity at constant pressure (C_p; discussed also in Chapters 3 and 11) is related to the heat capacity at constant volume (C_v) by the relationship.

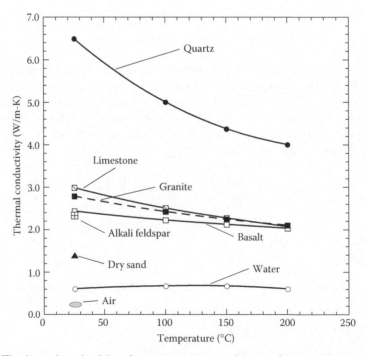

FIGURE 2.3 The thermal conductivity of some common materials, as a function of temperature.

$$C_p = C_v + \alpha^2 \left(\frac{VT}{\beta} \right) \tag{2.8}$$

where:

α is the coefficient of thermal expansion

β is the coefficient of compressibility

V is the molar volume

T is the absolute temperature (K)

Thermal diffusivity provides a quantitative measure for evaluating how quickly a material will change temperature in comparison with its volumetric heat capacity. Materials with a high thermal diffusivity will quickly change temperature. The thermal diffusivities for quartz, alkali feldspar, and most other minerals and rocks are in the range of approximately 1×10^{-6} to 10×10^{-6} m^2/s. For comparison, the thermal diffusivities of many common metals are in the range of 1×10^{-4} to 5×10^{-4} m^2/s. Thus, a metal will heat up or cool down ten to a hundred times faster than a mineral. This also suggests that the rock enclosing a magma body will behave as an insulating medium, transferring the heat away from the cooling molten rock at a relatively slow rate.

Complicating these relationships is the fact that rocks and soils are porous materials. The amount of porosity and its properties can vary significantly (see discussion in Chapter 4). In general, the greater the porosity, the lower the thermal conductivity. The extent to which thermal conductivity and diffusivity are diminished by porosity will depend upon what fills the pore space. Water and air are the most common pore-filling materials, and their respective thermal conductivities (Figure 2.3) are significantly different. Thus, for two rocks or soil samples in which all other things are equal, thermal conductivity will be higher in materials in which all of the pore space is water-filled (saturated), compared to materials in which some or all of the pore space is air-filled.

In Figure 2.4, the effect of saturation in quartz sand on the thermal conductivity is dramatically apparent. There are several points relevant for geothermal considerations that emerge from Figure 2.4. The first point is that the relationship between saturation and k_{th} is not linear. This results from the fact that the surface tension of water causes it to distribute itself primarily along contact points and junctions between sand grains, rather than evenly and uniformly within a pore space, in contrast to what a gas would do. As a result, the ability of thermal energy to be transmitted at grain contact points rapidly improves with the addition of a small amount of water. The rate at which thermal transmissivity improves quickly falls between 10% and 20% saturation.

A second point evident from Figure 2.4 is that the pore size also affects how saturation influences k_{th}. This effect is a consequence of the fact that the number of contact points between grains, per unit volume, depends directly on the size of the grains. Because there are more contact points per unit volume for a smaller grain size, k_{th} will increase more rapidly with increasing saturation in fine sand than coarse sand.

Both of these points make it clear that efficient use of geothermal energy must be based on thorough knowledge of the properties of the geological materials at the site that will be developed. Such knowledge should include thermal conductivity measurements that have been made in a laboratory on material that is as little disturbed from its natural state as possible. This is true for applications that are dependent on the thermal properties of near-surface materials, for example, geothermal heat pump installations (discussed in Chapter 11) as well as for applications involving power generation using geothermal energy from deep bedrock sources.

Although the thermal conductivity of minerals is not insignificant, minerals are poor conductors of heat compared to metals. Aluminum, for example, has a thermal conductivity of about 210 W/m-K at room temperature and iron has a thermal conductivity of about 73 W/m-K. Minerals conduct heat at rates that are, in general, 1–2 orders of magnitude less than those for common metals (Table 2.3). Although thermal conductivity has a strong influence on the local thermal properties of

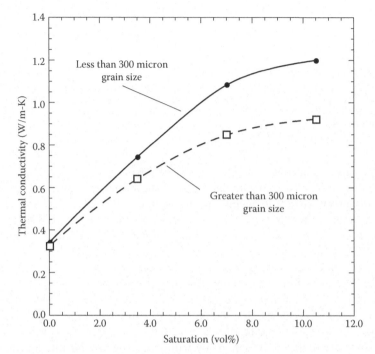

FIGURE 2.4 Thermal conductivity as a function of saturation for sands of different grain size. (From Manohar, K. et al., *West Indian Journal of Engineering*, 27, 18–26, 2005.)

a geothermal site, the amount of heat available at a site is a reflection of another heat transfer process in the earth that derives from the relatively low thermal conductivity of minerals. That heat transfer process is convection, which is the dominant mode of heat transfer in the earth.

CONVECTION

Conductive heat transfer occurs without movement of mass. However, a warm mass of any material flowing into a cooler region is also a means for accomplishing heat transfer. If, for some physical reason, the flow of the heated mass into the cooler region is not accompanied by any heat conduction, the process is called advection. Under most circumstances in the earth, however, the movement of the mass will occur as heat is simultaneously being conducted away. This combined process of heat being transferred by both mass movement and heat conduction is called convection.

In the presence of gravity, materials that have a lower density will tend to rise above materials of a higher density. In a planetary body composed of randomly distributed materials of differing density, an equilibrium state will eventually be achieved when all of the materials are ordered sequentially from the highest density material in the center of the body to the lowest density material at the outer edge of the body. This arrangement of densities is called *density stratification*. Density stratification will occur spontaneously over time provided some or all of the materials are capable of flow. The rate at which this will occur depends upon the viscosities and density differences of the materials involved. The early earth achieved this state of density stratification within a few tens of millions of years, as discussed in Section "Heat from Formation of the Core". However, this condition did not result in a static planet because it was not a stable configuration.

Viscosity is the resistance of a material to flow when stressed. Materials with a high viscosity are materials that, because of their molecular structure, possess high internal friction. Fluids, such as cold honey or molasses, have a high viscosity, compared to water or air. The resistance to flow is measured in units of pascal-seconds (Pa-s). One Pa-s is equivalent to $1 \text{ m kg s}^3/\text{m}^2$, which

TABLE 2.4

Dynamic Viscosities of Geological and Common Materials

Material	Temperature (°C)	Viscosity (Pa-s)
Water	20	0.001
Honey	20	10.0
Tar	20	30,000
Molten rhyolite[a]	~1400	~3.55×10^{11}
Upper mantle[b]	~1000	~1×10^{19}
Lower mantle[c]	~3500	~1×10^{21} to ~3×10^{22}

Sources: [a] Webb, S.L. and Dingwell, D.B., *Journal of Geophysical Research*, 95, 15695–15701, 1990

[b] Hirth, G. and Kohlstedt, D., *Geophysical Monograph*, 138, 83–105, 200

[c] Yamazaki, D. and Karato, S.-I., *American Mineralogist*, 86, 385–391, 2001.

is a measure of applied stress and the resulting deformation (or strain) it experiences. The range of viscosities materials possess is shown in Table 2.4. Note that many materials that seem incapable of flow at room temperature conditions (such as portions of the solid mantle of the earth) do, in fact, have viscosities at high temperatures and pressures that are sufficient to allow flow over geological timescales.

In the absence of an energy source, a density-stratified earth is a stable configuration for the distribution of materials that compose the earth. However, the hot, molten outer core is a very large energy source that continuously heats the base of the mantle. As previously noted, the minerals that compose the earth are poor thermal conductors. The mantle, therefore, is essentially a thermal insulator surrounding the core. As the base of the mantle heats, the minerals immediately adjacent to the core expand, thus becoming less dense. In addition, they also become less viscous. As this happens, portions of the thermally perturbed lower mantle begin to experience gravitational instability since they become relatively buoyant compared to the overlying, cooler and denser mantle. Eventually, the combined effect of decreasing density and viscosity overcomes the resistance to flow, and the heated lower mantle begins to rise buoyantly toward the surface. It is at this point that convection starts.

Qualitatively, the physical conditions that favor the onset of convection are low viscosity, significant thermal expansion, the presence of gravity (necessary for buoyancy effects), and low thermal conductivity (or large distance between heat source and heat sink). In addition, the temperature increase between heat source and sink must be greater than the temperature increase that would normally occur simply as a result of increasing pressure with depth (the "adiabatic gradient"; see Chapter 3).

The conditions that favor convective flow can be quantitatively represented by the ratio

$$\text{Ra} = \frac{(g \times \alpha \times \nabla T \times d^3)}{(\nu \times \kappa)} \tag{2.9}$$

This ratio evaluates the relative magnitudes of buoyancy and viscous forces. In this equation, g is the acceleration of gravity (9.8 m s^{-2}), ∇T is the vertical temperature gradient (K), α is the coefficient of thermal expansion (1/K), d is the depth interval (meters) over which the temperature difference occurs, ν is the kinematic viscosity (m^2 s^{-1}), κ is the thermal diffusivity (m^2 s^{-1}), and Ra is a dimensionless number, that is, it has no units. This can be demonstrated by substituting into the equation the units for each parameter and carrying out the appropriate reconciliation of units. Ra is called the Rayleigh number and provides an indication of the relative contribution of conduction and convection to heat transfer. For values of Ra greater than about 1000, convection is the dominant heat transfer mechanism, whereas lower values of Ra indicate that conduction is the dominant heat transfer mechanism.

For the mantle, estimated values of Ra range between 10^5 and 10^7, depending upon the model used for the mineralogy of the mantle, the temperature distribution with depth, and the scale of possible convective overturn (Anderson 1989). Clearly, heat transfer from the lower mantle to the earth's surface is strongly dominated by convective processes. It is this fundamental attribute of the earth that drives plate tectonics and accounts for the distribution of geothermal resources around the globe.

PLATE TECTONICS AND THE DISTRIBUTION OF GEOTHERMAL RESOURCES

In 1912, Alfred Wegener published his landmark paper on what became known as continental drift (Wegener 1912). That paper was followed in 1915 by his book *Die Entstehung der Kontinente und Ozeane* (*The Origin of Continents and Oceans*). Over the next 14 years, the book went through three revisions, culminating with his 1929 fourth edition, which provided the most complete discussion of his argument for movement of the continents (Wegener 1929). The hypothesis proved to be extremely controversial. It was not until the 1960s that the geological community overwhelmingly came to accept the view that the continents and ocean basins are, in fact, mobile.

One of the primary stumbling blocks to the acceptance of the concept of a mobile crust was the absence of a convincing mechanism for driving the movement. This issue was eventually resolved after the Second World War when oceanographic research vessels were outfitted with magnetometers that were originally intended to be used to detect submarines during the war. Evidence obtained from oceanic surveys documented the presence of a globe-encircling mountain chain that became known as the mid-ocean ridge system. As the magnetometer-equipped vessels cruised the world ocean, they discovered unexpected patterns of magnetic anomalies that paralleled the ocean ridge system and that extended for hundreds of miles on either side of them. It was quickly recognized that the anomaly pattern on one side of the ridge system was precisely mirrored by the anomaly pattern on the other side of the ridge. The only explanation for this symmetry was that ocean crust must be forming at the ridges and spreading away from it. For that to be the case, it was postulated that the mantle must be upwelling at the mid-ocean ridge system. That upwelling process was bringing hot, deep mantle rocks to the surface, which resulted in melting of the hot rock as it rose to shallower levels in the earth where pressures were lower. This process of hot, upwelling mantle was a classic example of convection. The places where the upwelling convection cells intersect the surface of the earth are called spreading centers because they define those places where crust forms and migrates away to either side of the ridge system.

To balance the upward flow of the hot, convecting mantle required that there exist a downward flow as well. Otherwise, the earth would be expanding, and conservation of mass arguments made it clear such could not be the case. It was quickly realized that most of the volcanoes on the planet were associated with deep ocean trenches and zones of very deep earthquakes, which were the likely locations for the downwelling portion of the convecting mantle system (Figure 2.5). These locations became known as subduction zones.

This spreading center–subduction zone couple defines the boundaries of the major tectonic plates of the earth. Each plate behaves as a rigid unit of crust that moves in response to forces from the underlying convection cells in the mantle, as well as in response to forces generated by interactions with adjoining plates. Figure 2.6 shows our present day understanding of plate boundaries and plate motions.

The combined convective transfer of heat and mass explains why our previously computed heat flow calculation from Equation 2.4 resulted in such a low value for the continental crust, namely, heat in addition to that coming directly from radioactive decay is being actively transferred from deep within the earth via convective processes.

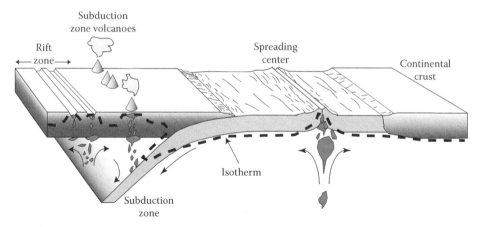

FIGURE 2.5 Schematic diagram showing the configuration of the main elements that compose plate tectonic structures. The arrows indicate local motion of convecting mantle material. The gray, irregular masses represent magma bodies as they ascend from the mantle into the crust. Note that the bulk of magma that occurs in the earth is found at spreading centers, subduction zone volcanoes, or rift zones.

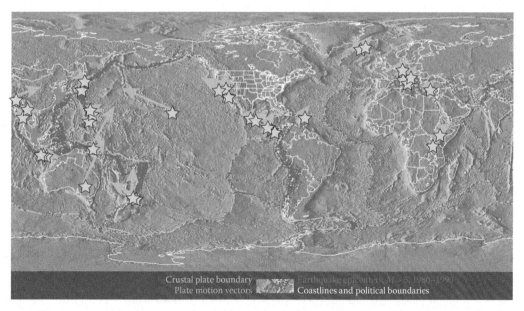

FIGURE 2.6 **(See color insert.)** Global map showing the locations of earthquakes (red dots) that indicate plate boundaries (yellow lines), political boundaries (in white), and the locations of the world's geothermal power plants (stars). The directions of some plate motions are shown by the green arrows, with the length of the arrow corresponding to relative velocity of plate motion. Note the strong correlation between power plant sites and plate boundaries. There are many more power plants than stars because many sites have several power plants. The global map, earthquake data, and boundaries are from the National Oceanic and Atmospheric Administration Plates and Topography Disc and the power plant sites from the International Geothermal Association website (http://iga.igg.cnr.it/geo/geoenergy.php).

As previously noted, the average global heat flow is 87 mW/m². However, recalling that convection is the transfer of heat by conduction and the movement of mass, spreading centers must be loci of very high heat flow. Indeed, computer models of heat flow at spreading centers suggest heat flow could be as high as 1000 mW/m² (Stein and Stein 1994). Heat flow measurements at spreading centers are generally lower (~300 mW/m²; see Figure 2.7) than predicted by computer models,

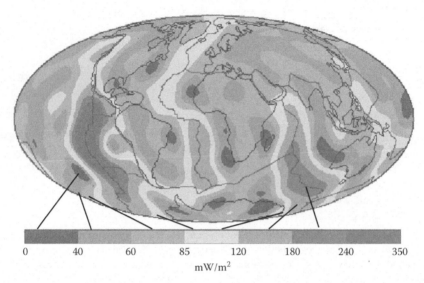

0   40   60   85   120   180   240   350

mW/m^2

FIGURE 2.7 **(See color insert.)** Low resolution global map showing the distribution of heat flow at the surface. Compare this figure with that in Figure 2.6 to see the relationship between plate boundaries, geothermal power plants, and heat flow. (From International Heat Flow Commission, http://www.geophysik.rwth-aachen.de/IHFC/heatflow.html.)

reflecting the effects of water circulation in the crust where spreading occurs, thus transferring large quantities of heat into the ocean water by advection and convection.

By analogy with classic models of convection cells, one would also assume, a priori, that subduction zones must be regions of very low heat flow, because they mark those places where the cooler, downwelling portion of a convection cell returns to the mantle. And yet, subduction zones are marked by the globes highest concentration of volcanoes, which bring tremendous volumes of molten rock (and heat) to the surface.

This apparent contradiction results from the fact that subduction zones transport water back into the mantle. Water is primarily contained in certain *hydrous* minerals that form during the alteration and metamorphism of the oceanic crust as it migrates away from spreading centers. These hydrated minerals are stable at relatively low temperatures, but recrystallize to new, less hydrated minerals at elevated temperatures. When the oceanic crust descends into the mantle at subduction zones, it heats up, eventually reaching temperatures at which the hydrated minerals begin to recrystallize to new mineral phases that do not accommodate water in their structure. As a result, the water molecules that are released form a separate fluid phase. This process of dehydration of the original mineral phase to form an anhydrous mineral and a coexisting water phase can be represented by the following reaction that schematically represents the dehydration of serpentine and brucite (common hydrated minerals in the oceanic crust) to form olivine:

$$Mg_3Si_2O_5(OH)_4 + Mg(OH)_2 \Leftrightarrow 2Mg_2SiO_4 + 3H_2O$$

Serpentine   Brucite   Olivine   Water

Dry rock, when sufficiently heated, begins to melt. Wet rock, when sufficiently heated, also melts, but does so at much lower temperatures than dry rock. The water released during subduction causes melting to occur in the hot mantle immediately above the descending oceanic crust. The resulting melt is less dense than the solid rock from which it formed and migrates upward. This process is, in essence, a secondary convecting system that brings molten rock and heat to the surface in the vicinity of subduction zones (Figures 2.5 and 2.8).

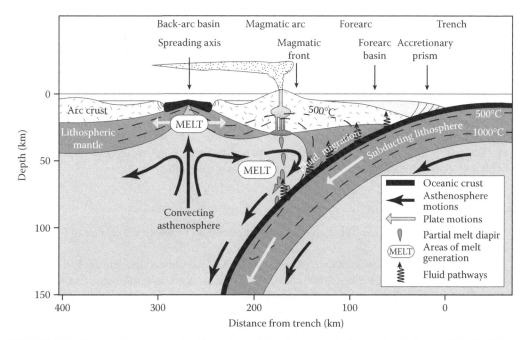

FIGURE 2.8 Schematic cross section through a subduction zone similar to that in Japan. (Modified from DuHamel, J., 2009. Wry heat—Arizona history Chapter 5: Jurassic time. http://tucsoncitizen.com/wryheat/tag/subduction/.)

Heat is also brought to the surface in the regions behind the volcanic front that forms at subduction zones. It is believed that the subduction process gives rise to small-scale convection cells above the descending slab (Figure 2.8; Hart et al. 1972). The upwelling part of these convection cells often causes the overlying crust to spread apart, forming rift zones and rift basins that can be in places where shallow-level magma chambers develop. The northern part of the North Island of New Zealand is such a place and also happens to be the site where the first large-scale geothermal power development was pioneered. The Basin and Range province of the western United States is also, at least in part, likely to reflect similar processes, but for which the extent of crustal separation is very small.

An additional tectonic setting that is commonly geothermally significant are locations that are known as hot spots. Hot spots are places that persist for tens of millions of years at which magma rises from the deep mantle to the earth's surface in a nearly continuous fashion. The geological reasons why hot spots form, and what maintains their persistent eruption histories, are a matter of considerable debate. Regardless of what powers them, they are prodigious sources of heat. Hawaii and Iceland are two prominent examples of hot spots, and both are well-known sites for geothermal energy.

Shown in Figure 2.6 are the locations of geothermal power plants around the world. Their correspondence with spreading centers, hot spots, and subduction zones is obvious. Shown in Figure 2.7 are the high heat flow regions of the earth. Their relationship to key plate tectonic elements is clear. This correspondence has been an influence in guiding exploration for geothermal resources suitable for electricity generation and continues to be important.

CLASSIFICATION OF GEOTHERMAL SYSTEMS BY THEIR GEOLOGICAL CONTEXT

The relationship between plate tectonics and heat at the earth's surface is fundamental. As described above, the locations of geothermal systems correlate with specific geological settings, when considered within a plate tectonics context. The principal structural elements of plate tectonics are spreading centers, subduction or convergence zones, and transform faults. Each of these elements, and how they interact, establishes a particular type of environment that has specific geological manifestations. These manifestations are mainly expressions of how the crust responds to stresses generated by the underlying driving forces that cause plate tectonics. In the subsequent paragraphs, the geological characteristics of these settings are considered in more detail, with relevant geothermal examples provided. By classifying the primary tectonic and geological characteristics that possess geothermal resources, the groundwork is established for more detailed classification schemes useful for exploration purposes, as outlined in Chapter 6. Numerous classification schemes for geothermal systems have been proposed over the years. The following discussion highlights key elements relating tectonic settings and geothermal systems, synthesizing material from White (1973), DiPippo (2008), Moeck (2014), and numerous references therein.

EXTENSIONAL ENVIRONMENTS

There are three distinct types of extensional environments, all of which are related to fundamental plate tectonics processes.

Spreading centers are one such extensional system. As diagrammed in Figure 2.5, they are sites where diverging limbs of uprising mantle convection systems move in opposite directions, transporting the crust laterally. The uprising mantle carries with it melts that are extruded at these spreading centers, forming hot, young new crust. The vast majority of the globe's spreading center extent is underwater in the world oceans (Figure 2.6). Extensive surveys of these systems by oceanographic research vessels have documented the pervasive presence of hydrothermal/geothermal systems at these sites. Recent estimates are that these systems have a heat flux of between 2 and 4×10^{12} W and a total hydrothermal flux of about 9×10^{12} W (Elderfield and Schultz 1996). Oceanic spreading centers are thus a huge, but currently untapped, potential source of geothermal energy. However, such systems are not restricted solely to the ocean basins. They extend on land in Africa (the East African Rift system) and the Imperial Valley of California. Both settings have extensive geothermal resources. In the case of the former, they are just beginning to be explored and developed, whereas in the case of the latter, ~1000 MW of geothermal power generation capacity has been installed.

Another extensional environment in which geothermal systems are likely to occur is the so-called back-arc basins (see Figures 2.6 and 2.8). These regions form behind island arcs in oceanic settings. The underlying motivating forces that control development of these features remain obscure, although they are likely related to mantle flow that occurs in response to subducting crust. As these basins form, spreading of the crust occurs, much in a manner similar to that of oceanic spreading centers, with the concomitant formation of melts and new, hot young crust. As with spreading centers, most back-arc basins are restricted to oceanic settings and are deeply submerged. However, in a few locations such systems extend on to major landmasses. The northern portion of the North Island of New Zealand is a location where back-arc spreading appears to extend onto land. It is within this area that one of the world's best-developed complexes of geothermal systems has been established.

A third example of extensional tectonics within which geothermal systems are present are intra-continental rift zones. Extension of the crust results from geological processes in which portions of the crust become thin and separate. This thinning reduces pressure on the underlying mantle, allowing it to rise. This decompression also allows the formation of melts to occur in the mantle, which rise through buoyancy effects. These melts invade the crust, providing extensive heat sources at shallow to deep levels. As with back-arc basins, the underlying global dynamics that cause such systems to form is not well known. The Great Basin of the western United States is an example of such a system. Extensive geothermal development is underway in this region, most of which is concentrated in Idaho and Nevada.

COMPRESSIONAL ENVIRONMENTS

These settings are regions where converging plates result in the subduction of one plate under another. As the subducting plate descends to depths in excess of 100 km, a variety of processes result in the formation of melts, as described in Section "Plate Tectonics and the Distribution of Geothermal Resources". The melts that are generated in this environment rise through the mantle, eventually erupting as volcanoes on the overriding plate. Because subduction generally lasts for tens of millions of years, these very hot volcanic systems are long lived, providing a persistent heat source that drives very extensive geothermal systems. Examples of such compressional settings that host geothermal systems are the volcanic chains in Indonesia, New Zealand, the Philippines, Japan, the Aleutians, the Pacific Northwest of the United States, Central America, and South America (the so-called Pacific Ring of Fire).

TRANSLATIONAL ENVIRONMENTS

An important structural element that defines a third type of plate tectonic boundary is a transform fault. These settings are places where tectonic plates move horizontally past each other. Perhaps the most famous of these is the San Andreas Fault in California. Other examples are the Anatolian Fault in northern Turkey, the Alpine Fault Zone in New Zealand, and the Dead Sea Fault that runs through Israel, Lebanon, Palestine, and Syria. Such zones are major ruptures in the crust that allow the circulation of fluids to great depths, often with the result that hot springs are present along them. In addition, evidence exists that such systems also allow the escape of mantle fluids (Kennedy et al. 1997).

In addition to the three basic plate tectonic settings in which geothermal resources are located, other settings can host geothermal systems as well. Some of these are, in fact, some of the largest geothermal resources on the planet.

HOT SPOTS

Among the most prodigious localized heat sources on the planet are hot spots. These are locations where magma persistently erupts in a localized region. The causes of these localized volcanic sources are hotly debated and may, in fact, be diverse. Whatever their underlying cause, they are important hosts for geothermal resources. Iceland and Hawaii are two classic examples of hot spots. Both of these have geothermal resources already being used for power production. Other hot spots include the Canary Islands, the Cape Verde Archipelago, the Galapagos Islands, the Cook Islands, and Yellowstone, Montana.

TRANSITIONAL SETTINGS

Plate boundaries can often be sites of complex interactions between the crust and mantle. This particularly holds true when a site is evolving from one type of boundary to another or is at the junction between two types of environments. One type of transitional setting that is an example of what happens as a plate boundary evolves is in the vicinity of The Geysers in California. The Geysers provides one of California's most important geothermal resources, as we will discuss in Chapter 10. The Geysers is located near the triple junction between the San Andreas transform fault, the Humboldt fracture zone (another transform fault), and the subducting Gorda plate (part of the greater Pacific plate system). The plate motions in this area, and the interactions with the underlying mantle, are complex and evolving. They appear to have resulted in the formation of a "window" that allows hot mantle to interact with the overlying crust, resulting in the generation of magmas that rise to relatively shallow levels. The result is a hybrid system that generates very hot, dry steam. It is this dry steam resource that provides energy for generating power in this region.

The Long Valley Caldera in eastern California and the Surprise Valley system in northeastern California are also examples of a transitional setting. Rather than being a location where plate boundaries are evolving because of complex motions, these sites are at the transitional boundary between the compressional environment of the Sierra and Cascade volcanic systems formed by past and recent subduction of the Pacific plate and the extensional setting of the Great Basin. As described in Chapter 4, the Long Valley Caldera is a relatively recent collapsed volcano, whereas

Surprise Valley is the location of recent volcanic activity, as well as the site of a series of faults that allow circulation of waters to deep levels. Both of these settings have proven geothermal potential, with the Casa Diablo power plant at Long Valley Caldera already providing nearly 50 MW of power.

A final type of system that lacks any direct link to plate tectonics settings is that which can form when rocks rich in radioactive elements occur within a geological setting that insulates them for long periods of time. One such system is that which is currently being developed as an *enhanced geothermal systems* (EGS) site in southeast Australia. This particular location has a granitic body with high abundances of potassium and other radioactive elements. Its long history of burial and associated radioactive heat production has made a good target for EGS efforts, as described in Chapter 13.

AVAILABILITY AND UTILIZATION OF GEOTHERMAL ENERGY

Geological controls on the near-surface (meaning less than 3 km depth to the resource) availability of geothermal energy has resulted in its concentration in specific regions. As Figure 2.6 makes clear, generating power from geothermal energy has consistently been pursued in regions where volcanic activity is present. As a result, power production has been restricted to 25 countries, with an installed capacity of over 11,000 MW (see Table 10.1). However, global resources far exceed that value. It is estimated by the International Energy Agency (IEA 2005) that global capacity exceeds 150 EJ/year, which is equivalent to more than 4.5 million MW. Global power production from all sources was estimated at 14.7 million MW. Clearly, by this estimate, geothermal power production could contribute more than a quarter of all power production in the world.

Figures 2.9 and 2.10 show in more detail the relationship between heat flow and power generation for the continental United States and Europe, respectively. In the case of the United States, the western half of the continent is a region that is rich with potential geothermal resource sites, with heat flow values locally exceeding 150 mW/m^2, reflecting the rift tectonics occurring in this region. However, it is also clear that even within this region the variability in heat flow is high. Although the eastern half of the continent generally has low heat flow (reflecting its geologically inactive character) in the range of 25–60 mW/m^2, there are regions where heat flow values exceeding 70 mW/m^2 occur. Some of these regions, particularly in the Gulf Coast area where oil and gas production have occurred, are being considered for power production using binary power generation technologies that can exploit moderate temperature resources. We will discuss these in more detail in Chapter 16 on the future of geothermal energy. Similarly, new approaches that rely on deeper drilling and reservoir engineering, collectively described as EGS (also discussed in Chapter 13), have the potential to access significant geothermal resources that occur at depths of 6–10 km even in the stable continental regions with low or moderate heat flow.

The map of heat flow in Europe (Figure 2.10) presents the same range of heat flow values as seen within the United States. However, Europe's geological history over the last 50 million years is complex, making it difficult to interpret heat flow in terms of plate tectonics processes. The standout exception to this is, of course, Iceland, which is a hot spot that sits directly atop the mid-Atlantic spreading system. Noted in the figure are some of the current sites where geothermal power is being generated or actively pursued. Note that they occur in close proximity to areas with high surface heat flow.

It is important to point out that geothermal resources are more widespread than that which is useful for power production. As we will discuss in Chapters 11 and 12, so-called direct-use applications, in which warm geothermal waters are used for heating and a variety of other applications, and the deployment of geothermal heat pumps can also be important ways in which to utilize geothermal energy. In that same analysis undertaken by the IEA (2005) as discussed in this section, it was estimated that more than 78 countries employed geothermal energy for such applications. The thermal energy utilized was equivalent to more than 50,580 MWt (Chapter 12; Lund et al. 2010).

Heat flow map of North America 2004

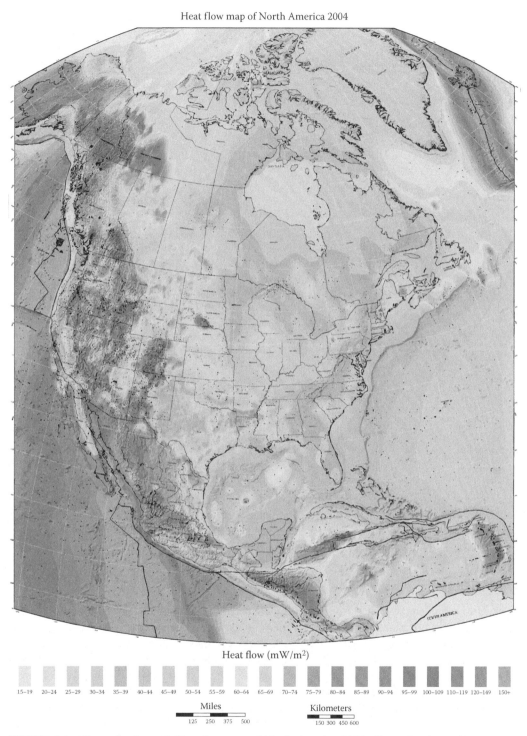

Heat flow (mW/m^2)

FIGURE 2.9 **(See color insert.)** Heat flow map of North America 2004. (From Geothermal Laboratory, Southern Methodist University. http://smu.edu/geothermal/2004NAMap/Geothermal_MapNA_7x10in.gif.)

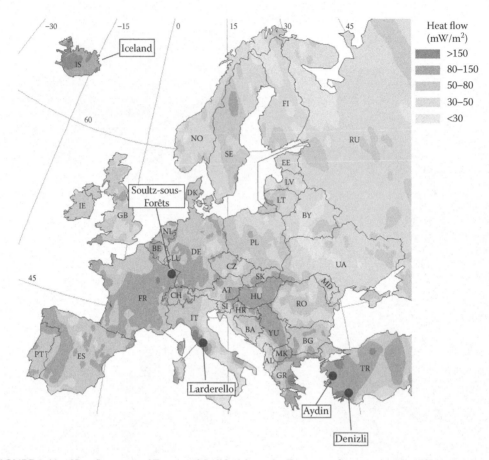

FIGURE 2.10   Heat flow map of Europe. (Modified from the European Community Nr. 17811.)

Using geothermal energy for such applications displaces electrical demand and improves energy efficiency. Although not always included in discussions of geothermal energy, such applications can have a very significant impact on global energy use and the sources employed for energy generation.

SYNOPSIS

The earth is a dynamic planet driven by heat. This heat, which can be accessed for power generation and other applications, comes from several sources. Formation of the planet involved a complex series of events including impacts from objects in the early solar nebula, heat by radioactive decay of short-lived isotopes, and the resultant melting of the planet. Density stratification then occurred, with the release of gravitational potential energy as liquid iron settled to the interior to form the core of the earth. Additional heat is generated by the decay of longer lived isotopes, especially U, Th, and K in the continental crust. The remnant heat from the early melting and slow cooling of the earth combined with the heat given off by radioactive elements has resulted in a heat budget that provides an average rate of heat escaping from the surface of about 87 mW/m^2. Transfer of heat in the earth is dominated by conduction on local scales but by convection on global scales. The convective processes, in particular, determine the distribution of heat sources in the near surface through the processes that manifest as plate tectonics. The largest heat sources are at spreading centers and regions associated with subduction zone volcanoes. These various processes have lead to a heterogeneous distribution of heat, with heat flow values on the continents ranging from less than 30 mW/m^2 to much more than 150 mW/m^2. Although power generation tends to occur in areas with heat flow values greater than about 80 mW/m^2, virtually any region, regardless of heat flow, can support the use of ground source heat pumps for heating and cooling of buildings. In addition, so-called direct-use applications in which warm waters are used for aquaculture, agriculture, and other purposes can be developed anywhere in the vicinity of springs or warm subsurface waters that can be found in a broad range of settings.

CASE STUDIES

Examples of power generation from several of the key plate tectonic settings are explored in more detail below. Many of these sites are also described in more detail in Chapters 10 and 13. An important consideration is that plate boundaries are often the primary locations for developing geothermal power plants. However, new developments and technological capabilities are making it possible to exploit geothermal power in regions far from plate boundaries. This new approach reflects two important developments. One is the ability to generate power from lower temperature resources than previously utilized, employing organic Rankine cycle technology, as described in Chapter 10. The other development is a new concept for geothermal power generation called EGS. EGS methodology is discussed in detail in Chapter 13. Applications of geothermal heat in nonpower settings is discussed in detail in Chapters 11 and 12.

EXTENSIONAL ENVIRONMENTS—SPREADING CENTERS

Salton Trough, California—The Gulf of California has its northern termination at the junction of Baja California and mainland Mexico. Extending northward from this location is a broad, fault-bounded valley that extends through that part of Mexico into Southern California. The Gulf of Mexico is a new ocean basin, formed by an extension of the main spreading center in the Pacific Ocean, the East Pacific Rise. Spreading along that extension caused a sliver of Mexico (i.e., Baja California) to split away from the mainland, beginning about 20 million years ago. The primary motion direction of that sliver is to the northwest and is accommodated by slip along the San Andreas fault. In Southern California and Northern Mexico, this spreading has resulted in the formation of a major depression in the crust, which is the Salton Trough. The San Andreas fault is a part of this dynamic system.

Separation of the crust and the formation of the trough has resulted in the development of an extensional environment that can accommodate the intrusion of hot magma that is associated with the spreading system. This magma has risen to shallow levels in the Salton Trough, causing the development of extensive geothermal reservoirs. Important power generation facilities exist in northern Mexico at Cerro Prieto, where a complex of power plants has a combined installed generating capacity of 720 MW. This single location accounts for nearly 75% of Mexico's total installed capacity of 953 MW.

In southern California, the Salton Trough is the location of more than 530 MW of installed generating capacity. Although the size of the geothermal potential in the region is a matter of considerable debate, the minimum capacity for the region is estimated to be an additional 2000 MW. This region contributes approximately a third of the total geothermal power production in California, which ranks as the world's largest producer of geothermally generated electricity.

Another example of an extensional geothermal complex is the East African Rift Valley. This is the globes' most extensive spreading center complex within a continental framework. The main branch of the rift runs from the Afar region, through Ethiopia, Kenya, Tanzania, Malawi, Mozambique, Uganda, Djibouti, Eritrea, Rwanda, Zambia, and Botswana. Although detailed resource assessments have only been carried out in a few locations, resulting in a few hundred megawatts of power generation being put in place, the consensus is that the potential for power generation in this region approaches 10,000 MWe or more (Gizaw 2008; Omenda 2008).

COMPRESSIONAL ENVIRONMENTS—SUBDUCTION ZONES

The Philippines—The Philippine Islands are a complex of over 500 volcanoes formed above the westward subducting Pacific plate. Of those 500, approximately 130 are active. Beginning in 1977, geothermal power production was initiated on the island of Leyte. As of June, 2008, that initial effort had grown to more than 1900 MW of electric power production distributed among the islands of Luzon, Negros, Mindanao, and Leyte. Geothermal power production now accounts for 18% of the country's electrical needs. The national goal is to achieve 3131 MWe of geothermal power by 2013. Worldwide, the Philippines rank second to the United States in producing geothermic energy.

HOT SPOTS

Hawaii—The Hawaiian Island chain tracks the movement of the Pacific plate over a mantle hot spot that has persisted for more than 80 million years. The island of Hawaii is the currently active volcano in the chain. On the eastern flank of the volcano is the East Rift Zone, a zone along which lava has repeatedly erupted. Geothermal exploration was started in this area during the 1960s, eventually leading to development of a small (3 MWe) power generating facility in 1981. Power production continued until 1989 when the plant was shut down. A newer facility was built in 1993, which has continuously produced 25–30 MWe of baseload power. The geothermal resource in the region is capable of providing about 200 MWe of power on a continuous basis.

Iceland—It is one of the world's largest islands. It sits directly on the mid-Atlantic ridge and is interpreted to represent the superposition of a hot spot on a spreading center. At that location on the mid-Atlantic ridge, the spreading rate is about 2 cm per year. Iceland has the greatest per capita use of geothermal energy in the world. In 2006, 26% of its electricity (322 MWe) was produced from geothermal electric generators. In addition, hot water from power plants as well as directly from geothermal resources accounts for 87% of the hot water usage in the country. This usage includes hot water for space heating using district heating systems, snow melting, and domestic hot water use.

PROBLEMS

2.1. What is the global average heat flow at the surface of the earth? What is its range?

2.2. What quantities do you need to calculate heat flow? What physical processes influence heat flow?

2.3. Assume a well is drilled in dry sand to a depth of 2800 m and the temperature at the bottom of the well is measured to be 200°C. Assume, too, that the thermal conductivity of dry sand is constant between 10°C and 250°C. Is there likely to be a geothermal resource in the area?

2.4. For a material that has a Rayleigh number of 10, what mode of heat transfer will dominate? For a value of 100? For a value of 10,000? Will this behavior have any influence on heat flow measured at the surface of the earth? If so, what?

2.5. What geological regions are most likely to have high heat flows?

2.6. What are the geological hazards that might influence whether a geothermal resource is developed? Discuss this in terms of the geological environments that host geothermal reservoirs.

2.7. Where on earth does the highest heat flow occur? Why is it unlikely it will be developed in the near future for power generation?

2.8. Which would be better as a geothermal resource—dry sand or fully saturated sand?

2.9. What regions in the United States are best suited for geothermal power facilities? Why? Which regions are the least well suited?

2.10. Considering Figure 2.9, what geological environment makes the Lesser Antilles islands in the Caribbean a promising place for geothermal energy development? Consider the same question for Nicaragua.

REFERENCES

Alfé, D., Gillian, M.J., and Price, G.D., 2007. Temperature and composition of the Earth's core. *Contemporary Physics*, 48, 63–80.

Allégre, C.J., Manhés, G., and Göpel, C., 1995. The age of the Earth. *Geochimica et Cosmochimica Acta*, 59, 1445–1456.

Anderson, D.L., 1989. *Theory of the Earth*. Boston, MA: Blackwell Scientific Publishing. p. 255.

Beardsmore, G.R. and Cull, J.P., 2001. *Crustal Heat Flow: A Guide to Measurement and Modeling*. Cambridge: Cambridge University Press. p. 324.

Burroughs, J., 1876. *Winter Sunshine*. New York: Hurd & Houghton. p. 221.

Canup, R.M. and Agnor, C., 2001. Accretion of the terrestrial planets and the Earth-Moon system. In *Origin of Earth and Moon*, eds. R.M. Canup and K. Righter. Cambridge: Cambridge University Press.

Chambers, J.E., 2001. Making more terrestrial planets. *Icarus*, 152, 205–224.

Clauser, C. and Huenges, E., 1995. Thermal conductivity of rocks and minerals. In *Rock Physics and Phase Relations*, ed. T.J. Ahrens. Washington, DC: American Geophysical Union, pp. 105–126.

DiPippo, R., 2008. *Geothermal Power Plants*, 2nd edn. Elsevier, Oxford, 493 pp.

DuHamel, J., 2009. Wry heat—Arizona history Chapter 5: Jurassic time. http://tucsoncitizen.com/wryheat/tag/subduction/.

Elderfield, H. and Schultz, A., 1996. Mid-ocean ridge hydrothermal fluxes and the chemical composition of the ocean. *Annual Review of Earth and Planetary Sciences*, 24, 191–224.

Gizaw, B., 2008. Geothermal exploration and development in Ethiopia. United Nations University, Geothermal Training Programme, Reykjavík, Iceland, 12 pp.

Göpel, C., Manhés, G., and Allégre, C.J., 1994. U-Pb systematics of phosphates from equilibrated ordinary chondrites. *Earth and Planetary Science Letters*, 121, 153–171.

Hart, S.R., Glassley, W.E., and Karig, D.E., 1972. Basalts and sea-floor spreading behind the Mariana island arc. *Earth and Planetary Science Letters*, 15, 12–18.

Hirth, G. and Kohlstedt, D., 2003. Rheology of the upper mantle and the mantle wedge: A view from the experimentalists. *Geophysical Monograph*, 138, 83–105.

International Energy Agency, 2005. IEA Geothermal Energy Annual Report 2005. p. 169. http://iea-gia.org/wp-content/uploads/2012/08/GIA-2005-Annual-Report-Draft-Wairakei-4Dec2006-Gina-5Dec06.pdf.

Kennedy, B.M., Kharaka, Y.K., Evans, W.C., Ellwood, A., DePaolo, D.J., Thordsen, J., Ambats, G., and Mariner, R.H., 1997. Mantle fluids in the San Andreas fault system, California. *Science*, 278: 1278–1281.

Kleine, T., Münker, C., Mezger, K., and Palme, H., 2002. Rapid accretion and early core formation on asteroids and the terrestrial planets from Hf-W chronometry. *Nature*, 418, 952–955.

Kortenkamp, S.J., Wetherill, G.W., and Inaba, S., 2001. Runaway growth of planetary embryos facilitated by massive bodies in a protoplanetary disk. *Science*, 293, 1127–1129.

Lund, J.W., Freeston, D.H., and Boyd, T.L., 2010. Direct utilization of geothermal energy: 2010 worldwide review. *Proceedings of the World Geothermal Congress*, Bali, Indonesia, pp. 1–23.

Manohar, K., Ramroop, K., and Kochhar, G.S., 2005. Thermal Conductivity of Trinidad "Guanapo Sharp Sand." *West Indian Journal of Engineering*, 27, 18–26.

Moeck, I., 2014. Catalogue of geothermal play types based on geologic controls. *Renewable and Sustainable Energy Reviews*, 37, 867–882.

Omenda, P.A., 2008. The geothermal activity of the East African rift. United Nations University, Geothermal Training Programme, Reykjavík, Iceland, 12 pp.

Sass, J.H., 1965. The thermal conductivity of fifteen feldspar specimens. *Journal of Geophysical Research*, 70, 4064–4065.

Stein, C.A., 1995. Heat flow in the Earth. In *Global Earth Physics*, ed. T.J. Ahrens. Washington, DC: American Geophysical Union, pp. 144–158.

Stein, C.A. and Stein, S., 1994. Constraints on hydrothermal heat flux through the oceanic lithosphere from global heat flow. *Journal of Geophysical Research*, 99, 3081–3095.

Thompson, A. and Taylor, B.N., 2008. *Guide for the Use of the International System of Units (SI)*. Washington, DC: National Institute of Standards and Technology (NIST) Special Publication, 78 pp.

US Energy Information Agency, 2008. International Energy Annual, 2006. Table E1 World Primary Energy Consumption, http://www.eia.gov/totalenergy/data/annual/archive/038406.pdf.

Van Schmus, W.R., 1995. Natural radioactivity of the crust and mantle. In *Global Earth Physics*, ed. T.J. Ahrens. Washington, DC: American Geophysical Union, pp. 283–291.

Weast, R.C., 1985. *CRC Handbook of Chemistry and Physics*. Boca Raton, FL: CRC Press, p. E-10.

Webb, S.L. and Dingwell, D.B., 1990. Non-Newtonian rheology of igneous melts at high stresses and strain-rates: experimental results for rhyolite, andesite, basalt, and nephelinite. *Journal of Geophysical Research*, 95, 15695–15701.

Wegener, A., 1912. Die Entstehung der Kontinente. *Geologische Rundschau*, 3, 276–292.

Wegener, A., 1929. *Die Entstehung der Kontinente und Ozeane*. 4th Auflage, Vieweg & Sohn Akt.-Ges., Braunschweig, Germany, 1–231.

Wetherill, G.W., 1990. Formation of the Earth. *Annual Review of Earth and Planetary Sciences*, 18, 205–256.

White, D.E., 1973. Characteristics of geothermal resources. In *Geothermal Energy: Resources, Production Stimulation,* eds. P. Kruger and C. Otte. Chapter 4. Stanford University Press, Stanford, CA.

Yamazaki, D. and Karato, S.-I., 2001. Some mineral physics constraints on the rheology and geothermal structure of Earth's lower mantle. *American Mineralogist*, 86, 385–391.

Yin, Q., Jacobsen, S.B., Yamashita, K., Blichert-Toft, J., Te'louk, P., and Albarede, F., 2002. A short timescale for terrestrial planet formation from Hf–W chronometry of meteorites. *Nature*, 418, 949–952.

FURTHER INFORMATION

Rudnick, R.L. (ed.), 2005. *The Crust*. New York: Elsevier, 683 pp.

This reference book is for the advanced student with a strong geological background. It contains numerous articles contributed by recognized scientists that provide an up-to-date scientific description of many of the processes described in this and other chapters.

United States Geological Survey (http://www.usgs.gov/).

The USGS website has numerous excellent pages describing the basic elements of plate tectonics and how they relate to volcanism and seismicity.

University of California Museum of Paleontology (http://www.ucmp.berkeley.edu/geology/tectonics.html).

The website of the museum has an excellent series of animations relating to motion of the tectonic plates.

SIDEBAR 2.1   Power, Energy, and Units

Throughout this book, the conventions of the International System of Units, or SI units, will be used. The SI system is employed by the scientific community worldwide to assure consistency and avoid confusion when measurements of any type are being done. The beginnings of a standardized decimal metric system began with the adoption by the French Assembly of a standard meter and kilogram in the late eighteenth century. Since that time, the international scientific community has developed an internally consistent and well-defined set of units and constants for key fundamental physical properties. The latest internationally accepted version of these standards can be accessed through the National Institute of Standards and Technology (NIST) website, http://physics.nist.gov/cuu/Units/ (see also Thompson and Naylor 2008). The discussion based on the information available through NIST is provided in the subsequent paragraphs.

The SI system is based on seven base units that are independent and from which all other units are derived. The base units and the quantities they relate to are as follows:

- Meter (m)—length
- Kilogram (kg)—mass
- Second (s)—time
- Ampere (a)—electric current
- Kelvin (K)—temperature
- Mole (mol)—amount of substance
- Candela (cd)—luminous intensity

For our considerations of geothermal energy, some of the key derived units we must use relate to heat energy, power, and work. Of these units, the joule, which is a unit of energy or work, is fundamentally important. A joule (J) is defined as the work done (or energy expended) when 1 newton (N) of force is applied to an object that is displaced 1 m. The newton is also a derived unit, defined as the force required to give a 1 kg mass an acceleration of 1 m/s^2. Thus, a joule can be defined as

$$J \equiv N \times m = (m \times kg/s^2) \times m = m^2 \times kg/s^2$$

Power is the rate at which energy is used to accomplish work. In SI units, the rate at which energy is used is measured in watts (W), which are defined as

$$W \equiv J/s$$

Throughout this book, power generation or consumption will be described as the amount of time over which energy is used at a specified rate. For example, if 50 J/s are consumed by a laptop computer and it is used for 1 hour, its power consumption would be expressed as 50 Wh. Similarly, a geothermal power generator that produced 5 million joules per second (a modest generator) for a 24-hour period would provide 120 MWh of power.

Other units are often used when discussing energy and power. Some useful equivalencies are given below:

- 1 joule (J) = 0.2388 calories (cal) = 0.0009478 British thermal units (Btu)
- 1 kWh = 3.6 × 10^6 J = 8.6 × 10^5 cal = 3412 Btu

The following table provides equivalencies for a variety of units used throughout this book.

SI Unit	Other Units
1 J	= 1 N-m
	= 0.2388 calories
	= 0.0009478 Btu
1 m	= 3.281 feet
1 m^3	= 35.714 cubic feet
1 megapascal (MPa)	= 9.869 atm
	= 10 bar
	= 145.04 lb/sq.in
1 kg	= 2.205 lb
	= 9.8066 N
1 K	= −272.15°C

Throughout this book, energy equivalences between different energy sources are considered. Understanding such equivalencies allows quantitative comparison of the *energy intensity* of a given source. The table below provides some of these energy equivalencies. Note that these values are averages, since each of these sources varies in quality.

Energy Source	Energy Content
One barrel of crude oil	= 5.8 × 10^6 Btu
(0.159 m^3 or 42 gallons)	= 6.1178 × 10^9 J
	= 1.7 MWh
	= 164.24 m^3 natural gas
	= 5800 cubic feet natural gas
1 m^3 of natural gas	= 37.7 × 10^6 J
	= 35,714 Btu
	= 10.8 kWh
1 kg of coal	= 24 × 10^6 J
	= 22.75 × 10^3 Btu
	= 6.7 kWh

3 Thermodynamics and Geothermal Systems

Effective use of geothermal energy requires the ability to move and convert heat efficiently. In some instances, heat is used to do work, as in the generation of electricity. In other cases, heat is either concentrated or dissipated. Regardless of the application, an understanding of the behavior of fluids and materials when heated or cooled, and the implications for energy balances, is the foundation for achieving an economically successful outcome for any geothermal application. This chapter provides an introduction to the elements of thermodynamics that are important for such considerations.

FIRST LAW OF THERMODYNAMICS: EQUIVALENCE OF HEAT AND WORK AND THE CONSERVATION OF ENERGY

CONSERVATION OF ENERGY

By the second half of the 1700s, the engineering community had become intrigued by the repeatedly observed fact that doing work on some materials generated heat. It was noted by Benjamin Thompson in 1798 that, in the process of making cannons, boring into metal resulted in the metal becoming very hot. A series of experiments he conducted (followed shortly by experiments performed by Humphrey Davy in 1799) demonstrated that mechanical work and heat are directly related. They and others eventually demonstrated that a given amount of mechanical work would result in the generation of a predictable amount of heat.

However, it was not until Julius Mayer published a seminal paper in 1842 that the concept of conservation of energy, as embodied in the equivalence of heat and mechanical work, was articulated. Although Mayer's paper was the first to present the concept directly, it lacked sufficient experimental grounding and mathematical rigor to be considered an adequate demonstration of the equivalence of work and heat. In 1847, Hermann von Helmholtz developed a mathematical basis for the concept. Then, in 1849, the thorough experimental and observational work of James Prescott Joule was presented to the Royal Society, in a paper entitled "On the Mechanical Equivalent of Heat." These achievements established the concept that mechanical work and heat are equivalent and that, invariably, energy is conserved. This principle became the first law of thermodynamics.

Simple statements of the first law of thermodynamics have been numerous, two of which follow:

Energy can neither be created nor be destroyed.
All forms of energy are equivalent.

INTERNAL ENERGY

The most rigorous description of the energy in a system relies on the concept of internal energy (E). E is a characteristic of a specific, defined system. A system can be a cylinder of gas, a bottle of water, a bar of steel, a rock—anything that can be physically described by parameters of state (such as temperature T, pressure P, volume V). If a system is completely isolated from its surroundings (i.e., it is a closed system, meaning no mass can move into or out of it), then at any given set of conditions (T, P), the internal energy (E) of the defined system is fixed and depends only on the properties of the materials of which the system is composed. The internal energy (E) will change solely in response to changes in the state parameters P and T. If the conditions to which the system is subject are changed, either by moving heat (q) into it or by doing work (w) on it, the internal energy E must also change. The mathematical statement of this process is

$$\Delta E = E_f - E_i$$

which states that the change in internal energy (ΔE) is equal to the internal energy of the system in its final state (E_f) minus the initial internal energy (E_i).

This simple equation is profoundly important. It establishes the significance of knowing the internal energy at the end points of a process that affects a system. It is the difference between these end points that determines how much energy is required to heat a space, generate power, or cool a room. It also emphasizes that the pathway that was followed to get from the initial to the final states has absolutely no significance for the change in internal energy. For example, imagine a volume of gas contained in a cylinder and that the cylinder is a perfect insulator—that is, it will not allow heat to be added to or removed from the gas it contains. Imagine, too, that one end of the cylinder is movable (Figure 3.1). There are an infinite number of ways to get from some specified initial state (the cylinder on the left of the figure) to the final state (the cylinder on the right of the figure), two of which are depicted in the figure. In the sequence A1 → A2, the gas undergoes simple compression in a single step. In this case, the pressure (P_i) and temperature (T_i) of the gas in the initial state are increased to P_f and T_f, respectively, and the internal energy change (ΔE) is equal to $E_f - E_i$. In the sequence B1 to B2 to B3 to B4 to B5, the gas undergoes a series of changes in pressure and temperature before reaching the same P_f and T_f as in the A1 to A2 sequence. Each step results in a change in the internal energy:

$$\Delta E = E_{B2} + E_{B3} + E_{B4} + E_{B5} - E_{B1}$$

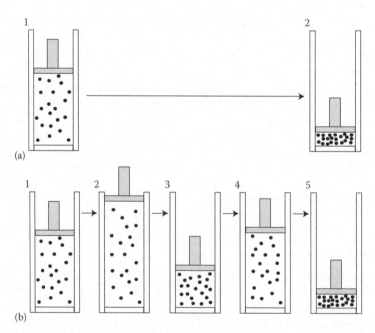

FIGURE 3.1 Diagrammatic representation of a cylinder of gas that follows two pathways (a and b) from one set of pressure and temperature conditions to another. Path (a) represents a single-stage compression, whereas path (b) involves two expansion and two compression stages. For both pathways, the change in the internal energy (E) is the same. The black dots in the cylinder schematically represent the gas molecules. In reality, there are an infinite number of pathways that could be followed that would achieve the same result.

If we were to measure the internal energy at the end of each step in B, we would find that

$$E_{B2} + E_{B3} + E_{B4} + E_{B5} = E_f$$

which results in

$$\Delta E = E_2 + E_3 + E_4 + E_5 - E_1 = E_f - E_i$$

Thus, regardless of the complexity of the path that is followed in getting from one set of conditions to another, or to put it another way, regardless of how much work one does to a system in moving it from one state to another, the internal energy will always be the difference between the initial and final states.

PRESSURE–VOLUME WORK

The conclusion that is inescapable is that any change in the internal energy of a system is solely the result of the work (w) done to the system, or that the system performs, and any heat (q) added to or taken from the system:

$$\Delta E = q + w \tag{3.1}$$

Mechanical work is performed when a force applied to a point, surface, or volume results in displacement of that point, surface, or volume. As an example, consider the changes in the cylinder in Figure 3.1. As the piston moves from its position in B1 to its position in B2, if there is an external force applied to the piston, mechanical work is performed by the gas as its volume increases and forces the piston to move. Hence, the element of mechanical work is defined by

$$dw = -P \times dV \tag{3.2}$$

By convention, mechanical work done *on* a system is positive, whereas work done *by* a system is negative. Work performed is, therefore, equivalent to the difference in volume between the two states:

$$w = -P \times (V_2 - V_1) = -P \times \Delta V \tag{3.3}$$

ENTHALPY

It follows from this discussion that in a process in which no change in volume occurs, no mechanical work is performed and any change in internal energy is solely related to heat added to or removed from the system:

$$\Delta E = q_v \tag{3.4}$$

The subscript v is used to indicate heat at constant volume; similarly, a subscript p would refer to a constant pressure condition. If, however, a change in volume occurs at constant pressure and heat is also added to or removed from the system, the change in internal energy is

$$\Delta E = q_p - (P \times \Delta V) \tag{3.5}$$

The heat added to or removed from the system at constant pressure is called the enthalpy (H) and the change in enthalpy (ΔH) that is realized when moving from one state to another is defined as

$$\Delta H = H_2 - H_1 = [E_2 + (P \times V_2)] - [E_1 + (P \times V_1)] = q_p \qquad (3.6)$$

Enthalpy is an important system property in geothermal power applications because it provides a means for establishing the behavior of a system in the subsurface and allows for evaluation of the useful energy that can be extracted from a working fluid. It has units of J/kg. We will consider this topic in more detail in Chapter 10.

SECOND LAW OF THERMODYNAMICS: INEVITABLE INCREASE OF ENTROPY

EFFICIENCY

At the same time as the concept of conservation of energy was being formulated, advances in steam engine technology were flourishing. By 1769, James Watt, building on early work of Thomas Savery, Thomas Newcomen, and others, had developed the steam engine to the point that it began to be the dominant driver of the Industrial Revolution. As the steam engine was adopted and modified, interest grew in the factors that determined the efficiency of steam engines.

The issue, at its core, was how much work could be done for a given amount of heat. The ideal, of course, would be a situation in which all of the energy contained in a given amount of heat would be converted to work with 100% efficiency. Mathematically, the efficiency of any situation involving heat and work can be expressed as

$$e = -w/q$$

In the above expression, w is the amount of work output (to reiterate, by convention we assign work done by a system a negative sign, and work done to a system a positive sign) for a given amount of heat input (q). In this expression, e is the efficiency. For the ideal case, $e = 1.0$ and $q = -w$.

CARNOT CYCLE

In 1824, Nicolas Léonard Sadi Carnot, a young French engineer, provided the definitive conceptualization that allowed efficiency to be rigorously determined. Carnot's conceptualization of the problem was further developed by Èmile Clapeyron and Rudolf Clausius, such that by the 1850s the concept was available in the form we generally use today.

To understand the fundamentals behind the Carnot cycle, the concept of equilibrium must be appreciated. From a thermodynamic perspective, a system has achieved an equilibrium state if it does not change spontaneously from the state it is in. An example of an equilibrium state would be a ball placed in a depression on a hill slope—as long as nothing perturbs the ball, it will not spontaneously roll out of the depression and it is thus in an equilibrium state. If, instead, the ball were placed on the slope uphill from the depression and released, it would spontaneously roll down the hill and would not be in an equilibrium state until it came to rest.

A Carnot engine is an imaginary engine that cycles through a series of four steps. At the end of the fourth step, the engine returns to its initial state. Each step must be carried out reversibly, meaning that equilibrium is achieved continuously throughout each step of the process. In reality, it is impossible to carry out a series of completely reversible (i.e., equilibrium) steps, because achieving complete equilibrium requires that no pressure or temperature gradients develop during the action of the engine. The only means by which such a state can be achieved is if each step is carried out infinitely slowly. Hence, the Carnot engine is an ideal that cannot be realized. But, as a means for understanding the relationship between heat, work, and efficiency, it is indispensable as a reference frame.

In its simplest form, assume the engine is composed of a gas-filled cylinder with a frictionless piston. The gas we will consider follows the ideal gas law, which states

$$P \times V = n \times R \times T \tag{3.7}$$

where:
  P is the pressure of the gas
  V is the volume of the gas
  n is the amount of gas (in moles)
  T is the temperature (K), which is the thermodynamic temperature scale in which absolute zero
      is equivalent to $-273°C$
  R is the universal gas constant (8.314 J/mol $\times$ K)

In the first step (Figure 3.2), the system is placed in contact with a very large thermal reservoir that adds heat to the gas as work is done on the gas. This step is carried out in such a way that the temperature of the gas never changes. This requires that the volume of the system increase and the pressure decrease, according to Equation 3.7. Because this has been done in a way that keeps the temperature constant, the step is an *isothermal* process. For the second step, work is done on the gas by increasing the volume without adding or removing heat, which also requires that the pressure decrease in the system. In addition, because we are not allowing heat to move into or out of the cylinder, the temperature must drop, according to Equation 3.7. This is an *adiabatic* step, which means the change in state parameters (P and T), and therefore the change in internal energy (ΔE) is only a function of the work done on the system. The third step is an isothermal compression, which is accomplished by having a heat sink into which any heat that might be generated during the compression is absorbed immediately. The last step is an adiabatic compression that returns the gas to its original pressure, temperature, and volume. Figure 3.3 graphically summarizes the pressure–volume path for the gas.

The net work performed by this engine in the cycle must be equal to the difference between the amount of heat put into the gas at step 1 and the (smaller) amount of heat removed at step 3. If no net work were performed, the system would oscillate along the isothermal line of step 1, and the

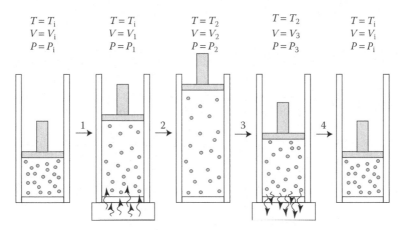

FIGURE 3.2 Diagrammatic representation of a gas cylinder following the steps in a Carnot cycle. Subscripts i and f represent, respectively, initial and final conditions. The arrows at the base of the cylinders in steps 1 and 3 indicate the direction of heat flow, relative to the heat reservoirs that are indicated by the boxes at the bases of the cylinders.

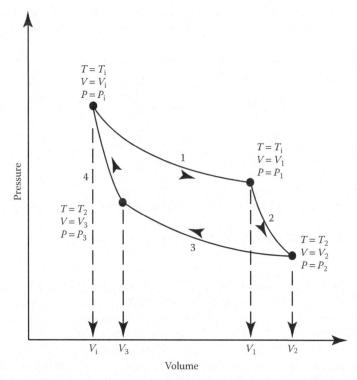

FIGURE 3.3 Pressure versus volume graph for the series of changes for the Carnot cycle depicted in Figure 3.2.

heat removed at step 3 would equal the amount of heat put in at step 1. The efficiency of the engine is then related to the amount of heat converted to work, which can be written as

$$e = -w/q = \Delta q/q_{in} \tag{3.8}$$

which indicates that an efficiency of 1.0 can be achieved only if there is no heat returned to the cold reservoir in step 3 (i.e., all of the heat is converted to work), and the system returns to its initial state.

HEAT CAPACITY

The behavior of this isolated, reversible system raises a question about the property of materials. It was noted as early as the mid- to late-1700s that different materials require different amounts of heat in order to achieve a specific change in temperature, say from 50°C to 60°C. It was also noted that, if the same amount of heat were added to two different materials, each material would reach a different temperature. If these two materials were placed in contact, the warmer material would cool down by transferring its heat to the cooler material and the cooler would warm up by absorbing the warmer material's heat content—the universal observation that heat always flows spontaneously from a warmer to a cooler material. The end temperature for both materials would be the same but their respective heat contents had to be different for the process to begin in the first place. Together, these observations led to an important conclusion, namely, that different materials will reach a state of equilibrium (i.e., a state at which no heat flows between them) only when they are at the same temperature. And yet, the amount of heat each material contains will not be the same.

These conclusions led to the concept that materials, for some physical reason that was unknown at the time, must have specific and unique internal characteristics that determine how much heat

must be added to (or removed from) the material in order for it to change its temperature by a specified amount. The quantity that represented this phenomenon ultimately became known as the heat capacity (*C*), which is expressed in joules per gram for each degree of temperature change (J/g-K). Eventually, it was recognized that changes in volume and changes in pressure affected the heat capacity, so it became a standard to specify heat capacity either at constant pressure (C_p) or at constant volume (C_v). The mathematical expression for the general relationship for heat capacity and heat is

$$C = \frac{dq}{dT} \tag{3.9}$$

where d*q* and d*T* are the differential changes in heat and temperature, respectively.

Rearranging Equation 3.9, we see that the amount of heat that can be taken from a system to do work is equal to the number of degrees the temperature changes multiplied by the heat capacity:

$$C \times dT = dq \tag{3.10}$$

As noted in Section "Enthalpy", for the general case the change in internal energy is Equation 3.5. From Equations 3.5, 3.6, and 3.10, it follows that, at constant pressure,

$$dH = C_p \times dT \tag{3.11}$$

and, at constant volume,

$$dE = C_v \times dT \tag{3.12}$$

which requires that

$$dw = C_V \times dT \tag{3.13}$$

ENTROPY

If we consider the Carnot cycle, we are confronted by the fact that we have started at some initial condition for which there is no history. The process that brought the Carnot cycle to the pressure, volume, and temperature at which it begins is not specified, nor does it matter for the overall evaluation of how heat and work relate to each other. As we move through the cycle from beginning to end, we add and subtract heat and the system does work or has work done on it. As a result, the enthalpy of the gas is changing. All of the changes in temperature are a function of the heat capacity of the gas in our cylinder, but the heat content at the beginning of our cycle is never involved in doing work and we can never use it. Furthermore, because the Carnot cycle is an idealized, reversible system that is unattainable in real life, there is a certain amount of heat that we simply cannot access—moving through a real-life cycle, we will inevitably lose some heat through friction and conduction that can never be used for useful work. A measure of this unattainable heat that is present at the initial state of our system as well as that lost in the process of moving through the cycle is called *entropy*.

The definition of entropy is

$$dS \equiv dq/T \tag{3.14}$$

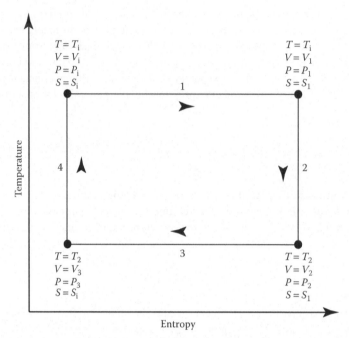

FIGURE 3.4 Temperature versus entropy graph for the same pathways indicated in the pressure versus volume graph in Figure 3.3.

which states that any differential change in the heat a system contains, at a given temperature, leads to a change in the entropy of the system. One way to conceptualize this relationship is to consider the temperature–entropy path our Carnot cycle followed (Figure 3.4). At the initial point in the cycle, the temperature and entropy are fixed (as are the pressure and volume). When heat is reversibly added to the system as work is done at constant temperature (dq/T), and, hence, entropy increase. In step 2, as the gas adiabatically expands, the temperature drops. Because there is no change in the heat content, dq is 0 and there is no change in the entropy. Steps 3 and 4 are the exact reverse of steps 1 and 2, respectively, and the system returns to the same entropy and temperature it initially had. In this reversible process, the system is now poised to continue the cycle again. In this case, the isolated Carnot engine has experienced no discernible change.

But, in reality, the entropy of the universe within which the engine exists has increased. This is apparent if one considers the fact that the addition and removal of heat was done using external heat reservoirs. The higher-temperature reservoir loses some heat (even though it is imperceptible) with each cycle, and the lower-temperature reservoir gains heat. If this process were carried out a very large number of times, the two thermal reservoirs would approach the same temperature. Once this happens, no more work can be done because $\Delta q/q_{in}$ approaches 0, which is the same thing as saying the efficiency approaches 0. Once the two thermal reservoirs have reached the same temperature, regardless of how that happens, they are no longer of any use for doing work, and the entropy of the system has reached its maximum state.

A key lesson in this phenomenon is that the entropy of a system can be manipulated, as is evident from Figure 3.4, but only at the expense of increasing the overall entropy of the surrounding environment. The scale upon which these changes occur can be large. If one considers, for example, carrying out the Carnot cycle experiment on a laboratory bench, the various stages of the cycle would presumably be done using electricity generated at a distant power station burning a fossil fuel. Although the entropy remains constant during the first step in our cycle engine, the electrical power required to carry out that step came from dramatically increasing the entropy of the fuel that

was burned to generate the power. It is for this reason that, when considering overall energy budgets, the energy generation and consumption milieu must be taken into account.

GIBBS FUNCTION AND GIBBS ENERGY (ΔG)

As discussed in Section "Carnot Cycle", when one defines the initial state of the Carnot cycle, there is no need to specify how that initial state was achieved. Indeed, it would be impossible to independently determine how the system achieved its current, initial state because the physical system contains no information that records the system's history. As a result, it is not possible to determine the actual, *absolute* internal energy of a substance or system—the best we can do is determine how the internal energy changes as the system evolves from its current initial state to some other state. We can also determine whether or not two systems (whether they be minerals, rocks, liquid, and gas, or any other pair of materials or substances) are in equilibrium, and what the absolute differences are in their respective heat contents. Two systems that have the same temperature, and are therefore in thermodynamic equilibrium, are incapable of doing work without some external action being taken. If, however, they are not at the same temperature, they can be a source of useful energy. To determine the amount of available energy that can be extracted for useful work, the difference in the heat contents of the systems must be evaluated.

To compare states, we must have a means of comparing the energy contained in one substance or system with that in another. Consider, again, the Carnot cycle and how the internal energy changes. It is evident that there are three fundamental attributes that contribute to the energy in that system at any point along the cycle—the energy that exists in the system at the initial set of conditions before the cycle begins, the energy the system acquires or gives up along an isothermal path, and the energy it acquires or gives up along an adiabatic path. In 1876, J. Willard Gibbs defined a function that mathematically described the energy contained in such a system and how that energy is affected by changes in temperature and pressure. The *Gibbs function* follows from the discussions above regarding the first and second laws of thermodynamics. Gibbs showed that the internal energy of a substance at some specified pressure (P) and temperature (T) is fully described through the following relationship:

$$\Delta G_{P,T} = \Delta H_{P_1,T_1} - T \times \Delta S_{P_1,T_1} + \int_{T_1}^{T} \Delta C_p \, dT - T \times \int_{T_1}^{T} (\Delta C_p / T) \, dT + \int_{P_1}^{T} \Delta V dP \qquad (3.15)$$

where:

$\Delta G_{P,T}$ is the *Gibbs energy* at P and T

$\Delta H_{P_1,T_1}$ is the enthalpy at some standard state, which is usually selected to be 1 bar (0.1 MPa) pressure and 25°C (298 K)

$\Delta S_{P_1,T_1}$ is the entropy at the standard state

ΔC_p is the constant pressure heat capacity

ΔV is the change in volume

The Gibbs energy of a substance is defined in terms of some reference components and compounds that are given explicit values at some reference condition, and all else is derived from that data. What the reference materials are (such as pure metals, pure elements, some suite of specified oxides or compounds), what the reference conditions are (such as a specified room temperature [e.g., 298 K] and pressure [e.g., 0.1 Mpa]), and what the energy content is for a material at the specified reference conditions (such as 0.0 J at the reference condition) are established on the basis of convenience and are usually determined on the basis of what the characteristics are of the system being considered. For many geological applications, the reference condition is taken as 298 K (25°C) at 0.1 MPa

(1 bar), that is, the so-called room temperature and pressure. Although the selection of reference conditions and initial energy values will determine the internal energy values that are computed for a substance at the set of conditions of interest, it is critical to realize that the *differences* in internal energy from one set of conditions to another set of conditions will be exactly the same, regardless of the reference frame one is using (see Section "The Standard State").

Gibbs developed his concept on the basis of reversible processes. Hence, what we call the Gibbs function represents the maximum energy a system possesses that can be used to do useful work.

STANDARD STATE

Note that Equation 3.15 is written in terms of the Δ's for the thermodynamic properties. That reflects the fact that we can compare states and properties only on the basis of some arbitrary reference system we define and studiously respect. For example, it is usually assumed that the Gibbs energy of formation of an element is 0 at 0.1 MPa and 298 K. Any measurements that are then made using those reference materials as the starting materials will produce data (heats of reaction, for example) that characterize the more complex compound. For example, consider the reaction

$$H_2 + \frac{1}{2}O_2 \Leftrightarrow H_2O$$

where:
  H_2 is a pure gas
  O_2 is a pure gas
  H_2O is a liquid or vapor

If this chemical reaction took place at standard conditions in a calorimeter, which is an instrument to measure the heat given off or absorbed in a reaction, the amount of heat given off in the formation of the water molecule would be, by definition, the energy of formation of water at standard temperature and pressure (STP; also known as the standard state). In principle, similar measurements could be made for all simple oxides or compounds of interest and the resulting data, along with Equation 3.15, would allow the thermodynamic properties for all minerals and fluids to be deduced.

Conducting similar experiments at conditions other than those of our defined standard state allows *equations of state* to be developed, which define how the energy content of a material changes with pressure and temperature. Equations of state can take many forms and are developed for substances to varying degrees of accuracy. As with the STP, we can use these relationships to determine the energy content of our systems at any set of conditions of interest.

The importance of this capability can be appreciated by considering the energy content of water as a function of pressure and temperature. From Equation 3.15, it is clear that if a substance such as water is heated at constant pressure, the change in the Gibbs energy will be equivalent to the enthalpy contributed by the heating process minus the product of the temperature times the entropy change:

$$G_{P,T} - G_{STP} = (H_{P,T} - H_{STP}) - T \times (S_{P,T} - S_{STP}) \tag{3.16}$$

This relationship is plotted in Figure 3.5 for liquid and vapor H_2O at 0.1 MPa and 1.0 MPa. This figure provides some important insights into material behavior that is predicated on the thermodynamic properties of the material in question. One obvious fact is that, at constant pressure, changes in temperature affect a vapor phase more severely than a liquid. This observation reflects the fact that the molecules in a fluid are more tightly bound by local molecular forces than are gas molecules. As a consequence, more thermal energy is required to affect the thermodynamic properties of a liquid material. The same is true for a solid phase.

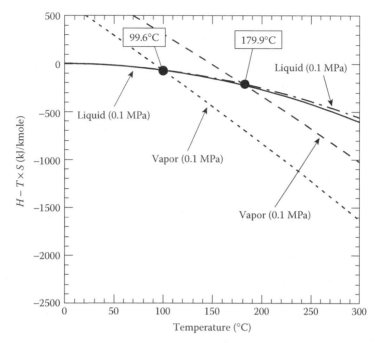

FIGURE 3.5 The isobaric changes in Gibbs energy at 0.1 MPa and 1.0 MPa as a function of temperature, for liquid and vapor H_2O. The temperatures at which the isobaric curves intersect are indicated.

An additional point that is clear from Figure 3.5 is that, for relatively modest pressures, pressure has minimal effect on changes in the thermodynamic properties of liquids (or solids) but can dramatically change the thermodynamic properties of gases. This point, too, reflects the molecular interactions that occur in the material as pressure and temperature change. Liquid water is nearly incompressible at near-surface conditions, whereas water vapor is highly compressible. Both of these effects imply that the energy content of gases will be sensitive to the physical conditions they experience, which will become an important point when we consider geothermal resources and how they can be used.

A third, critical aspect that emerges from Figure 3.5 is a simple explanation for why water changes phase, from liquid to vapor. Recalling the fact that all systems are driven to achieve their minimum energy state, note that at 0.1 MPa liquid, water has a lower value for the Gibbs energy change than does vapor at all temperatures less than 99.6°C. This means that water is more stable in the liquid state than in the vapor state at these temperatures. Beyond 99.6°C, however, the situation is reversed and the thermodynamic properties of the vapor phase make it the more stable phase. At higher pressures (e.g., 1.0 MPa), the same is true but the transition in phase takes place at a higher temperature.

THERMODYNAMIC EFFICIENCY

Importantly, Equation 3.8 is solely a function of temperature. Because steps 2 and 4 in our Carnot engine were conducted adiabatically, there is no heat added to or removed from the system in these steps. Hence, Equation 3.8 for this cycle can be written as

$$e = \Delta q/q_{in} = (T_i - T_2)/T_i \tag{3.17}$$

where:

T_i is the initial temperature of the gas

T_2 is the temperature of the cooled gas

This expression represents the *thermodynamic efficiency* of the engine (all temperatures are expressed in Kelvin). This relationship has profound implications for the thermodynamic efficiency of any geothermal application.

Equation 3.17 demonstrates the importance of achieving as great a temperature difference as possible between a working fluid and its cooled state in any cyclic process. Consider, for example, the several hypothetical geothermal heat sources used for power generation and their respective cooling systems represented in Figure 3.6. For the working fluid temperatures portrayed in the figure, there is a 5%–10% (absolute) increase in efficiency for every 50°C increase in the working fluid temperature. These are significant differences and are an important argument for careful site selection and site analysis for any geothermal project. Examples of the differences in efficiencies that can result from different operating conditions for geothermal power generation are shown in Table 3.1.

In real life, the actual thermodynamic efficiencies that are achieved are influenced by additional factors. One such factor is the depth at which the working fluid resides and the resulting

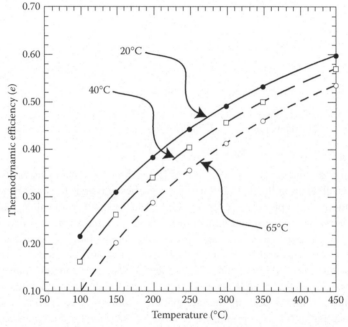

FIGURE 3.6 The thermodynamic efficiency that can be achieved for working fluids that begin at the temperature indicated on the horizontal axis and have an exit temperature as indicated by the arrowed lines. For example, a fluid that began at 200°C and had an exit temperature of 65°C would achieve an efficiency of 0.29, whereas that same fluid would achieve an efficiency of 0.39 if it had an exit temperature of 20°C.

TABLE 3.1

Relationship between Temperatures and Efficiencies for Some Hypothetical Geothermal Reservoir Systems

Reservoir	Reservoir Temperature (°C)	Cooled Temperature (°C)	Efficiency (e)
Low-temperature resource	100	25	0.20
Moderate-temperature resource (winter)	200	10	0.40
Moderate-temperature resource (summer)	200	35	0.33
High-temperature resource	300	25	0.48
High-temperature resource	450	25	0.59

pressure change that it experiences when it is brought to the surface and utilized for power generation (discussed in detail in Chapter 9; see also Sidebar 3.1). To understand the implications of this process for geothermal systems, the thermodynamic properties of water must be considered.

SYNOPSIS

Thermodynamic principles have long established that heat and work are directly related by simple functions. These functional relationships establish that all substances possess some quantity of heat; the availability of that heat to do work depends upon the temperature difference between the substance and its surroundings; that the maximum amount of heat that can be converted to work is independent of the pathway; and the only thermodynamically significant determinant of the amount of work that can be done is the temperature difference between the initial and final states of the system. The temperature difference also establishes the thermodynamic efficiency of the process being employed to extract the heat for work. The Gibbs function defines the available heat by considering the attributes and state of the system. The parameters that determine the Gibbs energy are enthalpy, entropy, heat capacity, and the temperature and pressure of the system. From these parameters, and through knowing the initial and final pressure and temperature states of a system, it is possible to completely characterize the thermodynamic properties of a material. Because all systems are stable only in their lowest energy state, the Gibbs function allows determination of the stable configuration of a system by providing the means to compare the Gibbs energy of the various possible combinations that a chemical system can take. Water, for example, can exist in solid, liquid, or vapor forms; the Gibbs function allows computation of the Gibbs energy at any pressure and temperature for each of the possible phase states, thus identifying the lowest possible Gibbs energy at any set of pressure and temperature conditions.

CASE STUDY: THERMODYNAMIC PROPERTIES OF WATER AND ROCK–WATER INTERACTION

Figure 3.7 shows the standard phase diagram for water. The pressure–temperature conditions for liquid water, water vapor, and solid water (ice) are shown, separated by their respective phase boundaries. Also shown, by the gray-shaded region, is the range of conditions normally encountered in the earth. Atmospheric pressure is equivalent to 0.1 MPa (1 bar) and is indicated by the thin horizontal line.

The phase diagram for water provides insight into the physical basis for the breadth of applications for which geothermal projects can be developed. Notice that for most conditions found on the surface of the earth, or deep within the earth, either liquid water or water vapor is the stable phase for water. Each phase has its own characteristic range of values for molar enthalpy, volume, heat capacity, and entropy. These relationships have been well established and tabulated for water and can be found in many references (e.g., Bowers 1995). There are several important aspects of how these parameters change with temperature and pressure that impact how geothermal energy can be used.

Consider, for example, the heat capacity of water. In Table 3.2, the heat capacity at constant pressure (C_p) for 1 kg of water is compared to that of air and potassium feldspar (a common mineral in rock and soil). The values in this table indicate how much heat must be added to a kilogram of the material being considered in order to raise its temperature 1 degree (K or C). Note that for each material C_p changes as temperature changes and that it changes by a different amount. Between 25°C and 300°C, for example, the heat capacity of liquid water decreases by about 50%, increases by a few percentage points for air, and nearly doubles for potassium feldspar. This contrast in behavior is a reflection of the atomic structure of each material.

For geothermal applications that make use solely of liquid water, such as ground source heat pumps (left hand box in Figure 3.7), the enthalpy of water in the liquid state at near-surface conditions (pressures less than 100 bars and temperatures less than 90°C) is approximately 200 kJ (Figure 3.8). This value for the enthalpy establishes the reference internal energy that the water contains. That amount of thermal energy per kg of water also applies to a volume of water of about

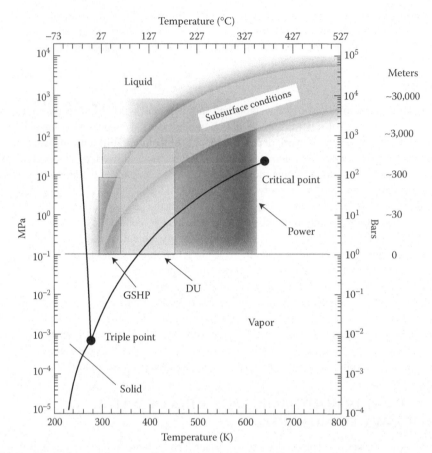

FIGURE 3.7 Phase diagram for water. Temperatures are indicated in degrees Celsius (upper horizontal axis) and Kelvin (lower horizontal axis), and pressures are indicated in megapascals (left vertical axis) and bars (right vertical axis). Also indicated on the right of the figure is the approximate equivalent depth, in meters, below the earth's surface, for the corresponding pressures. The gray band indicates the range of pressure–temperature conditions encountered with depth in the earth. The shaded boxes enclose those sets of conditions appropriate for ground source heat pump applications (medium gray), direct-use applications (light gray), and power generation (gray gradient).

1 liter. Given that the constant pressure heat capacity of water is approximately 4.18 kJ/kg × K under these conditions (Table 3.2), a ground source heat pump that removed 1000 J of heat energy from 1 liter of water would have changed its enthalpy by 0.5% and its temperature by

$$\frac{(1.0\,\text{kJ/kg})}{(4.18\,\text{kJ/kg} \times \text{K})} = 0.24\,\text{K}$$

Thus, whether water is pumped to the surface from a depth of 300 m and passed through a heat pump or the heat pump is installed at a depth of 300 m, the same result would be obtained.

That result is in striking contrast to the behavior of a system in which water moves from one set of physical conditions to another set of conditions and in the process crosses the phase boundary between vapor and liquid. Consider, for example, a water-saturated geothermal reservoir that is at a depth of 1500 m and a temperature of 250°C. If a production well extracts that water at a high rate, it is reasonable to assume that the fluid will lose an insignificant amount of heat to the surrounding environment as it ascends the well. Because no heat is removed from or added to the fluid, the ascent is adiabatic. Because the process is irreversible (as we have noted previously, no process in nature

---

TABLE 3.2

Constant Pressure Heat Capacity (C_p) of Some Common Materials Important for Geothermal Applications at Atmospheric Pressure (1 bar) and 25°C (273 K) and 300°C (573 K); Units Are kJ/kg-K

Material	25°C, 1 bar	300°C, 1 bar
Water[a]	4.18	2.01
Air[b]	1.00	1.04
Potassium feldspar[c]	0.66	1.05

Sources: [a] Bowers, T.S., *Rock Physics and Phase Relations*, ed. T.J. Ahrens, American Geophysical Union, Washington, DC, 45–72, 1995.

[b] Rabehl, R.J., Parameter Estimation and the Use of Catalog Data with TRNSYS. M.S. Thesis, Mechanical Engineering Department, University of Wisconsin-Madison, Madison, WI, 1997.

[c] Helgeson, H. C. et al., *American Journal of Science*, 278-A, 229, 1978.

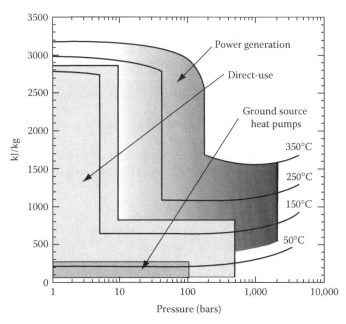

FIGURE 3.8 Enthalpy versus pressure diagram for water, with the corresponding regions for various geothermal applications, color coded as in Figure 3.7.

will take place reversibly when rapid changes are occurring under common geothermal conditions), the process occurs at constant enthalpy (i.e., isenthalpic) but not constant entropy (i.e., isentropic). During the ascent, the pressure continuously drops. At the point the pressure corresponds to the boundary between liquid and vapor, which is at approximately 40 bars pressure for fluid at 250°C (Figure 3.7), steam will begin to separate from the liquid, forming small bubbles. The process of vapor formation and separation from liquid water is called *flashing*.

When the phase boundary is encountered, the fluid temperature will migrate along the set of conditions defined by the phase boundary as the fluid ascends. This occurs because the change from the liquid phase to the vapor phase requires energy, the so-called heat of vaporization. As a result, the temperature of the fluids (liquid and steam) will decrease as the fluids ascend and steam continues to evolve from the liquid. Although this process occurs rapidly, it is not instantaneous. As a result, fluids that exit the wellhead will be a hot mixture of liquid water and steam. In the reminder of this section, we will assume the exiting temperature is 100°C.

Because neither heat nor mass is being added to or removed from our idealized system, the combined fluid (liquid + vapor) enthalpy is constant and the mass (liquid + vapor) is constant. This fact leads to the important concept of *heat and mass balance*, which is crucial when evaluating a geothermal system. Because the process is considered to be isenthalpic, we can write the following equation, which describes the enthalpies of the phase components in our system at the beginning and end point of the extraction and separation process:

$$H_{1,250°C} = x \times H_{1,100°C} + (1 - x) \times H_{v,100°C}$$

where:
  the subscripts l and v stand for liquid and vapor, respectively
  x is the fraction of the mass of the system that is liquid

Because the total mass fraction must equal 1.0, by definition, the mass fraction of the vapor phase must be $1 - x$.

This simple relationship is useful for understanding the characteristics of a geothermal system. If, for example, we have put in place a well that has reached a depth of 3 km with a bottom hole pressure of 1000 bars, and fluid is exiting the wellhead at a temperature of 100°C and 1 bar, and the fluid consists of 70% liquid and 30% steam, we can readily establish that the enthalpy in the reservoir is

$$(0.7 \times 419\,J/gm)_l + (0.3 \times 2676\,J/gm)_v = 1096\,J/gm$$

which would indicate that the reservoir working fluid temperature was about 252°C (enthalpy values for coexisting liquid and vapor are given in Table 3.3).

The heat and mass balance equation can be generalized to account for any component of the system that represents a concentration (mass or energy) that is conserved in the system:

$$C_{reservoir} = x \times C_l + (1 - x) \times C_v$$

Ideally, this relationship allows one to use chemical analyses and energy measurements made at the wellhead to establish characteristics in a reservoir. In other words, as long as a component or element of the system can be shown to be conserved, balance relationships such as this can be applied to determine the characteristics of the reservoir. In reality, however, a number of issues require thoughtful consideration in order to appropriately apply this relationship because a variety of changes occur in the fluid phase as it ascends from the reservoir to the wellhead. In Chapters 5 and 6, we will discuss in more detail the particular features that contribute to this issue, including nonideal behavior in the chemical components of the fluid and how best to account for them.

Diagrammatically, the behavior of the system can be portrayed in a pressure–enthalpy diagram, contoured for temperature (Figure 3.9). The arrow at 250°C and 1000 bars represents an initial fluid condition ($H_1 = 1113$ J/gm). As the fluid ascends the well bore, it encounters the liquid–vapor phase boundary at 40 bars and steam begins to separate from the liquid phase (points A). As the fluid ascends the well and pressure drops, the amount of steam increases and the amount of liquid decreases. The enthalpy of the liquid and steam follows the limbs of the two-phase region to point B, where the fluid exits the wellhead. From the heat and mass balance relationship, the amount of steam and liquid can be directly calculated at each point on the ascending pathway. For example, when the fluid exits the wellhead at 100°C, the respective enthalpies are as follows:

$$H_{1,100°C} = 419\,J/gm$$

$$H_{v,100°C} = 2676\,J/gm$$

TABLE 3.3

Temperature, Pressure, and Enthalpy of Coexisting Steam and Vapor along the Liquid–Vapor Saturation Curve

Temperature (°C)	Pressure (bar)	Enthalpy (J/gm) of Vapor	Enthalpy (J/gm) of Liquid
20	0.02	2538	83.96
25	0.03	2547	104.9
30	0.04	2556	125.8
35	0.06	2565	146.7
40	0.07	2574	167.6
45	0.10	2583	188.4
50	0.12	2592	209.3
55	0.16	2601	230.2
60	0.20	2610	251.1
65	0.25	2618	272.0
70	0.31	2627	293.0
75	0.39	2635	313.9
80	0.47	2644	334.9
85	0.58	2652	355.9
90	0.70	2660	376.9
95	0.85	2668	398.0
100	1.01	2676	419.0
110	1.43	2691	461.3
120	1.99	2706	503.7
130	2.70	2720	546.3
140	3.61	2734	589.1
150	4.76	2746	632.2
160	6.18	2758	675.5
170	7.92	2769	719.2
180	10.02	2778	763.2
190	12.54	2786	807.6
200	15.54	2793	852.4
210	19.06	2798	897.8
220	23.18	2802	943.6
230	27.95	2804	990.1
240	33.44	2804	1037.00
250	39.73	2802	1085.00
260	46.89	2797	1134.00
270	54.99	2790	1185.00
280	64.12	2780	1236.00
290	74.36	2766	1289.00
300	85.81	2749	1344.00
310	98.56	2727	1401.00
320	112.70	2700	1461.00
330	128.40	2666	1525.00
340	145.80	2622	1594.00
350	165.10	2564	1671.00

Source: Keenan, J.H. et al., *Steam Tables: Thermodynamic Properties of Water Including Vapor, Liquid and Solid Phases (International Edition–Metric Units).* John Wiley & Sons Inc., New York, 1969.

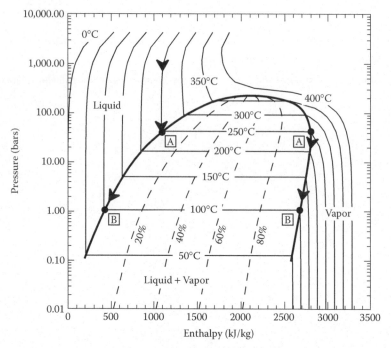

FIGURE 3.9 Pressure versus enthalpy diagram, contoured for temperature. The heavy black line encloses the region where steam and liquid coexist. The dashed lines are lines of constant mass percent steam coexisting with liquid water. The arrowed paths indicate the pressure–enthalpy path followed by a liquid at 250°C that ascends from 1000 bars and exits a wellhead at 1 bar and 100°C.

Thus,

$$1113\,\text{J/gm}_{l,250\,°C} = x \times 419\,\text{J/gm}_{l,100\,°C} + (1-x) \times 2676\,\text{J/gm}_{v,100\,°C}$$

Rearranging and solving for x,

$$(1113\,\text{J/gm} - 2676\,\text{J/gm}) = x \times (419\,\text{J/gm} - 2676\,\text{J/gm})$$

$$x = 0.69$$

Thus, exiting the wellhead is a mixture that is 69% liquid and 31% steam.

These considerations regarding water and its thermodynamic behavior are important for extracting reservoir information from geothermal fluids captured at the wellhead. We will discuss their use, and the limitations that must be recognized, when considering in more detail reservoir assessment (Chapter 8) and geothermal power production (Chapter 10).

PROBLEMS

3.1. In a Carnot cycle involving an ideal gas that has a molar volume of 40,000 cm³/mole, assume that the initial pressure of the gas phase is 1 bar and the volume is 1 m³. What is the initial temperature of the Carnot cycle?

3.2. In order to accomplish an isothermal expansion from the initial condition in Problem 3.1 to a pressure of 0.25 bars and a volume 3 m³, how much heat must be added (assume Cp = 1.02 kJ/kg-K and the density of the gas is constant at 1.2 kg/m³)?

3.3. What volume of water would be required to cool 1 m³ of a rock composed of 100% potassium feldspar from 300°C to 295°C, if the water is allowed to increase in temperature by only 1 degree (assume the heat capacities for water and potassium feldspar are their respective values at 25°C and 300°C)? The gram formula weight for potassium feldspar is 278.337 gm/mole and is 18.0 gm/mole for water. The molar volume for potassium feldspar is 108.87 cc/mole and is 18.0 cc/mole for water.

3.4. A borehole is drilled into a geothermal reservoir and encountered 300°C water at a depth of 3,000 m. As the fluid rises to the surface, at what depth would it flash to steam if a lithostatic pressure gradient (see Sidebar) were maintained throughout the column? What would be the depth of flashing if a hydrostatic pressure gradient existed in the well?

3.5. If an ideal gas isothermally expands from a volume of 1 m³ to 2 m³ against a pressure of 1 MPa, how much work is performed and what is the efficiency?

3.6. If 10 kg of liquid water completely flashes to steam at a pressure of 10 bars, what is the total enthalpy available for work in the steam?

3.7. How much of the enthalpy in problem 3.6 would be used for work if the end point for the cycle is 50°C?

3.8. Using the data in Table 3.3, make a plot of temperature versus enthalpy, contoured for constant pressure, similar to the pressure versus enthalpy plot in Figure 3.9. Discuss when one figure is more useful than the other, when considering enthalpy harvesting.

REFERENCES

Bowers, T.S., 1995. Pressure-volume-temperature properties of H_2O-CO_2 fluids. In *Rock Physics and Phase Relations*, ed. T.J. Ahrens. Washington, DC: American Geophysical Union, pp. 45–72.

Helgeson, H.C., Delany, J.M., Nesbitt, H.W., and Bird, D.K., 1978. Summary and critique of the thermodynamic properties of rock-forming minerals. *American Journal of Science*, 278-A, 229.

Keenan, J.H., Keyes, F.G., Hill, P.G., and Moore, J.G., 1969. *Steam Tables: Thermodynamic Properties of Water Including Vapor, Liquid and Solid Phases (International Edition–Metric Units)*. New York: John Wiley & Sons Inc.

Rabehl, R.J., 1997. Parameter Estimation and the Use of Catalog Data with TRNSYS. M.S. Thesis, Mechanical Engineering Department, University of Wisconsin-Madison, Madison, WI, Chapter 6.

ADDITIONAL SOURCES

Berman, R.G., 1988. Internal-consistent thermodynamic data for minerals in the system Na_2O-K_2O-CaO-MgO-FeO-Fe_2O_3-Al_2O_3-SiO_2-TiO_2-H_2O-CO_2. *Journal of Petrology*, 29, 445–522.

This reference provides a summary of the thermodynamic properties of minerals, as does the Helgeson et al. (1978) reference mentioned in Table 3.2. These and other tabulations of the thermodynamic properties of minerals are necessary starting points for calculating the available heat in subsurface reservoirs.

Klotz, I.M. and Rosenberg, R.M., 2008. *Chemical Thermodynamics: Basic Concepts and Methods*. Hoboken, NJ: John Wiley & Sons, Inc.

This is a thorough introduction to thermodynamics as applied to chemical processes. A good reference book for calculation methods.

The International Association for the Properties of Water and Steam.

A website useful for conducting calculations for the properties of water. Provides links to National Institute of Standards and Technology (NIST) and other organizations for obtaining relevant thermodynamic data. http://www.iapws.org/.

SIDEBAR 3.1   Lithostatic versus Hydrostatic Pressure

Pressure is a critical variable in geothermal systems. It has a profound influence on their performance and behavior. However, the pressure characteristics of geothermal systems reflect a complex interplay between lithostatic and hydrostatic effects. Understanding what these are and how they affect each other is important for analyzing geothermal systems.

The pressure exerted by a column of standing water is called the hydrostatic pressure (P_H). A column of standing water 1 m² in cross-sectional area and 1 m long will exert a force on the bottom of the column

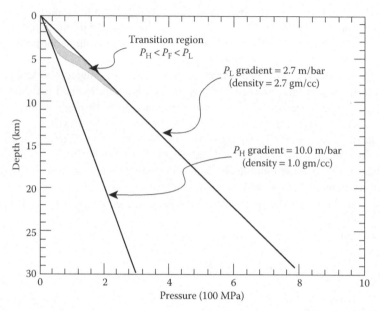

FIGURE 3S.1 The relationship between hydrostatic pressure (P_H), lithostatic pressure (P_L) and depth. The gray shaded region delineates the interval over which hydrostatic pressure transitions to lithostatic pressure.

of 1000 kg (water has a density of 1 kg/liter, and there are 1000 liters/m^3). This is equivalent to 1000 kg per 10,000 cm^2 or 0.1 kg/cm^2, which is equal to 0.1 bar or 1e4 Pa. If that same column were 3000 m tall, the pressure on the bottom of the column would be 300 kg/cm^2, or 300 bars or 3e7 Pa. In this case, the hydrostatic pressure (P_H), would be equal to 30 MPa.

The pressure exerted by a column of rock is called the *lithostatic* pressure (P_L). A column of rock has a density of approximately 2.7 gm/cc, which is equivalent to 2700 kg/m^3. This is equivalent to 2700 kg per 10,000 cm^2 or 0.27 kg/cm^2, which is equal to 0.27 bars (2.7e4 Pa) on its base. A 3 km column of rock 1 m^2 in cross-sectional area would exert a pressure of 810 bars (8.1e7 Pa) on its base. In this case, the lithostatic pressure would be 81 MPa. Figure 3S.1 compares how P_H and P_L change with depth.

At shallow levels (<1000 m) in the crust, fluid in pore spaces is primarily subject to hydrostatic pressure, and therefore, fluid pressure (P_F) is equal to P_H. This fact results from the ability of rock to support an open network of pore space that is not compressed by the overlying rock burden. The rock behaves as though it were a container in which fluid is stored, and the pressure the fluid experiences is mainly a function of the overlying mass of water. However, at increasing depth, the mass of the overlying rock burden becomes sufficient to exert a compressive stress on the rock that is directed vertically. This stress leads to deformation of the supporting geological material, resulting in a progressive reduction in the volume of pore space with depth.

Reduction in pore space volume exerts an increasingly significant pressure on the fluid in the pore space. As a result, pressure exerted on water in pores progressively increases above P_H until it becomes equivalent to the lithostatic pressure. The depth interval over which this transition occurs is dependent on the local geology and state of stress in the crust. Generally, however, P_F becomes equivalent to P_L by the time depths of 10 km are reached.

It is important to realize that P_F is a force acting on the enclosing rock framework, as well as being acted upon by the rock. As a result, it must be viewed as part of the structural architecture of the rock system. If fluid is removed from depth and not replenished (P_F), which was contributing to the support of the overlying rock, is diminished. When this happens, the rock will compress proportionately to the extent to which P_F is reduced. The manifestation of this effect is the phenomenon of land subsidence. Land subsidence can occur in regions where large volumes of fluid are pumped from depth and not adequately replenished. This effect is discussed in detail in Chapter 10.

4 Subsurface Fluid Flow
Hydrology of Geothermal Systems

The ability to use geothermal heat for any purpose, whether it be power generation, heating a building, or drying fruit, requires some means of transferring the heat from the subsurface to where it needs to be used. In locations where hot springs exist, a natural conduit has developed that provides a means for water to reach the surface from a geothermal reservoir. If the temperature of the water at the spring is sufficient for the application that is being considered, then the only other consideration is whether the flow rate is adequate for the intended application. If there is no surface flow, or higher temperatures are required, and if there is reason to believe a higher temperature resource exists at depth, accessing and using that resource requires understanding of how fluid movement occurs in the subsurface, and what the physical attributes are in the subsurface that determine whether a system is a useful geothermal resource. This chapter considers the principles that determine how fluids move through the subsurface, the natural constraints that limit water flow, and the basic principles that apply to enhancing water availability.

GENERAL MODEL FOR SUBSURFACE FLUID FLOW

Virtually anywhere on the planet where one might choose to drill a well, one will encounter water. The depth to which one would have to drill, and the amount of water encountered, will vary tremendously, from virtually nothing to artesian flow out of the wellhead. If the well were to be pumped, in some cases the water would be quickly exhausted, in others, the amount of water would seem to be limitless. Indeed, it is not uncommon for wells only a few hundreds of meters apart to exhibit completely different behavior—either the water is reached at dramatically different depths, or one well will rapidly go dry upon pumping, whereas the other will seem to provide a boundless water supply. And, wells that go dry when pumped will often recharge after some period of time if pumping is stopped. What controls this diverse behavior?

The fundamental determinant for fluid flow behavior is the structural character of the rock through which the fluid moves. There are three primary types of rocks—igneous, sedimentary, and metamorphic. Igneous rocks are rocks that were once molten but have cooled to temperatures well below their melting points. These rocks tend to be massive, but are also often riddled by fractures. Sedimentary rocks are rocks that were deposited either by or in watery environments (e.g., ocean basins, lakes, and streams) or as erosional material along steep scarps or other regions where there is significant topographic relief. Sedimentary rocks commonly are layered and have properties one would associate with mud, sand, and gravel (e.g., granular textures, lateral continuity, and heterogeneity on a scale of centimeters). But limestones are also sedimentary rocks, and they can be massive and quite homogeneous. Metamorphic rocks are any of the above that have been transformed by being recrystallized at elevated temperatures and pressures as a result of being buried to some depth below the surface. Metamorphic rocks can have a variety of textures that grade between those seen in igneous rocks to those approaching limestones. Figure 4.1 provides four examples of what such rocks can look like. These examples represent the type of rock often encountered in geothermal developments.

FIGURE 4.1 Examples of rock types in which geothermal systems occur. (a) Massive, fractured granite (igneous rock). Note the irregular form and diverse orientations of the fractures. (Width of image is 4 m. Sangre de Cristo Mountains, New Mexico.) (b) Porous sandstone that is cemented to different degrees. Note the diversity of grain sizes (fine sand to coarse cobbles). (Width of image is 1.5 m. Rio Grande Rift Basin fill, New Mexico.) (c) Metamorphic gneisses with parallel, planar fractures. (Width of image is 2.5 m. Sangre de Cristo Mountains, New Mexico.) (d) Fault zone in granite. The bracket bounds multiple, parallel, large-scale fractures that define the fault zone. (Width of image is 7 m. Eastern edge of the Rio Grande Rift, Sangre de Cristo Mountains, New Mexico.) (All photographs by the author.)

The fundamental control on subsurface water movement is the amount of space in the rock that is available for water to occupy and the physical characteristics of that space. The examples in Figure 4.1 emphasize that the primary distinction in flow pathways is between flow within and through the body of the rock (matrix flow) and flow within fractures that may be present (fracture flow). Sedimentary rocks commonly have few fractures, and flow is dominated by movement through pore space in the matrix. Igneous and metamorphic rocks have very little pore space but are commonly fractured. Hence, in these rock types fluid flow is primarily through the space within the fractures. The attributes of these flow regimes are discussed in detail in Sections "Matrix Porosity and Permeability" and "Fracture Porosity and Permeability".

MATRIX POROSITY AND PERMEABILITY

If one walks along a beach carefully observing what happens as waves lap against the shore, one will notice that, after the wave advances and begins to recede, some portion of the returning flow seeps into the sand. Often that process is accompanied by bubbles streaming up from below. The bubbles are air that is forced out of the pore spaces between the grains of sand in the subsurface by infiltrating water. The extent to which these actions can be seen depends upon the coarseness of the sand. Very coarse sand beaches will sometimes have virtually no return surface flow, because the advancing water washing up the beach will immediately disappear into the sand. Very fine sand beaches, however, often have a very strong backflow, with little infiltration of the wave wash into the underlying sand. This behavior exhibits the profound interplay between porosity and permeability and how they are affected by the characteristics of the pores.

Pores are the open spaces between grains in gravel, sand, soil, and rock. In coarse gravels and sands, pore spaces can be large (e.g., a significant fraction of a cm) and take up as much as 40% of the total rock volume, whereas in fine sands, muds, and rocks, the size of the pores can be quite small (on the order of a small fraction of a mm down to a 1×10^{-6} m or less) and the total pore volume may be only a few percent or less of the total rock volume.

Within a given sample of rock or sediment, the number of pores with a given size range will vary considerably. The variability of natural materials makes any mathematical representation of the pore size distribution little more than an approximation. Nevertheless, it is common to assume that the distribution of pore size follows, approximately, a log-normal distribution. Shown in Figure 4.2 is the log-normal probability density function (PDF) for three different samples with mean pore sizes of 4.5×10^{-5}, 5.9×10^{-3}, and 1.0×10^{-2} cm. A PDF is simply a description of what the probability is of finding a parameter of a certain value where the sum of all probabilities is 1.0, and the distribution follows some defined mathematical function (in this case, a log-normal form; Kosugi and Hopmans 1998). Also shown in the figure is the mean value and the region around the mean that accounts for 50% of the total pore size distribution.

Natural systems will not exactly follow any mathematical description of the pore size distribution because very few geological materials are sufficiently homogeneous to accommodate such regularity. Nevertheless, mathematical descriptions are useful because they provide a quantitative approximation that allows modeling of the behavior exhibited by natural systems. Such models allow flow rates to be calculated, thus allowing evaluation of the suitability of a rock material for development of geothermal applications.

The fraction of the pore space that is actually water-filled also varies considerably, depending primarily upon climate, elevation, and the ability of the soil or rock to drain water. The extent to

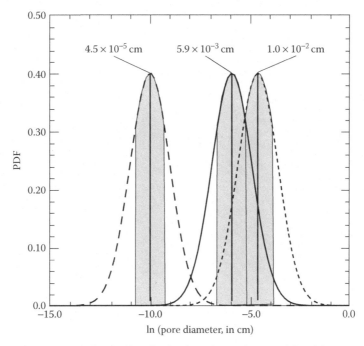

FIGURE 4.2 Examples of PDFs for the size distribution of pores in materials with mean pore diameters of 4.5×10^{-5}, 5.9×10^{-3}, and 1.0×10^{-2} cm. The pore size distribution is assumed to obey a log-normal form. The shaded region about the mean (solid vertical line) encloses 50% of the total distribution.

which pores are filled is called the *saturation*. In regions with high rainfall and low elevation, fully saturated conditions (saturation = 100%) are usually encountered within a few meters to a few tens of meters below the ground surface. The static water table is that point below which the rock is fully saturated.

Between the water table and the ground surface, the soil or rock is within the *unsaturated zone* or *vadose zone*. Within the unsaturated zone, the rock will not be dry, in the sense that water is completely absent from the pores, even though a sample of material from the unsaturated zone may appear to be absolutely dry. Water will be retained at points of contact between grains, and within small pores due to the surface tension of water. The extent to which this surface tension–pore geometry relationship affects the ability of rock to retain water is called capillary suction or *capillary suction potential*. The only conditions under which rock is completely free of liquid water in the subsurface are when the ambient temperatures are elevated significantly above the boiling point of water.

The ability of water to flow through rock pores depends upon a variety of factors. Obviously, any pores present in the rock must be interconnected to some extent to allow the passage of fluid. Two examples are shown in Figure 4.3, in which the volume of pore space is identical (in this case, 40%). In Figure 4.3a no fluid flow through the rock mass is possible because there is no interconnected, through-going porosity, whereas in Figure 4.3b, fluid can freely move through the rock via interconnected pore space. Other effects that influence fluid movement include how complex, or torturous, the path is, the size of the orifice between interconnected pores, and the viscosity of the fluid. In Figure 4.3b, the size of the orifices between pores is consistently much greater along the flow path to outlet point A than along the flow path to outlet point B, resulting in a preferential fluid flow path in the vertical direction. It is this type of porosity feature that results in observable flow anisotropy in intact (i.e., no fractures) rocks. As a consequence of this anisotropy, the total fluid volume exiting at A, for any given time period, will be greater than that exiting at B. This flow through a given cross-sectional area for a given period of time is called the *flux* and has units of volume (m^3) per cross-sectional area (m^2) per time (seconds), that is, $m^3/m^2/s$ or m/s.

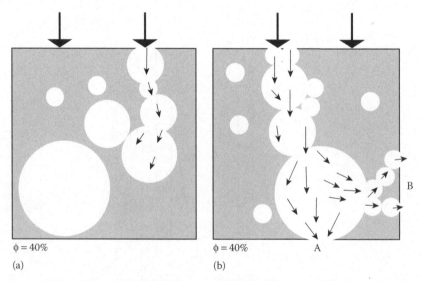

$\phi = 40\%$ $\phi = 40\%$ A

(a)   (b)

FIGURE 4.3 Schematic representation of the relationship between porosity (white regions) and flow (indicated by small arrows). Both (a) and (b) have equivalent porosities (40%). In (a), fluid flow is impossible because there is no connectivity of pores through the rock mass available to the fluid (represented by bold arrows at the top of the block). In (b), fluid flow occurs via interconnected porosity. Note that there is a preferred flow (flux through outlet A is less than that through outlet B) that results from greater pore cross-sectional area connectivity in the A direction as opposed to the B direction.

DEFINITION OF MATRIX PERMEABILITY

The quantitative description of fluid flow in porous media was first formalized by Henry Darcy in the mid-1800s when he developed the quantitative description of volumetric flow, Q (in m³/s) for a given cross-sectional area, A (in m²), which is known today as Darcy's law:

$$q = \frac{Q}{A} = -\left(\frac{\kappa}{\mu}\right) \times \left(\Delta \frac{P - \rho g z}{L}\right) \tag{4.1}$$

where:
  q is the flux (m³/m²-s)
  κ is the permeability (in units of area, m²)
  L is the length of flow regime over the pressure gradient
  ρ is the fluid density (kg/m³)
  g is the acceleration due to gravity
  z is the vertical distance of the system
  μ is the dynamic viscosity (kg/m-s)

The $\Delta (P - \rho g z)$ term is the gradient in pressure, including that due to gravity, that is, the specific weight of water. Strictly, this law only applies to very slow flow (i.e., no turbulence) of a single, homogeneous phase. It is often used as an approximation for more complex conditions, but its limitations need to be recognized. Non-Darcy flow, which is realized under conditions where fluid velocities are high, is commonly encountered in situations involving pumping of wells for geothermal applications. Flow in the subsurface under natural conditions commonly is slow enough so that Darcy flow can be assumed to be a reasonable approximation to the flow regime.

Permeability is a fundamental concept that underlies most considerations in which the flow of fluid in the subsurface is important. Consider, again, the various flow paths in Figure 4.3. In Figure 4.3a, despite the high porosity of the sample, fluid cannot pass through the material. The resulting flux (q, Equation 4.1) will therefore be 0, which also requires that the permeability (κ in Equation 4.1) be 0. In Figure 4.3b, flow can exit the sample in two locations, via a path to A and a path to B. Path A provides the most direct path and the least restrictive (i.e., widest) pore throats and the flux exiting via that route will be greater than that exiting the block via path B, even when the two paths available at B are considered. Path A must therefore have a greater permeability (κ) than path B.

The discrepancy between permeability values along the different paths in Figure 4.3b demonstrates an important aspect regarding permeability. First, permeability is often scale-dependent. Given that pores in rocks are often in the submillimeter size range, clearly the depiction in Figure 4.3 represents a very small piece of rock. If the depicted sample had been obtained from a much larger rock in which there was a random distribution of pore characteristics, it is possible that the permeability measured for the larger sample would average out the effects of different paths, such as A and B, and the resulting value for κ would be different from that obtained for either path individually. Second, the depiction in Figure 4.3b demonstrates that permeability can be directionally heterogeneous. For instances in which it would be important to maximize the fluid volume obtained from a well, for example, it would be important to know what the local permeability heterogeneity is, in order to assure that a borehole accesses the most favorable permeability field. In the example in Figure 4.3b, a horizontal borehole would see a greater fluid flux than would a vertical borehole, assuming that the flow fields at the borehole scale are the same as those shown in the figure.

The units of permeability (κ) reflect the means whereby it is measured in the laboratory. The most common unit used for permeability is the darcy. One darcy is defined as the volumetric flow rate of 1 cm³/s of a fluid with a viscosity of 1 cP through a cross-sectional area of 1 cm² under a pressure gradient of 1 atm per centimeter. As shown in Table 4.1, the range in permeability of geological materials is very large, spanning many orders of magnitude.

TABLE 4.1

Permeabilities for Some Representative Geological Materials

	Highly Fractured Rock	Well-Sorted Sand, Gravel	Very Fine Sand and Sandstone	Fresh Granite
κ (cm^2)	10^{-3}–10^{-6}	10^{-5}–10^{-7}	10^{-8}–10^{-11}	10^{-14}–10^{-15}
κ (millidarcy)	10^{8}–10^{5}	10^{6}–10^{4}	10^{3}–1	10^{-3}–10^{-4}

KOZENY–CARMAN EQUATION

The factors that determine permeability were formally quantified by Kozeny (1927) and later modified by Carman (1937, 1956). The final form of the equation they developed is

$$\kappa = \frac{\left[n^3/(1-n)^2 \right]}{(5 \times S_A)^2} \tag{4.2a}$$

where:

n is the porosity, as a fraction

S_A is the specific surface area of the pore spaces per unit volume of solid (cm^2/cm^3)

Equation 4.2a is known as the Kozeny–Carman equation. This equation allows the dependence of the permeability on the porosity of a porous sample to be determined. Implicit in this relationship are all of the factors discussed above regarding flow in the porous rocks. Of particular importance for permeability is the tortuosity of the flow path—the more tortuous the network of pores through which fluid must flow, the lower will be the permeability. Tortuosity can be accounted for by recasting Equation 4.2a as

$$\kappa = c_0 \times T \times \frac{\left[n^3/(1-n)^2 \right]}{S_A^2} \tag{4.2b}$$

where:

T is the tortuosity which is equivalent to the ratio of a straight path of length L connecting two points to the actual path followed along some tubular route L_t, that is, L/L_t

c_0 is a constant characteristic of the system

Generally, $c_0 \times T = 0.2$, thus reducing Equation 4.2a to 4.2b.

HYDRAULIC CONDUCTIVITY

A useful measure of the ability of a rock to allow fluid to flow is the *hydraulic conductivity* (K). The hydraulic conductivity is the proportionality constant in Darcy's law, (κ/μ) × *specific weight*, and is expressed in units of meters/second. It is defined as the volume of fluid flowing through a specified cross-sectional area under the influence of a unit hydraulic gradient. Examples of values for the hydraulic conductivity are given in Table 4.2 for various types of rocks.

The hydraulic conductivity, as with the permeability, can vary with direction, the scale over which it is considered and obviously with type of rock. For these reasons, laboratory measurements that derive a value for the hydraulic conductivity may be problematic when applied to a field site. Caution is warranted when using such measurements, and it is for this reason that it is common to have several measurements done on a suite of samples from a potential subsurface reservoir if a region is of particular interest for a geothermal application.

TABLE 4.2

Range of Hydraulic Conductivities (m/s) for Various Rocks

Material	Hydraulic Conductivity (Low)	Hydraulic Conductivity (High)
Clay	1.2×10^{-13}	1.2×10^{-7}
Fine sand	7.0×10^{-7}	3.0×10^{-6}
Coarse sand	5.8×10^{-6}	2.3×10^{-5}
Gravel	2.3×10^{-5}	7.4×10^{-4}
Granite	3.5×10^{-9}	3.5×10^{-7}
Slate	1.2×10^{-13}	1.2×10^{-10}

TABLE 4.3

Spacing of Natural Fractures in Different Types of Rocks

Material	Minimum Spacing (m)	Maximum Spacing (m)
Granite	1.2	33.5
Sandstone, shale	1.8	6.1
Gneiss	1.5	13.7
Slate	1.2	7.6
Schist	3.7	15.2

Source: Snow, D.T., *Journal of the Soil Mechanics and Foundations Division, Proceedings of American Society of Civil Engineers*, 94, 73–91, 1968.

FRACTURE POROSITY AND PERMEABILITY

As is obvious from Figure 4.1, many geological materials possess a population of cracks or fractures. The mechanical processes that cause fracturing are many—tectonic forces associated with movement along faults, slow uplift or burial that warps rocks, and cooling or heating, to name a few (see Sidebar 4.1 for discussion of the relationship between stress and fracture properties). The properties of the resulting fractures reflect the relationship between the mechanical properties of the rock mass and the tectonic regime within which the fracturing occurs. Table 4.3 gives the results of one attempt to quantify the range of spacings observed between fractures in the specified rocks. Such an analysis, however, needs to be viewed with appreciation for the complexities that are associated with fracture formation. For example, it is not uncommon for a rock to be traversed by several different sets of fractures, each set having its own characteristic time of formation, orientation, spacing, and properties. Each fracture set reflects, among other things, how the local stress field changed with time. In addition, it is often observed that conjugate fractures will develop in some rocks under specific conditions. Conjugate fracture sets are fractures in which two or more different fracture orientations develop at the same time, with systematic angular relationships between them. Such conjugate fracture sets usually have a specific relationship to the orientation of the local stress field and can be useful for mapping stress regimes. This is discussed more fully in Sidebar 4.1.

Fracture Permeability

Characterizing fractures in a way that adequately allows predictions to be made about flow properties has proven to be difficult. This difficulty arises from the fact that a thorough description of fractures must take into account the dimensions of any open space within a fracture (also called the *aperture*), the fracture orientation, the extent to which there is any interconnectedness of multiple fractures, the length and surface roughness of each set of fractures, the planarity of a fracture set, and the properties at the intersections of different fracture sets. In addition, the variability of each of these properties can be very large within a given fracture set. Examples of some of the ways in which fractures can vary are portrayed by the fracture sets shown in Figure 4.4.

A variety of approaches have been proposed to deal with this complex of relationships, but each has its limitations. In general, however, it appears that the properties that have the largest effect on the flux of fluid moving through any given fracture set in a volume of rock are the fracture aperture and fracture spacing.

The hydraulic conductivity in a fracture is defined as (Bear 1993)

$$K_{fr} = \left(\frac{\rho \times g}{\mu} \right) \times \left(\frac{a^2}{12} \right)$$ (4.3)

where:

K_{fr} is the fracture hydraulic conductivity (m/s)
ρ is the fluid density (kg/m^3)
g is the acceleration due to gravity (m/s^2)
μ is the dynamic viscosity (m/s^2)
a is the aperture (m)

Because the hydraulic conductivity and permeability are related by

$$K = -\left(\frac{\rho \times g}{\mu} \right) \times \kappa$$ (4.4)

the fracture permeability is then defined as

$$\kappa_{fr} = \frac{a^2}{12}$$ (4.5)

FRACTURE TRANSMISSIVITY

The transmissivity of a fracture, which is the discharge through a fracture at some velocity across a given unit aperture, can now be defined as

$$T_{fr} = \left(\frac{\rho \times g}{\mu} \right) \times \left(\frac{a^2}{12} \right) \times a = \left(\frac{\rho \times g \times a^3}{12 \times \mu} \right)$$ (4.6)

This relationship is often referred to as the cubic law because of the dependence of the transmissivity on the cube of the aperture. Hence, the overall movement of fluid through a fracture set in a rock volume can be characterized primarily by the fracture aperture and fluid properties.

However, as is evident from the fractures shown in Figure 4.4, the distance between fracture walls can be highly variable, both within a single fracture and within the fractures of a given fracture set. As a result, the definition of aperture becomes a challenge. In addition, the roughness of the surface of a fracture wall, when combined with the variability of the fracture aperture, can often lead to the development of preferential flow paths. Hence, the total volume of open space in a fracture is not necessarily representative of the volume of flowing fluid that could be moving in a fracture at any particular instant in time. This has lead to the concept of *effective aperture*, in which

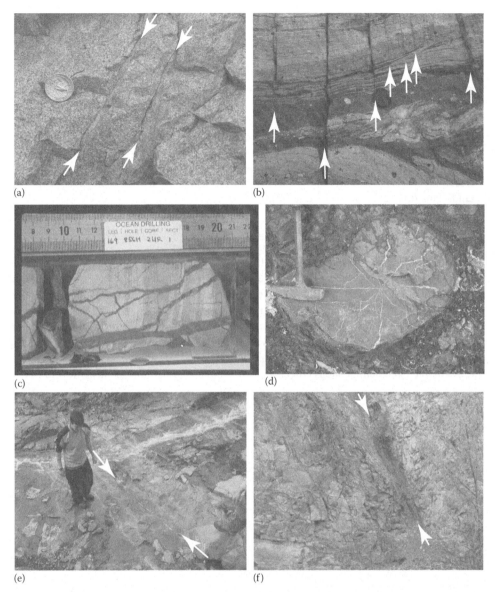

FIGURE 4.4   (See color insert.) Examples of the diversity of fracture forms in various rock types. (a) Two parallel, planar fractures (indicated by arrows) cutting granite. Note the change in color and texture along the fractures, compared to the bulk rock. Note, too, that the fracture has some open space. This change indicates fluid movement and chemical alteration took place along the fractures. (Coin for scale. Photograph by the author; Sangre de Cristo Mountains, New Mexico.) (b) Planar, parallel fractures in marble, indicated by arrows. Note that some fractures cut through all of the different compositional layers, whereas some fractures terminate at the boundary between one compositional layer and another. (Lens cap for scale. Photograph by the author; West Nordre Strømfjord, West Greenland.) (c) Irregular, bifurcating, and cross-cutting filled fractures in hydrothermally altered turbidite. Turbidite is a sedimentary rock, but in this instance, it has been slightly metamorphosed by hot fluids circulating through it. Recovered from a borehole drilled by the Ocean Drilling Project in the northeast Pacific Ocean. (Photograph courtesy of Robert Zierenberg; Middle Valley, Northeast Pacific Ocean.) (d) Radial, filled fractures in a pillow basalt. (Photograph courtesy of Robert Zierenberg; Clear Lake, California.) (e) Fracture with alteration halo from an exhumed ~5–6 million-year-old geothermal system in Geitafell, Iceland. The arrows indicate the location of the planar, linear fracture. Note the green alteration halo that extends for more than a meter around the fracture. (Photograph courtesy of Peter Schiffman.) (f) Small fault zone (arrows) with fault gouge. (Width of image is 5 m. Photograph by the author; Sangre de Cristo Mountains, New Mexico.)

the flow through a fracture set is a function of the integrated effects of multiple segments along a flow path through fractures, where each segment of length l has a common aperture (Wilson and Whiterspoon 1974). Given that fracture systems vary in their characteristics in three dimensions, predicting the effective aperture of a system is often impossible, but the measured flux in a system allows one to calculate what the effective aperture is.

Measured permeability of fractured materials provides a clear demonstration of the importance of knowing fracture properties for any geothermal application in which significant fluid flow rates must be obtained. In Figure 4.5, fluid flux is shown as a function of permeability and pressure gradient. The permeability ranges from Table 4.1 are also shown for reference. It is clear from the graph that the fluxes that can be obtained from highly fractured rocks are, at a given pressure gradient, at least 2–5 orders of magnitude greater than that which can be obtained from matrix flow in fine sand. This difference has profound importance for many geothermal applications where flow rate, and hence the amount of heat that can be made into work at a given rate, determines the economics or efficiency of an application. The requirements for flow rates are discussed in more detail in Chapters 10 through 12 dealing with specific applications.

In Figure 4.6, the effect on permeability of fracture aperture and spacing is modeled. Note that for a given fracture spacing, an order of magnitude increase in the fracture width results in a 2 orders of magnitude increase in the permeability. However, for a given fracture width, a decrease in the fracture spacing by an order of magnitude only results in an order of magnitude increase in permeability. The important point in this representation of fracture properties on flow properties is that permeability is very sensitive to the effective aperture. For this reason, and as will be emphasized later when discussing power generation using geothermal systems, thorough understanding and analysis of the fracture characteristics at a geothermal site is crucial for achieving and sustaining adequate flow rates.

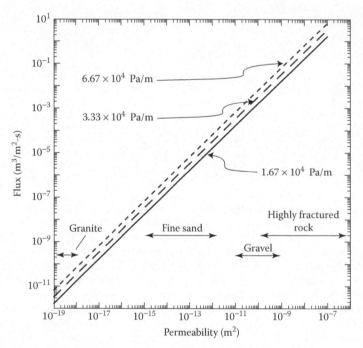

FIGURE 4.5 Flux (m^3/m^2-s) as a function of permeability (cm^2), computed using Darcy's law (Equation 4.1), for pressure gradients of 1.67×10^4, 3.33×10^4, and 6.67×10^4 Pa/meter. Also shown in the figure are the permeability ranges from Table 4.1 for various types of geological materials encountered in geothermal systems.

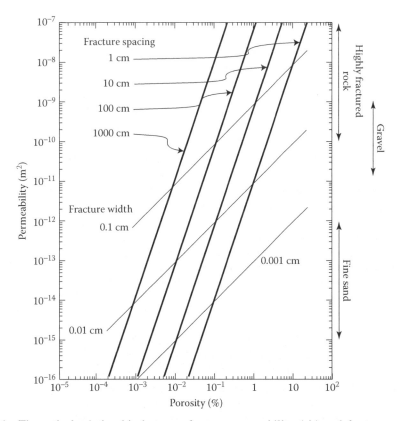

FIGURE 4.6 Theoretical relationship between fracture permeability (air) and fracture porosity. The bulk porosity and permeability for a given fracture width (or aperture) and spacing of those fractures is found by locating the intersection of the width and spacing of interest. It is clear that permeability is a function of both fracture width and spacing, both of which affect bulk porosity. (Modified from Reservoir Characterization Research Lab, University of Texas, Austin, TX, available at http://www.beg. utexas.edu/indassoc/rcrl/rckfabpublic/petrovugperm.htm; Lucia, F.J., *American Association of Petroleum Geologists*, 79, 1275–1300, 1995.)

EFFECT OF DEPTH ON POROSITY AND PERMEABILITY

Although there is considerable variability from one location to another in the type of rocks that are present, it is a general observation that porosity decreases as depth below the surface increases, regardless of rock type. The extent to which this happens depends upon the local geology and the depth being considered. This general behavior is a consequence of compaction and recrystallization. The deeper a rock or sediment is buried, the greater the lithostatic pressure of the overlying rock column. As a result, the rock responds by compressing. The magnitude of this effect depends upon the rock strength. For homogeneous crystalline rocks, such as granites and related rocks, decrease in porosity and permeability due to compression requires greater pressure than for an unconsolidated sediment. The magnitude of this compaction effect on unconsolidated sandstones can be seen in the reduced porosity as a function of depth for some North Sea sandstones in Figure 4.7.

Estimating how permeability changes with depth has primarily relied on models developed from empirical relationships. For depths less than a kilometer, approximations have been developed that

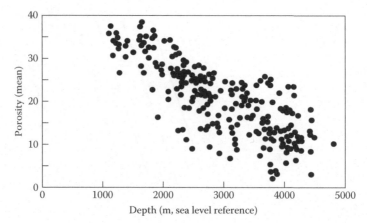

FIGURE 4.7 Porosity as a function of depth in sandstones from the Norwegian Shelf. The zero depth is sea level, and all depth intervals indicate distance, in meters, to mean sea level. The measured porosities plotted in the figure represent the 75th percentile value. (Modified from Ramm, M. and Bjørlykke, K., *Clay Minerals*, 29, 475–490, 1994.)

rely on how fracture aperture changes and utilize a concept called the "conducting" aperture (a_c), as opposed to the "real" aperture (e.g., Lee and Farmer 1990). These models are based on numerous measurements and provide a means to generate an estimate of fracture flow properties. Generally, the result is that fluid flux changes as the cube of the conducting aperture, as in Equation 4.6, which can be recast as

$$q = C \times a_c^3 \times \nabla P \qquad (4.7)$$

where C is an empirical constant that is characteristic of the material under consideration. The values for a_c are strongly dependent upon the roughness, tortuosity, and other properties of the dominant fracture set and must be estimated.

For geothermal resources at depths greater than about a kilometer, permeability is much more difficult to estimate. A variety of studies have used empirical data and model results that have greatly improved our understanding of how permeability changes with depth. In Figure 4.8, these results are presented to a depth of 35 km (modified from Manning and Ingebritsen 1999). At depths less than 5 km, the effective permeability can vary by up to 6 orders of magnitude, making it virtually impossible to develop predictive models that are sufficiently precise to allow useful estimates of how well a site may perform, without some reliable data set from sampling in the field. At deeper levels, the range in uncertainty diminishes, primarily because the elevated temperatures and pressures are sufficient to overcome the effects of heterogeneous distribution of the intrinsic rheological properties of the rocks, which ultimately determine the extent to which a rock will compress. As pressures increase, weak fractures are closed and pore space is reduced by compression, resulting in a more restricted range of permeability values.

Also shown in the figure is the depth reached by the deepest well yet drilled. This well was drilled offshore Sakhalin Island in 2012 and reached a depth of more than 12,376 meters. The deepest onshore borehole drilled reached 12,289 meters in Qatar in 2008. Geothermal wells generally are shallower than 5 km, although the future development of enhanced geothermal systems (EGS; see Chapter 13) will likely achieve greater depths. Clearly, for the foreseeable future, geothermal applications will require field tests and measurements, rather than a priori estimates, in order to establish the actual permeability that can be achieved at any given site.

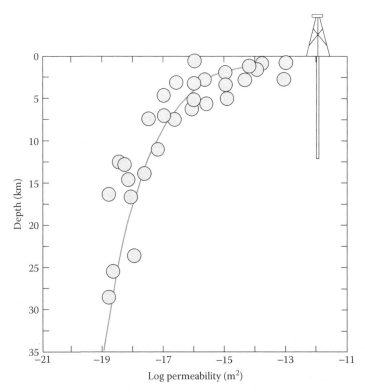

FIGURE 4.8 The variation of permeability as a function of depth. For reference, the depth of the deepest drilled oil well in the world is also portrayed. (Modified from Manning, C.E. and Ingebritsen, S.E., *Reviews of Geophysics*, 37, 127–150, 1999.)

HYDROLOGIC PROPERTIES OF REAL GEOTHERMAL SYSTEMS

The fluid flow properties of real geothermal systems have been summarized by Björnsson and Bodvarsson (1990) who surveyed published data on geothermal power plants. Their results (Figure 4.9) highlight several key points regarding experience in the power generation industry. Most obvious is that systems in which fluid flow is dominated by matrix porosity invariably require higher porosity to achieve a given permeability than systems that are dominated by fractures. For this reason, exploration for geothermal resources that are focused in sedimentary basins benefits from access to previous geological studies in which the hydrological properties of subsurface units have been established. Knowing which geological units have high matrix porosity and acquiring information regarding the subsurface distribution of those units can substantially save resources and reduce risk.

A second key point from the Björnsson and Bodvarsson (1990) survey is that low fracture porosity need not be the limiting factor for power generation. Fracture porosity as low as 0.2% may still be suitable for producing power, provided sufficient fluid flow rates can be obtained. This suggests that in a fractured rock mass, a single fracture set, with modest fracture spacing, may be sufficient for achieving adequate transfer rates of heat from the subsurface. In such a case, the primary drilling targets will be discrete fracture sets with high fluid flow, which may be difficult to locate. This is an important reason why high-quality, high-resolution subsurface data on fracture properties are essential in fracture-dominated systems.

Finally, the variability of porosity and permeability in natural systems can be quite high, spanning several orders of magnitude at a given site. For that reason, it is important to conduct an extensive exploration program for identifying porosity–permeability anisotropy. Such a program must also establish the extent to which properties are heterogeneously distributed in the subsurface.

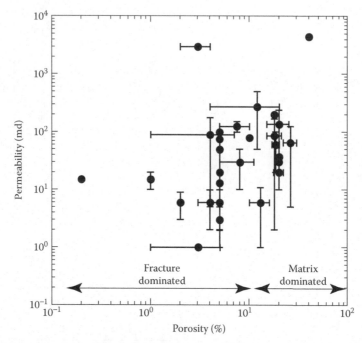

FIGURE 4.9   The variation between porosity and permeability for a suite of geothermal systems, plotted on a log–log scale. Permeability is in millidarcies. The principal source of the porosity (fracture vs. matrix) is indicated. (From Björnsson, G. and Bodvarsson G., *Geothermics*, 19, 17–27, 1990.)

Because stress fields commonly change orientation with depth, fracture orientations observed in one region may be different from that in another region or at a different depth. Such variability emphasizes the importance of obtaining high-quality subsurface information. This topic is discussed in more detail in Chapters 10 and 13.

SYNOPSIS

Flow of water in the subsurface depends upon the characteristics of the pores in the rock (i.e., the porous medium) and the properties of fractures (i.e., the fractured medium). In both of these possible flow pathways, the permeability controls the volumetric rate of fluid flow that can be accommodated by the rock. This in turn directly controls the rate at which energy can be transferred to the surface for geothermal uses. The variables that control flow in the porous medium are the extent to which the individual pores are interconnected, tortuosity and surface area affecting flow. For fractures, the primary variables are the aperture and number of fractures per rock volume. In both instances, the pressure gradient is an additional influence that affects flow rate. Both porosity and permeability are affected by the lithostatic load on the rock and thus are also a function of depth. The sophistication of hydrologic modeling efforts has greatly improved over the last two decades, to the point where quantitative analysis allowing predictive capabilities is now possible. However, the accuracy of modeling results is directly dependent on the availability of high-quality analytical data for *in situ* rock properties.

CASE STUDY: LONG VALLEY CALDERA

Long Valley Caldera has been the site of geothermal power generation since 1985. Currently, a 40-MWe power plant (Casa Diablo) produces electricity from a relatively shallow geothermal reservoir with a working fluid temperature of about 170°C. Flow rates of water from fractured

and porous volcanic rocks are about 900 kg/s. The geologic setting for this system and the information gained from the subsurface by a drilling program undertaken to evaluate the existence
of a near-surface magma body provide an excellent example of the complexity to be expected
in a subsurface flow regime. A series of papers published in the *Journal of Volcanology and
Geothermal Research* (2003, volume 127) provide a summary of the current state of knowledge
about this system.

Long Valley Caldera is a volcanic feature that lies on the boundary between the Sierra Nevada
mountain system and the Basin and Range tectonic province (Bailey 1989). It is part of a volcanic
system that began erupting over 3.5 million years ago along a north–south trend on the eastern edge
of the Sierra Nevada mountain range.

For more than 2.5 million years, the volcanic system poured out various lavas and other volcanic
rocks, while magmas accumulated at depth. At 760,000 years ago, a catastrophic eruption occurred
that spewed more than 600 cubic kilometers of rock into the atmosphere, spreading ash and other
volcanic debris as far east as Kansas and Nebraska and as far west as the Pacific Ocean. The eruption partially emptied the underlying magma chamber, causing collapse of the overlying volcanic
complex. This collapse process resulted in the formation of the Long Valley Caldera, a 17 km ×
32 km depression that is elongate east–west (Figure 4.10). During the eruption, a large volume
of the ejected volcanic ash fell back into the caldera, forming a thick layer of rock known as the
Bishop Tuff that only partially filled the huge depression.

Within 100,000 years after this major eruption, the central part of the caldera rose to form
a "resurgent" dome, apparently in response to pressure from rising magma some kilometers
at depth. The caldera thus took on the form of a large depression with a central highland. The
deep ring surrounding the resurgent dome slowly began to accumulate the eruption products
of later, smaller volcanic events. The volcanic activity within this "moat" has continued to
the present day, the most recent activity being the movement of magma in the subsurface
that caused extensive seismic activity and release of CO_2 that killed thousands of trees in
1989–1990 (Farrar et al. 1995).

The state of the thermal regime within the resurgent dome and the moat have been the subject
of recent studies. In Figure 4.11, the temperature profiles (from Farrar et al. 2003) in several of
the boreholes are shown. All of the selected boreholes are either from the moat, which is the same
geological complex as that in which the geothermal power station is located, or in close proximity
to the site of the power station. Also shown in the figure are the temperature profiles that would be
expected from purely conductive heat transport for thermal gradients of 20°C/km, 30°C/km, and
40°C/km, assuming a ground surface temperature of 10°C.

The most striking aspect of these temperature profiles is their strong deviation from the
linear gradient expected for systems controlled solely by conductive heat transfer. The very
high temperatures at shallow levels (>50°C at depths <1 km), the large (>10°C) temperature
reversals, and the substantial depth intervals (>100 m) over which the temperature remains
constant are indicative of domains in which heat transfer is dominated by a complex fluid
flow pattern. In this instance, the interpretation of the thermal profiles is complicated by the
geology, as well.

Within the moat, the first few hundred meters of rocks encountered in drilling is a complex
interlayering of volcanic lava flows and other eruptive rocks, some of which are likely to have
quite high permeabilities and low thermal conductivity (k_{th} = 0.8–1.2 W/m-K; Pribnowa et al.
2003). Below this series of rocks is the Bishop Tuff. Tuff is a volcanic rock formed when ash
particles settle out of the atmosphere after an eruption. It is mainly composed of particles the
size of very fine dust up to coarse sand-sized particles, but can also contain rock fragments as
large as boulders. When it settles, if it is very hot, it can weld together into a coherent, brittle
rock. If it is cold when it settles, it will have the properties of unconsolidated dust or sand. The
Bishop Tuff has a lower section that is unwelded and relatively unconsolidated, and a middle,
densely welded section. Recent faulting has caused the development of high-permeability fracture

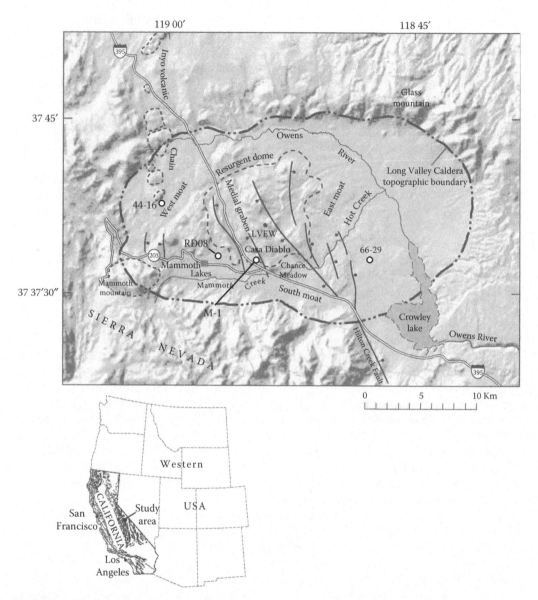

FIGURE 4.10 Location map and topography of the Long Valley Caldera. The heavy, dash–dot line marks the topographic definition of the boundary of the caldera. The lighter dashed lines enclose post-caldera volcanic centers and the resurgent dome. Faults are indicated by the solid black lines. The side of the fault that has dropped down is marked by the dots connected to the faults. The Casa Diablo power generation facility is labeled. (From Farrar, C.D. et al., *Journal of Volcanology and Geothermal Research*, 127, 305–328, 2003.)

zones that are vertically oriented. This has resulted in an anisotropic permeability distribution that favors vertical fracture flow (Evans and Bradbury 2004). The Bishop Tuff is deposited on a sequence of earlier breccias and other volcanic rocks that overly the crystalline, very low porosity crystalline metamorphic basement rocks. The latter have relatively high thermal conductivities (k_{th} = 3.0–3.8 W/m-K; Pribnowa et al. 2003).

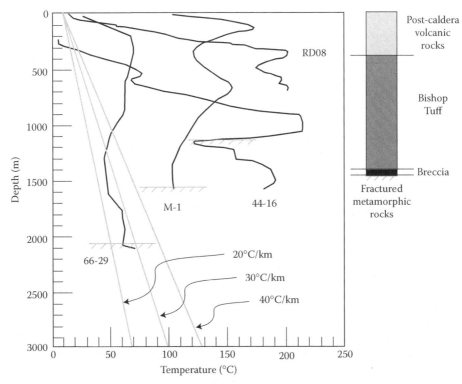

FIGURE 4.11 Temperature profiles from four wells in the Long Valley Caldera. The hachured lines indicate the depth at which basement rock was encountered in three of the wells. On the right is the approximate depth and thickness of the local geology. For reference, the temperature gradient that would be anticipated if heat transfer was solely via conduction in a uniform medium is shown for temperature gradients of 20°C/km, 30°C/km, and 40°C/km.

Farrar et al. (2003) interpret the shallow-level thermal spikes to be the result of meteoric water flowing down an hydraulic gradient that is maintained by recharge of groundwater in the Sierra Nevada Range to the west. This eastward flow encounters relatively recent hot intrusions west of the boreholes that heat the groundwater, resulting in an east-directed thermal plume, as depicted diagrammatically in Figure 4.12a. The shallow high temperature ground water system is maintained within the post-caldera volcanic rocks. Because this rock complex is very heterogeneous, in terms of rock type, porosity, and permeability, multiple preferential flow paths are possible, thus leading to the multiple thermal spikes seen in some of the borehole temperature profiles.

The shallow-level flow zone must be hydraulically isolated from the deeper thermal regime, probably by relatively impermeable rocks composing the bulk of the Bishop Tuff. Where there is little fracture permeability, heat transfer in the Bishop Tuff is primarily via conduction, and portions of the thermal profiles that approximately parallel the conduction isotherms drawn in Figure 4.11 are likely to be regions of low fluid flow. However, the virtually isothermal sections over substantial depth intervals that occur in boreholes 66-29 and the lower portion of M-1 must be maintained by vertical fluid flow in a fracture-dominated system in which convective fluid movement is driven by a heat source at depth, which is diagrammatically represented in Figure 4.12b. Temperature reversals in these sections are likely due to interaction with a small volume of infiltrating meteoric water.

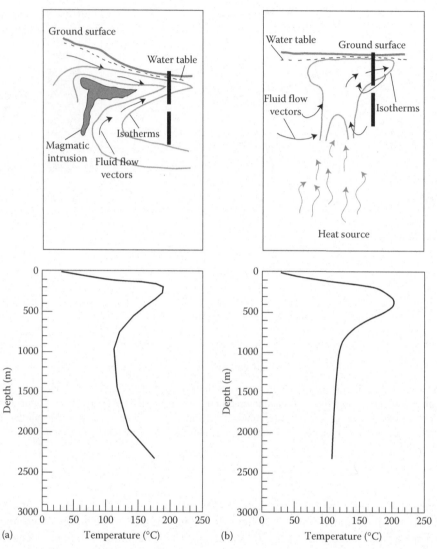

FIGURE 4.12 (Top) Two interpretive geological cross sections and the (a and b) respective temperature profiles that would be encountered if wells penetrated the geology at the bold vertically dashed lines. In (a), the thermal perturbation at a depth of about 200 m is caused by heat from an underlying intrusion that is affecting flow of meteoric waters from the highland to the left. In (b), the thermal perturbation at 300–500 m is caused by the convective flow of water heated from a deep heat source at depth greater than 2500 m. Both of these geological scenarios give rise to qualitatively similar temperature profiles.

PROBLEMS

4.1. What would be the exposed surface area in 1 m³ of rock that was cut by planar fractures with an aperture of 10 microns and which had a total fracture porosity of 10%?

4.2. What would be the fracture permeability of this rock?

4.3. What would be the fracture transmissivity of that same rock?

4.4. What would be the total volume of water that would move through a conducting fracture over an hour if the aperture were 1 cm and the pressure gradient was 1 kPa per meter?

4.5. What would be the permeability of a sandstone with a cross-sectional area of 0.1 m^2 if 0.01 m^3/s of fluid with a dynamic viscosity of 0.001 Pascal-seconds flowed through it under a pressure gradient of 1 MPa per meter?

4.6. In Figure 4.9, the maximum and minimum fracture permeabilities differ by more than three orders of magnitude, at a porosity of 30%. Using Figure 4.6, determine the range of respective fracture aperture and spacing for these conditions.

4.7. Develop two hypotheses for the temperature distribution shown in Figure 4.11 for well M-1. Which would be the most likely?

REFERENCES

Bailey, R.A., 1989. Geologic map of Long Valley Caldera, Mono Inyo volcanic chain and vicinity, Eastern California. US Geological Survey Miscellaneous Investigations Series Map I-1933.

Bear, J., 1993. Modeling flow and contaminant transport in fractured rocks. In *Flow and Contaminant Transport in Fractured Rock*, eds. J. Bear, C.-F. Tsang, and G. de Marsily. New York: Academic Press.

Björnsson, G. and Bodvarsson, G., 1990. A survey of geothermal reservoir properties. *Geothermics*, 19, 17–27.

Carman, P.C., 1937. Fluid flow through a granular bed. *Transactions of the Institute of Chemical Engineering London*, 15, 150–156.

Carman, P.C., 1956. *Flow of Gases through Porous Media*. London: Butterworths.

Evans, J.P. and Bradbury, K.K., 2004. Faulting and fracturing of non-welded Bishop Tuff, Eastern California. *Vadose Zone Journal*, 3, 602–623.

Farrar, C.D., Sorey, M.L., Evans, W.C., Howle, J.F., Kerr, B.D., Kennedy, B.M., King, C.-Y., and Southon, J.R., 1995. Forest-killing diffuse CO_2 emission at Mammoth Mountain as a sign of magmatic unrest. *Nature*, 376, 675–678.

Farrar, C.D., Sorey, M.L., Roeloffs, E., Galloway, D.L., Howle, J.F., and Jacobson, R., 2003. Inferences on the hydrothermal system beneath the resurgent dome in Long Valley Caldera, east-central California, USA, from recent pumping tests and geochemical sampling. *Journal of Volcanology and Geothermal Research*, 127, 305–328.

Kosugi, K. and Hopmans, J.W., 1998. Scaling water retention curves for soils with lognormal pore-size distribution. *Soil Science Society of America Journal*, 62, 1496–1505.

Kozeny, J., 1927. Über kapillare Leitung des Wassers im Boden. Sitzungsber. *Akademiie Wissenschaft Wien*, 136, 271–306.

Lee, C.H. and Farmer, I.W., 1990. A simple method of estimating rock mass porosity and permeability. *International Journal of Mining and Geological Engineering*, 8, 57–65.

Lucia, F.J., 1995. Rock-fabric/petrophysical classification of carbonate pore space for reservoir characterization. *American Association of Petroleum Geologists*, 79, 1275–1300.

Manning, C.E. and Ingebritsen, S.E., 1999. Permeability of the continental crust: Implications of geothermal data and metamorphic systems. *Reviews of Geophysics*, 37, 127–150.

Pribnowa, D.F.C., Schutze, C., Hurter, S.J., Flechsig, C., and Sass, J.H., 2003. Fluid flow in the resurgent dome of Long Valley Caldera: Implications from thermal data and deep electrical sounding. *Journal of Volcanology and Geothermal Research*, 127, 329–345.

Ramm, M. and Bjørlykke, K., 1994. Porosity/depth trends in reservoir sandstones: Assessing the quantitative effects of varying pore-pressure, temperature history and mineralogy Norwegian Shelf data. *Clay Minerals*, 29, 475–490.

Snow, D.T., 1968. Rock fracture spacings, openings, and porosities. *Journal of the Soil Mechanics and Foundations Division, Proceedings of American Society of Civil Engineers,* 94, 73–91.

Wilson, C.R. and Witherspoon, P.A., 1974. Steady state flow in rigid networks of fractures. *Water Resources Research*, 10, 328–335.

FURTHER INFORMATION SOURCES

Batu, V., 1998. *Aquifer Hydraulics*. New York: Wiley. 727 pp.
    This book is a good reference source for hydraulic principles that are relevant for understanding fluid flow in the subsurface.

Bear, J., 1979. *Hydraulics of Groundwater*. New York: McGraw-Hill. 569 pp.

This book is a standard reference for groundwater research. It thoroughly presents the concepts and quantitative considerations that facilitate understanding movement of water in rocks.

Sharp, J.M. Jr., 2007. *A Glossary of Hydrological Terms*. Austin, TX: The University of Texas. 63 pp.

This glossary provides a useful compilation of terms, along with appropriate equations and definitions, for hydrological properties and processes. It is available at http://www.geo.utexas.edu/faculty/jmsharp/sharp-glossary.pdf.

SIDEBAR 4.1   Stress and Rock Fractures

Deformation of a material occurs when a force is applied to the material. The magnitude and nature of the deformation, however, is very material-specific—push a finger against a car fender and there will barely be any visible indication that the fender temporarily yields (i.e., deforms), but apply that same amount of pressure to a lump of wet clay and the deformation will be obvious and permanent. The lesson is that all materials resist deformation, to one degree or another, but if a force is applied to a surface, a measureable (even if microscopic) deformation occurs.

In Newtonian mechanics one of the fundamental equations defines force as an acceleration acting on a mass:

$$F = ma$$

where:
   F is the force
   m is the mass
   a is the acceleration

The unit of force (in the SI system) is a Newton (N), which has units of (kg × m)/s^2 that, obviously, is mass (kg) times acceleration (m/s^2).

If a force (F) is applied to an object, the object experiences stress. When 1 N of force is applied to a specific area in an object, such as one square meter, the result is

$$1\,\text{N/m}^2 = 1(\text{kg} \times \text{m})/\text{s}^2/\text{m}^2 = 1\,\text{kg}/(\text{m} \times \text{s}^2) \equiv \text{Pascal}\,(\text{Pa})$$

The Pa is the SI unit of stress.

Stress can be normal stress or shear stress. *Normal* stress results when a force is applied perpendicularly to a surface, whereas *shear* stress results when the force is applied parallel to the surface. No matter how a force is applied to a body, it is always possible to resolve that stress into three stress components that are perpendicular to each other (Figure 4S.1). The large arrow in the figure indicates a hypothetical stress applied to the light gray face of the block. The stress direction is inclined at some arbitrary angle to the face. The x-, y-, and z-axes shown in the figure are an arbitrary set of perpendicular axes that allow the applied stress to be resolved into stress components σ_x, σ_y, and σ_z, with σ_x being the maximum stress, σ_z being the minimum, and σ_y being the intermediate. It can be demonstrated that there is one unique orientation of these reference axes for which the maximum and minimum stress components are parallel to two of the three axes. The third axis defines the orientation of the intermediate stress. These stress directions, defined as $\sigma_1 = \sigma_x$, $\sigma_2 = \sigma_y$, and $\sigma_3 = \sigma_z$, define the orientations of the maximum, intermediate, and minimum *principal* stresses to which the rock is subject. A figure can be drawn in this three-dimensional reference frame that encloses all possible values of stress in all orientations. Such a figure is called the stress ellipsoid. Two-dimensional sections drawn along the two axes of that ellipsoid are stress ellipses, the dimensions of which are defined by the principal stresses σ_x, σ_y, and σ_z along the respective axes (see Figure 4S.1).

For real rocks in the earth's crust, the state of stress is a complex function of several variables. In the simplest case, the only stress a rock experiences is due to gravitational forces. Under such circumstances, the stress a rock experiences only depends upon how deeply buried it is, and the source of the stress comes from the mass of rock above it. In that case, the vertical stress is the maximum principal stress, which is also called the lithostatic pressure. An approximate rule-of-thumb is that the lithostatic pressure in the crust increases at the rate of approximately 3.33×10^7 Pa, or 33.3 MPa (mega-Pascals) for every kilometer of depth.

For a homogeneous rock with sufficient internal strength to prevent plastic flow, the lithostatic pressure is also equivalent to the confining pressure the rock experiences in all directions imposed on it by its

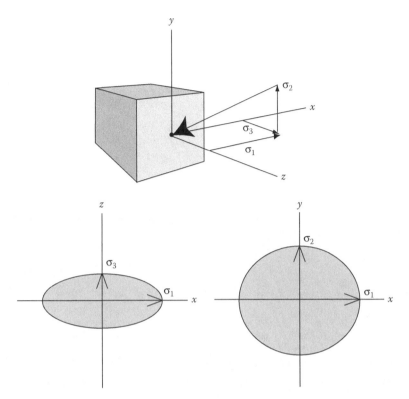

FIGURE 4S.1 Representation of a randomly oriented force (bold arrow) acting on a cube face. The force can be resolved into three orthogonal components, one acting perpendicularly to the cube face (σ_x), and two (σ_y and σ_z) acting parallel to the face. These components can be oriented such that they are parallel to the maximum and minimum stress directions, thus allowing the stresses to define the stress ellipse for the force acting on the face. Note that the stress ellipses are drawn at different scales; if scaled equally, σ_x ($= \sigma_1$) would be the same length in both ellipses.

neighboring rock mass. As a result, the stress ellipsoid is a sphere and σ_x, σ_y, and σ_z are equal. In such a case, there will be no fractures in the rock, and the only available space to accommodate fluid flow would be the intrinsic rock matrix porosity.

Failure of a rock by fracturing occurs when the difference between the maximum principal stress and the minimum principal stress (also called the stress differential, usually expressed in MPa) exceeds some value characteristic for that rock. For this to occur, some force must be applied to the rock in addition to that resulting from gravity. Many geological processes are capable of imposing such forces, such as uplift, settling, downslope mass movement, differential movement of tectonic plates, magma intrusion, and so on. Because these conditions change over geological time, it is common for a given rock to have experienced a long history of changes in both the orientation and magnitude of stresses it experiences. As a result, it is likely that a rock will, at some point in its history, experience conditions where its characteristic strength is less than the stress differential, and it will fail by fracturing. For some rocks, this situation may have happened multiple times in its history, and multiple fracture sets may develop.

The resulting fractures at the time of failure will be oriented in specific and approximately predictable directions, relative to the orientation of the principal stress directions. Fractures that form during such an event is called a fracture set. The characteristics of the fractures, such as how long they are in any particular direction, their planarity and the spacing between the fractures that form, will depend upon the characteristics of the rock, the pressure and temperature at the time of fracturing, the magnitude of the stresses, and the rate at which the stress is applied. Rocks that have experienced multiple failure events are likely to have multiple fracture sets, each with characteristic properties and orientations.

5 Chemistry of Geothermal Fluids

Geothermal waters exhibit a broad range of chemical compositions (Tables 5.1 and 5.2), from very dilute (a few hundred parts per million, by weight, of total dissolved constituents) to very concentrated (solutions containing tens of percent, by weight, of dissolved constituents). This dissolved load can provide important information about the characteristics of a reservoir including its temperature, mineralogy, and history. However, the dissolved load can also impact the performance of machinery in a geothermal power plant that is used in heating and cooling, as well as in generating power. This chapter considers the basic chemical processes that influence the chemical properties of geothermal fluids and how to use chemical analyses to gain information about the geothermal resource. Later, in Chapters 6 and 7, we will apply these principles to evaluate specific aspects of geothermal resources. In Chapter 15, we will consider the environmental aspects of the compositions of geothermal fluids and how the environmental consequences of the chemical components can be mitigated.

WHY THE GEOCHEMISTRY OF GEOTHERMAL FLUIDS MATTERS

Water occurs virtually everywhere in the subsurface, as we previously noted in Chapter 4. For geothermal applications that require temperatures of a few tens of degrees centigrade or less, such as heat for direct-use applications or for heating, ventilation, and air conditioning (HVAC) purposes, meteoric waters that occur in shallow groundwater systems are usually sufficient to provide the needed energy. Heat pumps, as will be discussed in Chapter 11, can efficiently move heat in such systems. Because such systems lead to a relatively small temperature drop in the fluids, and because the chemistry of the fluid is of little direct consequence for these uses, it is usually not necessary to pursue detailed knowledge of the fluid chemistry for such applications. That is not the case for instances in which higher temperature fluids are required.

High-temperature fluids used in geothermal power production are often associated with magma bodies or are in regions where igneous activity has occurred in the recent past. In those cases where recent igneous activity is not part of the geological history, high-temperature geothermal fluids can form where deep fluid circulation is facilitated by faulting and fracturing, allowing water to reach depths where heat is available to raise the temperature sufficiently to be useful for power production (usually >120°C). When water temperatures increase to such a degree, the water will interact with the surrounding rocks, taking on chemical characteristics that are influenced by the local geology. These interactions impart a chemical signature to the water. Deciphering the implications of that signature for the quality of the resource and the potential economic and environmental impacts that may need to be addressed depends upon a detailed understanding of the aqueous geochemistry (Brook et al. 1979).

The geochemistry of natural waters is also important for a different, but just as crucial aspect of geothermal energy considerations, namely, exploration for the resource. In many instances, surface evidence that a high-temperature resource is present at depth may not be obvious, thus making the resource difficult to find. These "hidden" resources are often detectable through chemical signatures in surface waters. Some of these signatures can be specific chemical constituents that indicate the presence of a heat anomaly at depth; in other cases, it may be shifts in the distribution patterns of certain elements, isotopes, or compounds, or changes in the ratios of elements. Understanding the processes that control these geochemical signatures provides the ability to assess the value and quality of a resource.

TABLE 5.1

Properties of Some Common Aqueous Species

Species	Atomic/Molecular Weight (g/mole)	Ionic Radius (angstroms)
H^+	1.008	0.25
Li^+	6.941	0.1
Na^+	23	1.1
K^+	39.1	1.5
Mg^{2+}	24.31	0.7
Ca^{2+}	40.08	1.2
Fe^{2+}	55.85	0.7
B^{3+}	10.81	0.1
SiO_2	60.09	–
H_2S	34.082	–
F^-	19	1.2
Cl^-	35.45	1.6
Br^-	79.9	1.8
HS^-	33.074	–
HCO_3^-	61.017	–
O^-	16	1.2
S^-	32.06	1.7
SO_4^-	96.064	–

Source: Compiled from the data in Whittaker, E.J.W. and Muntus, R., *Geochimica et Cosmochimica Acta*, 34, 945–966, 1970.

Note: Weights are in grams/mole and radii are in angstroms (10^{-10} m). The radius is that of the ion in an ionic crystal.

WATER AS A CHEMICAL AGENT

Water, as with any other chemical compound, reacts with materials with which it comes in contact. For example, solids that are placed in direct, intimate contact with each other will exchange atoms at their interface via diffusional processes. These interactions, however, are of limited extent and are slow to occur. Liquids, within which the constituent molecules are highly mobile, will relatively quickly interact with whatever compounds they are in contact. The extent of interaction is determined by the molecular characteristics of the fluid. In the case of water, the reactivity is significant. This is a reflection of the fact that water molecules are polar, meaning they possess electrical polarity due to the orientation of the hydrogen atoms covalently bound to the central oxygen atom of a water molecule (Figure 5.1). Compounds such as minerals, which invariably have exposed atoms and their associated electrical charges on their surfaces, will interact with water because of its electrically polarized nature. As a result, water is an excellent *solvent*, meaning it has the ability to dissolve constituents (*solutes*) and carry them as a dissolved load.

Another consequence of this charge polarity is that the water molecules will link together to form a weak polymeric structure. The water molecules will tend to orient themselves such that the side of a water molecule that has the net positive charge will be in proximity to the net negative charge portion of adjacent water molecules, as depicted in Figure 5.1. This structural linking is ephemeral, with water molecules aligning with different neighbors very quickly. This property has a strong influence on the way in which the internal energy of a volume of water changes with temperature. At low temperature, the vibration rate is relatively low, and the length and lifetime of the polymeric chains is relatively long. As temperatures increase, the vibration rate increases, and the

TABLE 5.2

Chemical Composition of Waters from Diverse Geothermal Systems

Location	pH[a]	Na	K	Ca	Mg	Cl	B	SO$_4$	HCO$_3$	SiO$_2$
Wairakei, New Zealand (W24)	8.3	1,250	210	12	0.04	2,210	28.8	28	23	670
Tauhara, New Zealand (TH1)	8.0	1,275	223	14	–	2,222	38	30	19	726
Broadlands, New Zealand (BR22)	8.4	1,035	224	1.43	0.1	1,705	51	2	233	848
Ngawha, New Zealand (N4)	7.6	1,025	90	2.9	0.11	1,475	1,080	27	298	464
Cerro Prieto, Mexico (CPM19A)	7.27	7,370	1,660	438	0.35	13,800	14.4	18	52	808
Mahia-Tongonan, Philippines (103)	6.97	7,155	2,184	255	0.41	13,550	260	32	24	1,010
Reykjanes, Iceland (8)	6.4	11,150	1,720	1,705	1.44	22,835	8.8	28	87	631
Salton Sea, California (IID1)	5.2	62,000	21,600	35,500	1,690	191,000	481.2	6	220	1,150
Paraso, Solomon Islands (A3)	2.9	136	27	51	11.1	295	5	300	–	81
Paraso, Solomon Islands (B4)	2.8	9	3	17	10	2	2	415	–	97

Sources: Henley, R.W. et al., *Fluid-Mineral Equilibria in Hydrothermal Systems*, vol. 1. Reviews in Economic Geology. Chelsea, MI: Society of Economic Geologists, 1984; Solomon Islands from Giggenbach, W.F., *Proceedings of the World Geothermal Congress*, Florence, Italy, 995–1000, 1995. The paranthetical expressions are the identifiers for the wells from which the analyses were obtained.

Note: All concentrations are in mg/kg.

[a] This is the pH measured in the laboratory at 20°C and is not the pH of the fluid in the reservoir.

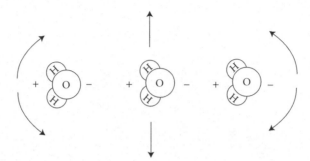

FIGURE 5.1   The distribution of charge for a water molecule. The positive and negative charge signs indicate the distribution of electrical charge concentrations on the water molecule. The three molecules are weakly bound by their charges, resulting in short-lived polymeric chains, in this case a trimer. The arrows indicate the vibration directions of the molecules in the trimer.

length and lifetime of the polymeric chains decrease. These effects directly influence solubilities, as we will see in the below paragraph.

Many solutes are themselves charged (Table 5.1), in which case they are called *ions*. Positively charged ions are called *cations* and negatively charged ions are called *anions*. Because it is a matter of universal human experience that water is not electrically charged, it is clear that the dissolved load is exactly electrically balanced. In other words, the positive electrical charge of the total dissolved load of cations exactly neutralizes the negative electrical charge of the total load of dissolved anions. This charge balance requirement, along with a variety of other chemical constraints that are discussed in Section "Components and Chemical Systems", determines the outcome of water interacting with rock. The end result is recorded in the broad range of compositions of geothermal waters found around the world (Tables 5.1 and 5.2). In the remainder of this chapter, we will examine the concepts of saturation, reaction rates, equilibrium, and other processes that determine what is contained in an aqueous solution and how those concentrations both reflect the characteristics of the geothermal resource and are affected by the use and development of geothermal resources. But first, we must develop the means for defining and describing a chemical system.

COMPONENTS AND CHEMICAL SYSTEMS

Any collection of compounds, whether they be metals, individual elements, mixed gases, mixed liquids, condensed solids, or any combination of these, is composed of chemical constituents. The minimum number of these chemical constituents that are needed to fully describe the collection of substances in a *system* is defined as the *components* of the system. How these components are identified depends upon the way in which the system is defined and how it is to be analyzed. In developing our list of components, it is important that they describe the system completely and are not themselves composed of two or more entities that occur in our system. The entities that make-up the system are the *phases*, which can be minerals, gases, or liquids.

As an example, consider the minerals quartz, tridymite, and chalcedony, all of which have the chemical formula SiO_2. These three phases are the system we will consider. The difference between them is how the atoms in the mineral structures are arranged. Three chemical reactions can be written that completely describe the possible interactions of the minerals:

$$Quartz \Leftrightarrow tridymite$$
$$Tridymite \Leftrightarrow chalcedony$$
$$Quartz \Leftrightarrow chalcedony$$

There are two ways in which the components of this system can be defined. One way is to note that the elements Si and O compose all of the minerals in this system and can therefore completely describe the chemical properties. In this case, the system would be treated as a 2-component system. It is also possible, however, to use SiO_2 as a chemical component because it, too, provides a complete description of this system. If our interest is in understanding the behavior of the minerals in the system (and not the elements or isotopes of the system), using SiO_2 as the chemical component is the preferable and necessary way to describe this system because it allows the smallest number of components to be used for defining the system. If, however, we were interested in the dissociation of these minerals into their respective atomic constituents or wanted to understand isotopic behavior, then we would have to use Si and O as the components, rather than SiO_2.

CHEMICAL POTENTIALS AND GIBBS ENERGY

Once the components are identified for a system, it is possible to describe such things as which mineral or collections of minerals will be the stable phases in the system under a given set of pressure and temperature conditions. Each component in a mineral has a *chemical potential* (μ), which has units of J/mole. Chemical potentials are analogous to gravitational potentials or electrical potentials. They are a measure of the tendency of a chemical component to change from one energy state (for a mineral, this state would correspond to a specific way in which the atoms in a structure are arranged) to another. Hence, in our example, we are considering the tendency of the component SiO_2 to exist in either the quartz, cristobalite, or chalcedony configuration. As noted in Chapter 3, the stable configuration of a system is that which provides the lowest energy state. Hence, in our example, the criterion for identifying the stable phase is to determine for which mineral phase the chemical potential of SiO_2 is the lowest. Of course, this will depend on the pressure and temperature conditions to which the system is subject.

The transition from one phase to another such as

Quartz ⇔ tridymite

is a chemical reaction that will occur at some set of physical (i.e., pressure and temperature) conditions. If all pressure and temperature conditions at which this reaction occurs are taken into account, a boundary in pressure–temperature space will be defined, on one side of which quartz, the *reactant*, will be stable and on the other side tridymite, the *product*, will be stable (by convention, reactants are those phases that occur on the left side of the written reaction and products occur on the right side of the written reaction). Along that boundary, the chemical potential of SiO_2 in both phases is equal.

$$\mu_{quartz} = \mu_{tridymite}$$

The various reactions that are possible in the example system are easily represented as expressions between components because there is only one component in the system and that component is the only one present in the phases we are considering. However, in more complex systems, such as those usually encountered in geothermal systems, keeping track of how all of the components are individually changing as the physical conditions evolve is simply too cumbersome. Instead, the following relationship is used to account for these changes:

$$\Delta G_j = \sum \mu_j^i \tag{5.1}$$

where:
ΔG_j is the Gibbs energy function of phase j
μ_j^i is the chemical potential of component i in phase j

In the summation, it is convention that the products are taken as positive and the reactants are taken as negative.

Recalling the definition of the Gibbs function in Chapter 3, it is evident that the sum of the chemical potentials of the components of a phase expresses the changes in enthalpy, entropy, and pressure–volume (PV) work that occur when a phase is affected by evolving physical conditions.

Equation 5.1 is an important general expression for all phases. It indicates that all phases that make up a physical system will respond to changes in their environment through the effects those changes have on the chemical potentials of the components composing the phase. Those changes will be expressed in a variety of ways, as will be discussed in more detail in this chapter and in Chapter 6. But the most dramatic change is that implied by the chemical reactions we are considering. Sufficient change in the chemical potentials of the components composing a phase will ultimately be sufficient to make that phase unstable, relative to some other arrangements of the components, and a reaction will occur, forming a new assemblage of phases. This then implies that Equation 5.1 can be generalized to represent the behavior of a collection of phases,

$$\Delta G_{rx} = \sum \Delta G_j \tag{5.2}$$

where ΔG_{rx} is the Gibbs energy function of the reaction we are considering, at some specified pressure and temperature.

Again, the same convention regarding the sign of the reactants and products is used in the summation as in Equation 5.1.

ACTIVITY

The way in which the chemical potentials of components change in response to changes in pressure and temperature or other state variables is represented by the *activity*, a. Activity is a thermodynamic measure of the difference between the chemical potential of a component in a phase from one set of conditions to another. This change in chemical potential follows the form

$$\mu_i^{T,P,x} = \mu_i^0 + R \times T \times \ln(a_i) \tag{5.3}$$

where:
   $\mu_i^{T,P,x}$ is the chemical potential of component i at some pressure, temperature, and composition (x) of the phase being considered
   μ_i^0 is the chemical potential of component i at some standard state, including a reference composition

Then, by analogy with Equations 5.1 and 5.2,

$$\Delta G_j = \sum \mu_j^{iT,P,x} = \sum \left[\mu_j^{i0} + R \times T \times \ln\left(a_j^i\right) \right] \tag{5.4}$$

$$\Delta G_{rx} = \sum \Delta G_j \tag{5.5}$$

These relationships provide the means to determine the chemical state of a system from the chemical compositions of the phases in the system. They also provide a means to predict the behavior of a system if it is subjected to conditions that are different from those that determined its current properties. Conversely, and importantly for considerations regarding geothermal systems, the chemical composition of phases in a system contains information about the temperature and pressure the system has experienced. We will exploit this fact when we consider exploration for geothermal systems in Chapter 6.

SATURATION AND THE LAW OF MASS ACTION

Let us return to that pot of boiling water we considered in Chapter 2. Assume that once the pot is boiling, we immediately remove it from the stove and place it in a perfectly insulated container. Just before we close the container and isolate the pot from the rest of the world, we pour in exactly 10 g each of table salt (i.e., halite, NaCl) and quartz sand (SiO_2), both of which have been ground to a particle size in which every particle has a diameter of exactly 10 microns. If we open the container ten minutes later, what would we find? An hour later? A day later? If we opened it a year later, would it be any different?

EQUILIBRIUM CONSTANTS

Experience would say that the salt would dissolve relatively quickly, eventually completely disappearing. But would it be gone after 10 minutes? And, if we kept adding 10 g aliquots of salt, how much salt could be added to the 100°C water before no more would dissolve? Would the quartz eventually dissolve completely, too? Would it do so in the same amount of time?

The answers to these questions are determined by the so-called law of mass action and reaction rate laws. The law of mass action relates to chemical reactions that have achieved equilibrium. Take, for example, the dissolution of halite, which can be written as a chemical reaction in which the halite is the reactant and the ions in solution are the products,

$$NaCl \Leftrightarrow Na^+ + Cl^-$$

Note that electrical neutrality is perfectly maintained when the halite is dissolved; all complete reactions must conform to this necessity. As the halite dissolves in the water, the overall reaction proceeds from left to right. If we could actually see the individual salt crystals at the atomic scale, we would see that as the atoms on the crystal faces separate from the crystal structure and enter the solution as ions (i.e., as dissolution occurs), we would also see a few ions in the solution reattach themselves to the crystal face (i.e., precipitate). The net effect, however, is that the rate at which atoms are removed from the crystal and enter the water as ions is much faster than the rate at which they reattach to the crystal surface, thus resulting in dissolution. At the point at which equilibrium is achieved, the reaction is proceeding in both directions at exactly same rate, and thus, there would be no net change in the amount of solid halite in the pot or the amount of ions in solution. In this state, the solution would be "saturated in halite." The law of mass action states that, at any given set of conditions, the following will hold true

$$\frac{[(a_{Na^+}) \times (a_{Cl^-})]}{(a_{NaCl})} = K$$

where:
  a is the activity
  K is the equilibrium constant, which is a function of temperature, pressure, and the composition of the solution

In Figure 5.2, the variation of the log of the equilibrium constant for the halite dissolution reaction and the quartz dissolution reaction is shown as a function of temperature up to 300°C.

By convention, the activity of solid minerals that are pure is equal to 1.0. For dissolved species, intuition would suggest that the activity would be equal to the molality of the species in solution, and as a result, the equilibrium constant would change linearly as temperature changed. However, the curvature of the line defining the measured values for the log K for halite in Figure 5.2 makes it clear that a more complex process is affecting the behavior of the ions in solution. This *nonideal behavior* is due to the interactions of the charged ions in solution with each other and with the polar

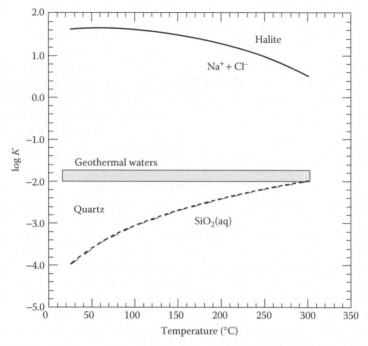

FIGURE 5.2   Variation of the equilibrium constant for halite and quartz dissolution, as a function of temperature. The gray box encloses the range of aqueous SiO_2 compositions for the geothermal waters listed in Table 5.1.

water molecules. Because sodium and chlorine ions have different ionic radii (Table 5.1), they will also have different charge densities, different frequencies with which they vibrate in the solution, and different number and orientation of water molecules that would be loosely bound to them by electrical charge interactions. In addition, adding ions to water modifies the polymeric structure of water, which in turn affects the internal energy of the solution.

ACTIVITY COEFFICIENTS (γ)

Together, these effects combine in nonlinear ways to influence how changes in the dissolved concentration of solutes in water affect the internal energy of a solution. It is this behavior that is considered nonideal and has led to the concept of activity coefficients. An activity coefficient can be defined as a factor which, when multiplied by a specie's molality, corrects the measured concentration in a solution so that the mass action law for a given reaction is satisfied. For the halite reaction, this means the mass action expression takes the form

$$[(\gamma_{Na^+} \times m_{Na^+}) \times (\gamma_{Cl^-} \times m_{Cl^-})] = K$$

where γ_i is the activity coefficient for species i.

For dilute solutions, γ_i is close to 1, and for most purposes, the measured molal concentrations can be used in calculations. Dilute solutions are those that have total dissolved solute loads equal to or less than that of sea water which has a dissolved load of about 35 parts per thousand. For more concentrated solutions, the effects of nonideality can be significant and need to be taken into account. The available data for activity coefficients that are experimentally determined are relatively limited, especially for highly concentrated solutions. Compilations of activity coefficients have been published or discussed by Harvey and Prausnitz (1989) and papers in Palmer et al. (2004).

AFFINITY

Establishing whether a chemical reaction such as the halite dissolution reaction will actually take place is accomplished by comparing the composition of the solution to the value of the equilibrium constant using the following expression:

$$A = R \times T \times \ln\left(\frac{Q}{K}\right) \tag{5.6}$$

where:
   A is called the affinity (J/mole)
   R is the universal gas constant (8.314 J/mole-K)
   T is the temperature (K)
   Q is the activity quotient for the relevant species in the applicable reaction
   K is the equilibrium constant for that same reaction

If the measured solution composition results in an activity product that is equal to K, the affinity will be 0, indicating that the solution is in equilibrium with the solids, and no net dissolution or precipitation will occur. If the affinity were greater than 0, the concentration of the solutes exceeds the equilibrium value and the solution would begin to precipitate solid. This situation is one in which the solution is *supersaturated* in the solid involved in the reaction. For affinities less than 0, the reactant (in this case, our halite solid phase) would continue to dissolve until it is completely dissolved or the activity product of the solutes becomes equal to K, and equilibrium would be achieved.

This situation with respect to the values of affinity suggests that, mathematically, A must be equivalent to the Gibbs energy function of a reaction as indicated in Equation 3.16 (given below)

$$A = G_{\text{products}} - G_{\text{reactants}} = (H_{\text{products}} - H_{\text{reactants}}) - T \times (S_{\text{products}} - S_{\text{reactants}})$$

This relationship leads to the fundamental expression

$$\Delta G = -R \times T \times \ln(K) \tag{5.7}$$

which is a statement of the relationship between the activities of the species involved in a reaction and the conditions under which they will coexist in thermodynamic equilibrium.

At 100°C, the value for log K for the halite dissolution reaction is about 1.6 (Figure 5.2). On a molar basis, and assuming ideal behavior of the ions in solution, this means that a little more than 6 moles, each, of Na^+ and Cl^- must be dissolved in solution to achieve saturation. Given that the molecular weight of sodium is about 22.99 g/mole and that of Cl is about 35.39 g/mole, the respective masses of Na and Cl in a saturated solution will be a little more than 145 g and 220 g. Clearly, our original 10 g of salt added to the pan would completely dissolve, the solution would remain strongly undersaturated, and the resulting affinity and Gibbs energy would be much less than zero.

For the quartz in the pot, the situation is different. The value for K at 100°C is about 0.001 (Figure 5.2), which translates into a total SiO_2 (aq) concentration of about 0.06 g at equilibrium, or approximately 60 parts per million. Clearly, in this case, we have added to the pot of water much more quartz than would ever dissolve. In fact, so little would dissolve that it would not be noticeable unless we carefully weighed it before and after our experiment. In other words, quartz has a low *solubility* under these conditions.

SOLUBILITY

Solubility is the physical manifestation that combines all of the processes we have discussed thus far in this chapter. Solubility is the maximum concentration of a solute in a solvent that can be maintained at a given temperature and pressure. It is measured in units of mass per volume or mass per mass. Common units are part parts per million (ppm) by weight and moles of solute per kilogram of solvent (molality, M).

The two examples of dissolving minerals, halite and quartz, both exhibit changes in solubility as temperature increases, but the pattern is different. Halite solubility increases until about 65°C, at which point solubility begins to decrease slightly over the temperature range to 300°C. Quartz solubility, however, increases throughout that range, although it drops precipitously beyond about 350°C. These behaviors reflect interactions that take place in the solvent between the solute components and the water molecules and demonstrate that even in simple chemical systems the process of dissolution and precipitation are the consequences of not-so-simple interactions at the atomic scale.

Part of this contrast in behavior is a reflection of the nature of these reactions. The halite and quartz reactions are representative of dissolution and precipitation reactions that will occur in geothermal systems. The quartz dissolution reaction represents a process in which a neutral molecule detaches from the mineral structure and enters solution as a neutral aqueous species, in this case SiO_{2aq}. The interaction of this electrically neutral component with the polar water molecules is one in which the aqueous species is surrounded by water molecules in a coordinated structure in which the number of water molecules surrounding the SiO_{2aq} is a function primarily of the temperature.

In the case of halite, dissolution occurs by individual ions on the surface of the crystal lattice breaking the ionic bonds with which they are bound to the structure and entering the solution as charged anions and cations. Because of their electrical charge, they interact more strongly with the individual water molecules than the neutral SiO_{2aq} species in solution. This leads to more complex behavior in solution, because any other charged solute components will also be interacting with the water molecules. As a result, solubility of halite in water can be strongly influenced by other solute components.

A third type of solubility behavior can be demonstrated by considering the dissolution and precipitation of the mineral calcite. Calcite is calcium carbonate, $CaCO_3$. The dissolution of calcite can be written as

$$CaCO_3(s) \rightarrow Ca^{2+} + CO_3^{2-}$$

This reaction is coupled to the hydrolysis of the carbonate ion to form bicarbonate via the reaction

$$CO_3^{2-} + H_2O \rightarrow HCO_3^- + H^+$$

This carbonate–bicarbonate equilibrium is obviously affected by the activity of the hydrogen ion in solution, which is also the determinant of the acidity of the solution. It is through this mechanism that pH becomes an important control on carbonate dissolution and precipitation.

The calcite dissolution exhibits an important characteristic that certain salts express when they dissolve, namely, *retrograde solubility*, which is defined as decreasing solubility with increasing temperature. Figure 5.3 compares the solubility of quartz and calcite as a function of temperature and, for the case of calcite, also as a function of the concentration of other solutes in solution. The retrograde solubility for calcite is obvious. But, as well, calcite solubility is also sensitive to other components in solution. At 150°C, for example, the calcite solubility can differ by more than 300%, depending upon the solution composition.

These examples demonstrate an important aspect of how fluid chemistry may evolve along a flow path. For example, assume a geothermal fluid is at 250°C and is migrating to a cooler region

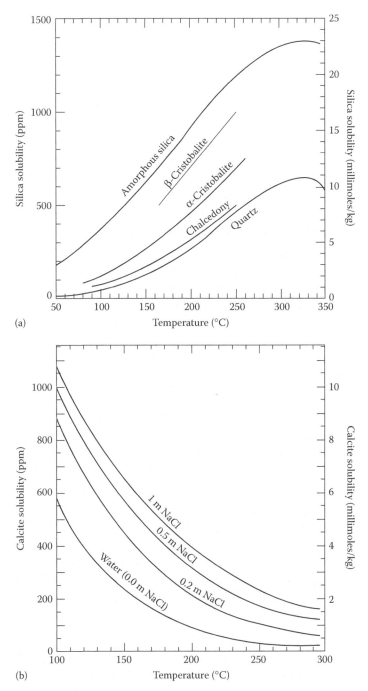

FIGURE 5.3 Comparison of the solubility of (a) quartz and (b) calcite, as a function of temperature and composition of the solvent (for calcite).

where the temperature is 100°C. Assume, also, that the fluid is pure water saturated with calcite and quartz. That means the solution will begin with a silica solubility of about 6.6 mM and a calcite solubility of less than 0.5 mM. As the fluid migrates and cools, quartz must be precipitated because its solubility will decrease, but the solution will become richer in calcite components, if calcite is in the rock the water is migrating through, because the calcite solubility increases with decreasing temperature. By the time the solution has equilibrated with the 100°C environment it finds itself in, the solubility of quartz will have dropped to about 1 mM, whereas that of calcite will be nearly 6 mM.

The halite, quartz, and calcite dissolution and precipitation reactions indicate the types of mechanisms that affect how fluids migrating through a porous or fractured medium will react with the rock framework. If the fluid is undersaturated with respect to the minerals in the rock, it will dissolve the minerals and thus increase the porosity and permeability along its flow path. If the fluid happens to become supersaturated with respect to one or more minerals, it will precipitate those minerals and decrease the porosity and permeability. Because most rocks are composed of a suite of minerals, a wide range of reactions are possible, thus leading to the possibility that the solution could become undersaturated in some minerals and supersaturated in others, with the consequence that precipitation and dissolution could occur simultaneously. Added to this is the possibility, which is often realized, that the solution can become supersaturated in a mineral phase not present in the rock. In this case, growth of a new mineral phase along the flow path can occur. If the solution composition evolves over time, a sequence of different minerals may grow on fracture surfaces or in pore spaces, resulting in mineralogically zoned features. Given the millions of years over which geothermal systems can persist, it is possible that the mineralogy seen today could be completely different from that when the geothermal system first formed. Geothermal fluids thus can be powerful agents of change. These changes can be important indicators of the properties of a geothermal resource, allowing them to be useful when exploring for geothermal reservoirs, as will be discussed in Chapter 6.

Ion Exchange

Another important reaction process that has only minimal impact on porosity and permeability evolution, but which has important implications for assessing geothermal systems, is ion exchange. Although some minerals, such as quartz, have a fixed chemical composition (i.e., SiO_2), other minerals can accommodate a variety of ions in their structure. One example of this is alkali feldspar, one of the most common minerals in the Earth's crust. There are two main end member chemical formulae for alkali feldspar, $NaAlSi_3O_8$ and $KAlSi_3O_8$. These two pure end member compositions are occasionally found in nature, but more commonly the mineral is found to be a mixture of the two end member molecules, with the chemical formula $Na_xK_{1-x}AlSi_3O_8$, where x is less than or equal to 1.0.

The relationship between the end members can be written as an exchange reaction:

$$NaAlSi_3O_8 + K^+ \Leftrightarrow KAlSi_3O_8 + Na^+$$

where Na^+ and K^+ are ions in an aqueous solution coexisting with the feldspar.

In Figure 5.4, the log of the equilibrium constant for this exchange reaction is plotted as a function of temperature.

The large variability of log K over a relatively small temperature interval raises an important point. Namely, an aqueous solution in contact with and in equilibrium with alkali feldspar at a given temperature should have a specific $Na^+:K^+$ ratio. In other words, the ratio of Na^+ to K^+ in a solution is temperature sensitive and has the potential to be a *geothermometer* (see, e.g., Giggenbach 1988, 1992; Fournier 1992). There are numerous caveats to applying this principle, but a variety of exchange reactions involving numerous minerals have been evaluated for their potential as geothermometers. We will consider these geothermometers in much more detail in Chapters 6 and 7.

Although the chemical processes we have been discussing provide a means for evaluating the dissolution and precipitation controls on minerals for a given set of conditions, and how those

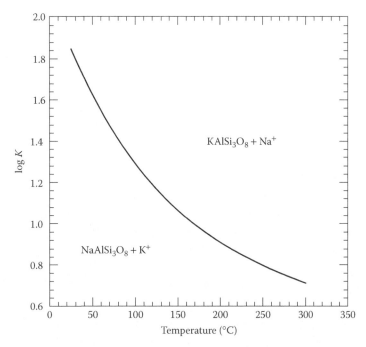

FIGURE 5.4   Variation of the equilibrium constant for the exchange reaction between sodium and potassium feldspar, as a function of temperature.

reaction processes will influence fluid and mineral compositions, they tell us nothing about the rate at which these processes will proceed. To evaluate how quickly, or even whether equilibrium will be achieved, and to know how fast dissolution or precipitation will occur, we must consider reaction kinetics.

KINETICS OF GEOTHERMAL REACTIONS

How quickly materials dissolve in or precipitate from a solution, and therefore how quickly equilibrium will be approached, is determined by the temperature and pressure, the effective surface area exposed to the solution, the chemical composition of the solution, and how readily mineral components can be removed from or added to the exposed surface of solids with which the fluid is in contact (for a detailed discussion of reaction kinetics see Laidler 1987). The rate at which such a process proceeds is called the *kinetics* of the reaction.

Experimental studies over the years have shown that mineral dissolution can be described by equations that take into account the variables described above. A number of versions of *reaction rate laws* have been proposed. Many of these laws work quite well for the specific types of minerals from which they were derived, but often cannot be successfully applied to other families of minerals that have strongly disparate structures or chemistries from the reference mineral system. As a result, care must be taken when applying rate laws when one is attempting to understand the time-dependent evolution of natural geological systems. With that caveat in mind, we will discuss generalized reaction rate laws as they are often applied to geological systems.

To quantitatively capture the overall behavior of a reacting system, the functional relationship must take into account how reaction kinetics are affected by the various parameters enumerated above. This can be accomplished by noting that each variable has an influence on the Gibbs energy function of a reaction. At equilibrium, by definition, the value of the Gibbs function is 0. Far from equilibrium, however, the reaction rate is a large positive or negative number.

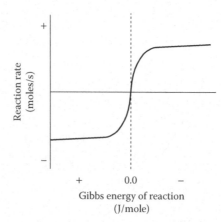

FIGURE 5.5   Variation of the reaction rate plotted as a function of the Gibbs energy of a reaction.

Graphically, this relationship appears as shown in Figure 5.5. Hence, any equation describing this behavior must be able to account for the far-from-equilibrium behavior as well as the behavior close to equilibrium.

One such general reaction rate equation, written in a form that uses the measured dissolution rate for reference, is given by

$$R = S_A \times k \times T_{fac} \times \alpha \times \phi \times \prod a_i \times \left(\frac{1-Q}{K} \right)^{\omega} \tag{5.8}$$

where:

R is the rate (moles/s)

S_A is the effective surface area exposed to the fluid (cm²)

k is the far-from-equilibrium rate constant (moles/cm²-s)

T_{fac} is the temperature correction factor for the rate constant k (usually an Arrhenius function)

α is a power function that accounts for changes in the rate close to equilibrium conditions

ϕ is a function that modifies the rate for precipitation relative to that for dissolution

a_i accounts for the dependence of the rate on the activities of specific components in solution (often this is mainly a reflection of the activity of the hydrogen ion, H^+)

Q/K is the same as in Equation 5.6

ω is the power dependence that accounts for the particular dissolution or precipitation mechanism (for details see discussions in the works of Lasaga et al. [1994] and Glassley et al. [2003])

Given the number of moles present, the time required to dissolve a material or the time required to precipitate a given amount of material can be calculated from Equation 5.8.

Rarely, however, are there sufficient data available to apply Equation 5.8 in a rigorous way to natural systems and processes. High precision and well-determined rate constants, for example, are known for only a few tens of minerals, and these can differ by many orders of magnitude (see the compilation of Wood and Walther 1983). In addition, the dependencies on α, ϕ, $\prod$, and ω are rarely known. As a result, various simplifications to rate laws have been proposed to account for the behavior of particular minerals or suites of minerals for a limited range of mineral–fluid conditions (e.g., Aagaard and Helgeson 1977; Lasaga 1981, 1986; Wood and Walther 1983; Velbel 1989; Chen and Brantley 1997), or simplifying assumptions are made to allow approximations to be computed for the extent to which reaction progress occurs.

Even though these various treatments for reaction kinetics in geological systems remain approximations, substantial insight can be gained by considering how rates of reactions influence what can be observed in natural systems. As an example of the insight that can be gained by evaluating reaction progress, consider the question of how reaction time is affected by grain size and temperature. Assume that we are considering a system in which a mineral has a known dissolution rate constant that was experimentally measured for a given set of conditions. Assume, also, that we know how that rate constant changes as temperature rises beyond the conditions for which Q/K equals 1.0.

Plotted in Figure 5.6 are the required times for such a reaction to go to completion, as a function of exposed surface area, and the extent to which the temperature of the system exceeds the equilibrium temperature for the reaction. The approach used is that of Lasaga (1986) and is specific to a particular reaction and reaction rate.

Figure 5.6 illustrates several key points relevant for understanding the geochemistry of geothermal systems. Foremost is that the geochemistry of any particular geothermal system is not likely to be

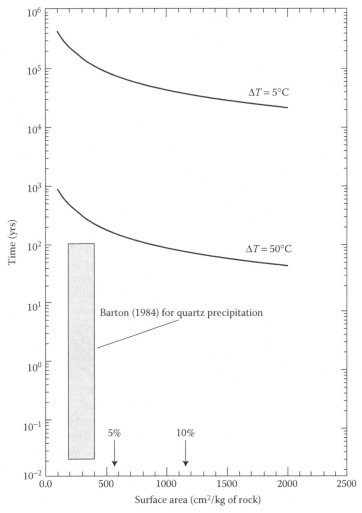

FIGURE 5.6   The time required for a reaction to go to completion, as a function of the exposed surface area, using the approach of Lasaga (1986). The ΔT values indicate the temperature overstepping, relative to the equilibrium temperature for the considered reaction, in this case the dehydration reaction for muscovite. Also shown is the range of times suggested by Barton (1984) for quartz to precipitate from hydrothermal fluids in the temperature range of about 100°C–300°C. The arrows indicate the total surface area, per kg of rock, of a spherical mineral with a radius of 0.1 cm, if the mineral composes 5 vol% and 10 vol% of the rock.

a reflection of an equilibrium state. The time it takes to achieve equilibrium for any specific reaction can vary from seconds to millions of years, depending upon the conditions at a specific location. Given the extremely heterogeneous distribution, on the meter scale, of mineralogy, grain size, pore volume, permeability, exposed surface area, and fluid composition, many competing reactions will be occurring simultaneously, each approaching an equilibrium condition at a different rate. This will be just as true for individual minerals distributed in a rock, as it will be for a multitude of reactions that involve multiple mineral phases. Hence, when considering the equilibrium condition of a specific geothermal site, it is wise to view those conditions as simply points along a multitude of coexisting evolutionary pathways.

The figure also illustrates the relative sensitivity of reaction progress to perturbations in temperature and changes in surface area along a flow path. Note that a few tens of degrees difference in the actual temperature, relative to the equilibrium temperature of a reaction, can change the time it takes to achieve complete reaction by thousands of years. Changes in the exposed surface areas will also affect the time required to achieve complete reaction, but the effect is less pronounced than for temperature, reflecting the fact that reactions scale approximately linearly with surface area, but exponentially with temperature, as is evident from Equation 5.8.

Although the results in Figure 5.6 are specific to the methodology outlined by Lasaga (1986), other approaches and other reactions would give qualitatively similar results, in the sense that larger surface areas and higher temperatures (relative to the equilibrium temperature) result in shorter reaction times. However, the absolute changes in reaction times will depend on the amount of mineral present, its surface area, and the reaction rate for the assemblage of minerals involved in the reaction at the temperature of the system. Clearly, to obtain an accurate understanding of the chemical characteristics of the fluids sampled in a specific geothermal system requires knowledge of the mineralogical characteristics throughout the flow path the geothermal fluid has traversed.

GASES IN GEOTHERMAL FLUIDS

Geothermal fluids always contain a dissolved gas component that can play an important role in establishing the chemical characteristics of the fluid, as well as providing insights into the properties of the geothermal reservoir. The source and influence of that component are multifaceted.

As noted in Chapter 2, the heat source for geothermal systems often is an underlying magmatic system that has perturbed the local geology. When magmas are generated, usually many kilometers underground, they include components of the rocks that melted to form them. Because the melting process involves large volumes of many different minerals, the melts that are generated are complex chemical systems that contain virtually every element in the periodic table. As the melts migrate to shallow levels in the crust, cool, and start to crystallize, a fractionation process occurs in which some elements are preferentially incorporated into the newly forming minerals, and others are excluded because they do not fit into the crystal structure of the new minerals. Many of the excluded elements and compounds have a tendency to form volatile compounds, including H_2O, CO_2, H_2S, O_2, CH_4, and Cl_2. As the magma crystallizes, the diminishing volume of molten material that contains these compounds as a dissolved load eventually reaches saturation in them and they begin to exsolve, migrating as gases and fluids from the crystallizing magma into the surrounding rock. As these volatiles migrate upward, they interact with geothermal fluids and groundwater. Consequently, geothermal fluids can contain elevated concentrations of these constituents.

The host rocks through which heated geothermal fluids migrate also can contribute to the development of a dissolved gas component. If, for example, the rocks contain calcite ($CaCO_3$), dissolution of that mineral will increase the amount of carbonate (CO_3^{2-}) and bicarbonate (HCO_3^-), which ultimately will increase the dissolved CO_2 content via the reactions listed below.

The influence of these components is expressed in the aqueous chemistry that occurs in solution. Some of the reactions in which these compounds are involved, and their respective log K values at 25°C, are the following:

$$\begin{array}{lll}
H_2S(aq) & \Leftrightarrow H^+ + HS^- & \log K = -6.9877 \\
H_2SO_4 & \Leftrightarrow H^+ + HSO_4^- & \log K = -1.9791 \\
HSO_4^- & \Leftrightarrow H^+ + SO_4^{2-} & \log K = 1.0209 \\
HCl(aq) & \Leftrightarrow H^+ + Cl^- & \log K = 0.67 \\
CO_2(aq) + H_2O & \Leftrightarrow H^+ + HCO_3^- & \log K = -6.3447 \\
HCO_3^- & \Leftrightarrow H^+ + CO_3^{2-} & \log K = -10.3288
\end{array}$$

Notice that all of these reactions involve H^+, which strongly influences the acidity of a solution (see Sidebar 5.1 for discussion of acidity).

In Figure 5.7 are plotted the pH values of the solutions listed in Table 5.3, as a function of the chloride content. The very strong correlation between acidity and chloride concentration reflects the fact that addition of chloride to a solution will make hydrochloric acid (HCl), and HCl strongly favors high activities of hydrogen ion in solution, as implied by the positive $\log K$ value shown above.

GAS PARTITIONING BETWEEN LIQUID AND VAPOR

In geothermal systems used for power production, the dissolved gases can have an important impact on the engineering and management of the power generation complex. Dissolved gases will stay in solution unless the pressure is decreased, which is exactly what happens as geothermal fluids are brought to the surface and used to turn turbines. Consider again the pressure–enthalpy figure we previously discussed in Chapter 3 (Figure 3.9). In this case, however, assume that the solution is one of the fluids listed in Table 5.3, in which there are dissolved gases. When the fluid reaches point A, steam begins to separate. Because dissolved gases have different thermodynamic

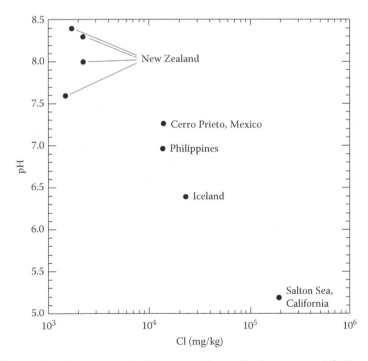

FIGURE 5.7 The relationship between chloride content (in mg/Kg) in geothermal fluids and the pH of the solution for the geothermal waters in Table 5.1.

TABLE 5.3
Chemical Composition of Gases from Diverse Geothermal Systems

Location	Enthalpy (J/g)	CO_2	H_2S	CH_4	H_2	NH_3
Wairakei, New Zealand (W24)	1135	917	44	9	8	6
Tauhara, New Zealand (TH1)	1120	936	64	–	–	–
Broadlands, New Zealand (BR22)	1169	956	18.4	11.8	1.01	4.85
Ngawha, New Zealand (N4)	968	945	11.7	28.1	3.0	10.2
Cerro Prieto, Mexico (CPM19A)	1182	822	79.1	39.8	28.6	23.1
Mahia-Tongonan, Philippines (103)	1615	932	55	4.1	3.6	4.3
Reykjanes, Iceland (8)	1154	962	29	1	2	–
Salton Sea, California (IID1)	1279	957	43.9	–	–	–
Paraso, Solomon Islands (A3)	–	966	19.8	2.8	2.5	0.02
Paraso, Solomon Islands (B4)	–	968	20.4	4.7	3.0	<0.01

Sources: Henley, R.W. et al., *Fluid-Mineral Equilibria in Hydrothermal Systems*, vol. 1. Reviews in Economic Geology. Chelsea, MI: Society of Economic Geologists, 1984; Solomon Islands from Giggenbach, W.F., *Proceedings of the World Geothermal Congress*, Florence, Italy, 995–1000, 1995.

Note: Expressed in millimoles of component per mole total gases.

properties (which is documented by the differing log K values above), some will have begun to exsolve from the solution prior to that point, whereas others will still be in solution. Once steam forms, however, a new process begins to influence the behavior of the dissolved gases, namely, the necessity to partition the total mass of each dissolved species between liquid and gas (steam). Thermodynamically, this partitioning process reflects the driving force for all substances to achieve a thermodynamic equilibrium condition, which can be expressed as (using H_2S as an example species)

$$\mu H_2S(aq) \Leftrightarrow \mu H_2S(g)$$

where $\mu H_2S(aq)$ and $\mu H_2S(g)$ represent the *chemical potential* of the hydrogen sulfide dissolved in the aqueous solution and in the gas phase, respectively.

Chemical potential is the sum of all the attributes of the system that determine the thermodynamic energy of a substance, for example, enthalpy, entropy, and PV work, and is the ultimate driving force for any chemical reaction. At equilibrium, all the chemical potentials of the participating substances in all phases must be equal, by definition. All compounds must obey this principle. For our considerations, the question we must address is how to determine how much of the compound must enter the gas phase in order for equilibrium to be achieved.

Currently, reliance is placed on experimental data to establish the partitioning relationship for any species, because the theoretical, quantum mechanical calculations are quite daunting. The approach used here is that of Alvarez et al. (1994), who developed the following function for describing partitioning behavior:

$$\ln K_D = (-0.023767 \times F) + \left\{ \frac{E}{T} \times \left[\left(\frac{\rho_l}{\rho_{cp}} \right) - 1 \right] \right\} + \left(F + \left\{ G \times \left[1 - \left(\frac{T}{T_{cp}} \right)^{2/3} \right] \right\} \right.$$

$$\left. + \left\{ H \times \left[1 - \left(\frac{T}{T_{cp}} \right) \right] \right\} \exp \left[\frac{(273.15 - T)}{100} \right] \right)$$

(5.9)

TABLE 5.4

Correlation Parameters for Use in Calculating Distribution Coefficients According to Equation 5.9

Solute	E	F	G	H
H_2	2286.4159	11.3397	−70.7279	63.0631
O_2	2305.0674	−11.3240	25.3224	−15.6449
CO_2	1672.9376	28.1751	−112.4619	85.3807
H_2S	1319.1205	14.1571	−46.8361	33.2266
CH_4	2215.6977	−0.1089	−6.6240	4.6789

Source: Data are from Fernandez-Prini, R. et al., Aqueous solubility of volatile non-electrolytes. In *Aqueous Systems at Elevated Temperatures and Pressures: Physical Chemistry in Water, Steam and Hydrothermal Solutions*, eds. D.A. Palmer, R. Fernandez-Prini, and A.H. Harvey, Elsevier, Boston, MA, 73–98, 2004.

where:

K_D is the mass ratio between the gas and liquid phases of the species of interest

T is the temperature (K)

T_{cp} is the temperature at the critical point of water (647.096 K)

ρ_l is the density of water at the temperature of interest

ρ_{cp} is the density at the critical point of water

$E, F, G,$ and H are correlation parameters that are derived from fits to the experimental data (Table 5.4)

Figure 5.8 shows the variation of the log K_D with temperature for H_2, O_2, CO_2, and H_2S. The figure demonstrates the strong partitioning into the gas phase of these species and the strong temperature

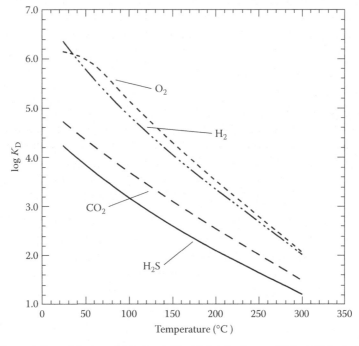

FIGURE 5.8   The variation of the log of the steam: liquid distribution coefficient (K_D) of selected gases with temperature.

dependence of the partitioning. Notice that, at high temperatures, the partitioning for these gases is within an order of magnitude of each other. As the gases cool, the differences in thermodynamic properties result in widely different partitioning behavior such that, at 100°C, the difference in partitioning is more than two and a half orders of magnitude. Among other things, this means that gas analyses carried out at low temperature must be corrected for this contrast in partitioning behavior if reconstruction of the water composition within a geothermal aquifer is to be obtained. The implications for gas compositional changes for generating equipment will be discussed in more detail in Chapter 10.

FLUID FLOW AND MIXING IN NATURAL SYSTEMS

The primary limitation of the approaches described thus far for understanding the geochemistry of reservoirs and geothermal systems comes from the fact that we have not quantitatively considered the complicating effect that flowing water has on the evolving geochemical system. In natural geothermal systems, fluid flow occurs. Flow is driven by gravitational effects and thermal effects, as described in Chapter 4. The rates of flow between different geothermal systems and within a given geothermal system can vary by many orders of magnitude, from millimeters per year to kilometers per year. In an engineered system where fluid is extracted and reinjected, flow rates can be hundreds of cubic meters per hour where extraction and injection occur, but can be much lower than that at some distance from the well. If fluid is moving through a porous or fractured medium, it is problematic whether fluid will remain in contact with a specific mineral for a period of time sufficient for chemical equilibrium to be achieved. Because rock systems can be highly variable in their mineralogy and permeability over distance scales of meters to kilometers, the environment a fluid experiences as it migrates along a flow path is not likely to be constant or uniform. The fluid, as a result, will evolve in response to competing and changing chemical reactions that occur along a flow path. It is thus likely that a fluid flowing through a geothermal system will, if sampled and analyzed at different locations along a flow path, possess detectably different compositions.

Conceptually, it is relatively straightforward to imagine the evolutionary history recorded in the chemistry of an aqueous fluid migrating through a uniform sequence of rock. As noted in Chapter 4, the hydrology of geothermal systems is strongly influenced by the local meteoric plumbing system. If recharge of an aquifer occurs primarily in response to precipitation in a nearby highland region, the hydraulic head that results drives the fluid flow. If our geothermal system is located within a basin, the fluid flow into the basin interacts with the thermal regime generated by a local heat source at depth, and the fluid is heated. The result will be an increase in reaction kinetics that will cause the migrating fluid to more aggressively interact with the surrounding and enclosing mineralogy. In general, the concentration of dissolved constituents will increase along the flow path. If the mineralogy along the flow path is known, the rate of fluid movement is known and estimates can be made of the exposed surface area of the minerals, a reasonably good approximation can be developed of how the fluid composition and mineralogy would evolve along the flow path. Mathematical formalisms useful for describing such a simple system have been developed and are relatively easy to apply (e.g., Steefel and Lasaga 1994; Steefel and Yabusaki 1996; Johnson et al. 1998).

To achieve realism, however, our conceptual model must allow for mixing of fluids from a variety of sources, as well as changes in the porosity and permeability of the geological framework as the solution interacts with it, dissolving the original minerals in the rock and precipitating new ones. Fluid mixing reflects the situation that the thermal regime that supplies the geothermal energy increases the temperature of deep fluids, thus causing them to become more buoyant. As a result, they rise from deep levels and interact with meteoric fluids at shallower levels, resulting in a complex mixing regime that modifies both the chemistry and flow pathways of the system.

The consequences of fluid mixing on the energy content (enthalpy) of geothermal fluids can be evaluated using the energy and mass balance equation we previously considered in Chapter 3.

Imagine, for example, that a meteoric fluid at 15°C mixes with a geothermal fluid rising from depth that is at 250°C. Assuming that the meteoric water has an enthalpy of 70 J/g and the geothermal fluid has an enthalpy of 1085 J/g, if the fluids mix in the ratio of 9:1 (meteoric to geothermal), the mixed fluid would have an enthalpy of (m indicating meteoric water and g indicating geothermal water)

$$(0.9 \times 70 \text{ J/gm})_m + (0.1 \times 1085 \text{ J/gm})_g = 171.5 \text{ J/gm}$$

Such a fluid would have a temperature of about 25°C. Hence, although not useful for power generation, such a thermal perturbation could be readily distinguished from the background ground-water temperature of 15°C and thus indicate the presence of a thermal anomaly, or heat source at depth. For nonpower applications (see Chapters 11 and 12), such a temperature may, in fact, be quite suitable.

Recent developments in reactive transport modeling allow rigorous description of the evolution of such systems over time. In Chapters 6 and 13, we will describe how such simulators can be useful for detecting and characterizing flow and mixing in geothermal systems, and how such capabilities also allow better management of reservoirs. In Section "Simulating Reactive Transport," we consider a very brief overview of the state-of-the-art of reactive transport simulators.

SIMULATING REACTIVE TRANSPORT

Rigorous representation of the time-dependent evolution of complex geological systems requires the ability to account for changes in the physical geological framework as chemical processes unfold in a flowing system that is experiencing changes in temperature and pressure. Within the past 15 years, the ability to undertake such modeling efforts has rapidly evolved. A number of computer codes currently exist that are useful for modeling coupled hydrological, geochemical, and physical processes (Reed 1997, 1998; Glassley et al. 2001, 2003; Xu et al. 2004; Steefel et al. 2005; Kühn and Gessner 2006). These "reactive transport" simulators, to varying degrees, rigorously represent the complex chemical interactions as well as the hydrology of the flowing systems in complex thermal regimes that possess, in three dimensions, varying hydrological and mineralogical properties.

The requirements for carrying out simulations of the evolution of geothermal systems using these computer codes are daunting. The basic inputs for defining the initial conditions of the model are as follows:

- The number of rock units and their three-dimensional distribution in space
- The mineralogy of the rock units and the initial reactive surface areas for each mineral
- The fracture and matrix porosity and permeability in each rock unit
- The temperature distribution
- The pressure distribution
- Initial compositions of the original fluid(s), including multiple liquid and gas phases

In addition, a database that contains the possible minerals that can form over time as the system evolves and their respective thermodynamic and kinetic properties must also be available in order to model the reacting system.

These initial constraints, as well as the adequacy of the thermodynamic and kinetic databases, have a very strong influence on the accuracy of any simulation that is conducted. Errors in any of the initial conditions will propagate through the simulation, resulting in mismatches between the model and what would happen in the real world. It is for this reason that it is important to have as complete a description as possible of the real-world system that is being modeled. As well, it is critical that numerous test runs be conducted to evaluate deficiencies and inaccuracies in the model by comparing the preliminary results to simple and well-understood systems before large, complex simulations be undertaken.

An example of an application of a reactive transport simulator is shown in Figure 5.9. In this instance, there is no flow of fluid considered. Rather, the primary variable is the relative proportion of water to rock. The physical significance of this variable can be viewed as analogous to a fracture in a rock through which fluid is flowing. At the edge of the fracture, the ratio of fracture water to rock is very high. But as the distance to the fracture edge increases, that is, as one moves from the edge of the fracture toward the interior of the rock, the amount of fracture water present will decrease. Hence, high values for the water-to-rock ratio would represent the interior of an open fracture where fluid is almost exclusively present, and low water-to-rock ratios would represent regions in the interior of the rock where little fracture water has penetrated. A real-world example of this variability is shown in Figure 4.4e, where the alteration around the fracture/vein results from interaction of fracture fluids with the rock surrounding the fracture.

Although there is substantial detail that can be derived from this treatment, the key points in this particular simulation, and which can be generalized to many geothermal systems, are the following:

1. The fluid-to-rock ratio is a key parameter in understanding water chemistry. The smaller the volume of water that interacts with a given mass of rock, the stronger will be the imprint of the rock chemistry on the water chemistry. Conversely, the larger the volume of water that interacts with the rock, the smaller will be the chemical modification of the water by the rock. In other words, the resulting chemistry of the water will either be rock dominated (small water-to-rock ratio) or water dominated (large water-to-rock ratio).

2. The extent of rock alteration to new or secondary minerals is strongly influenced by the amount of water that has interacted with the rock. Small water-to-rock ratios will lead to small degrees of rock alteration, whereas large water-to-rock ratios will result in extensive rock alteration.

3. Incremental changes in physical variables can result in abrupt changes in mineralogy and fluid chemistry. An example of this can be seen at a value of the log of the water-to-rock ratio of about 0.55. Such changes reflect the consequences of differences in the dependency of the Gibbs energy function of minerals on physical parameters—at one set of values for a range of variables, a particular set of minerals (a *mineral assemblage*) will have the lowest Gibbs energy, whereas a small shift in the values of those variables will result in another mineral assemblage having a lower Gibbs energy. It is this phenomenon that results in the often observed occurrence of *mineral zones* within and around fractures (e.g., the green zone around the white fracture in Figures 4.4e).

Similar tools allow modeling of migrating fluids through porous and/or fractured rock media, with the results of the simulations including changes in heart distribution and evolution of permeability due to mineral dissolution and precipitation, as well as the ability to track the changes shown for the simulations above. We will discuss applications of tools such as these in more detail in Chapters 6 and 13.

SYNOPSIS

Water acts as a solvent when it interacts with rock systems. As a result, it contains chemical signatures of processes, conditions, and compositions that reflect the history of its migration through the subsurface. This information is useful for evaluating the properties and conditions of geothermal reservoirs. The chemical processes that determine the signatures contained in a solution, and which must be considered when interpreting analytical data, include the chemical potentials of the components in a system of phases, the phases that are present, and the activities and Gibbs energies of components and phases. Also important are the activity coefficients and the influence they have on the equilibrium constants and resulting affinities and ion exchange processes.

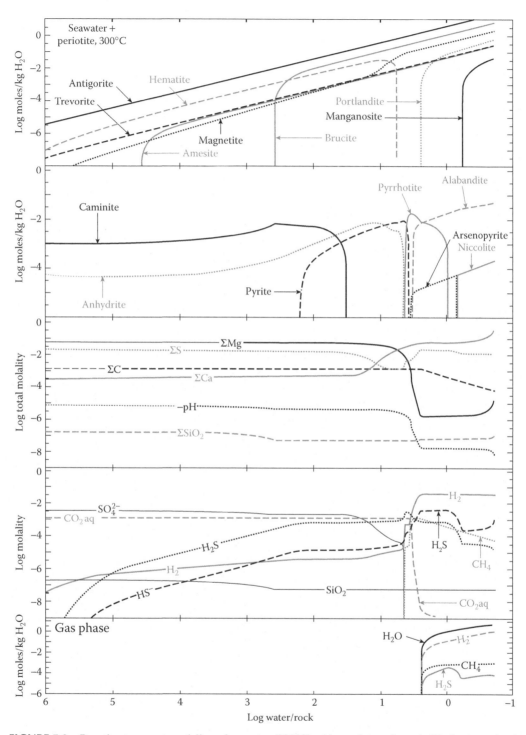

FIGURE 5.9 Reactive transport modeling of seawater (300°C) with an ultramafic rock. The horizontal axis is the mass ratio of water to rock. The upper two panels show the changing abundances of minerals (in log of the moles of solid per kilogram of solution). The third and forth panels show the molality of the aqueous dissolved species, and the bottom panel shows the gas phase species abundances (in log moles of the species per kilogram of liquid). (Modified from Palandri, J.L. and Reed, M.H., *Geochimica et Cosmochimica Acta*, 68, 1115–1133, 2004)

Together, this collection of basic chemical attributes of a system establishes the chemistry of the solution and allows the extraction of such information as reservoir temperature, mineralogy, gas compositions, and reaction paths.

CASE STUDY: SILICA SYSTEM

The mineral quartz is ubiquitous in geological systems. The properties of this system provide an example of the range of issues, and their complexity, that must be borne in mind when working with geological systems that make up geothermal sites.

The chemical composition of the silica system is SiO_2. SiO_2, however, occurs in several different forms or *polymorphs*. The minerals α-cristobalite, β-cristobalite, α-quartz, β-quartz, and tridymite are some of the mineral forms that can occur in geothermal systems. The differences between them are in how the individual atoms that form the crystal lattice for each mineral are arranged. Opal, a form of α-cristobalite that contains water in microscopic pores, is also relatively common. As well, amorphous silica, which is a form of SiO_2 that is glass-like with no crystalline structure on the scale of microns or larger, can also occur in a variety of settings where rapid precipitation occurs.

Each polymorph has specific conditions within which it is stable. Under dry conditions, α-quartz is stable from surface conditions to 573°C, at which temperature it transforms to β-quartz. Above 867°C, tridymite is the stable polymorph of SiO_2. In the presence of water or water vapor, SiO_2 can form as α-cristobalite up to about 220°C, beyond which β-cristobalite forms. This transformation is reversible in the presence of water vapor (Meike and Glassley 1990).

The different *P–T* fields reflect the different thermodynamic properties of the various polymorphs. By implication, therefore, each polymorph must also have its own rate constant for dissolution and precipitation and its own saturation value as a function of *T*. Plotted in Figure 5.10

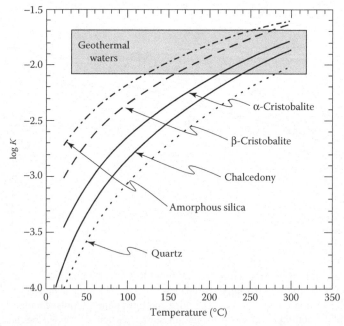

FIGURE 5.10 Log of the equilibrium constant as a function of temperature for the indicated silica polymorphs. The shaded region near the top of the figure encloses the compositions of the waters in Table 5.2.

are the saturation values for amorphous silica, α- and β-cristobalite, and quartz, as a function of temperature. Under nearly all conditions appropriate for geothermal systems, the geothermal waters listed in Table 5.2 are supersaturated with respect to quartz. At temperatures below 100°C, all of the waters are supersaturated with respect to all of the polymorphs shown in Figure 5.10. As a result, geothermal generating equipment that is in contact with geothermal fluids has the potential to acquire deposits of silica (silica scale). Indeed, many geothermal power facilities have in place programs for remediating this problem. This issue is discussed in detail in Chapters 10 and 14.

The rate at which silica precipitation occurs is not readily predicted, despite the work that has been done to evaluate silica polymorph precipitation and dissolution kinetics. Carroll et al. (1998) have shown that the precipitation rate of amorphous silica in New Zealand geothermal systems is influenced by, among other things, the concentration of aluminum in solution. This and other chemical properties of the solution can be a factor influencing reaction rates and argues for the importance of acquiring high-quality chemical analyses for geothermal fluids in order to fully understand the properties of any specific geothermal site.

PROBLEMS

5.1. The dissolution of calcite into water is represented by the reaction $CaCO_3 + H^+ \Longleftrightarrow Ca^{++} + HCO_3^-$. The log of the equilibrium constant for this reaction is -0.5838 at 200°C. Write the equilibrium constant expression for this reaction.

5.2. If calcite is dropped into a solution at 200°C that has a pH of 2.0 and an $HCO3^-$ concentration of 0.0001 molal, will it dissolve?

5.3. If the concentration at 200°C of Ca^{++} is 0.02 molal and the pH is 8.0, what is the concentration of HCO_3^-?

5.4. If the value of the equilibrium constant for a reaction was equal to -3.7, what would be the Gibbs energy of the reaction at 300°C?

5.5. What would be the concentration of $CO_2(aq)$ in water at 25°C if the water was in contact with calcite, and the same pH and bicarbonate concentrations were present as in problem 5.2?

5.6. What would be the amount of CO_2 in a coexisting gas phase? Assume that the density of water at its critical point is 0.32 g/ml.

5.7. Using the data in Figure 5.2, determine whether any of the fluids listed in Table 5.1 are in equilibrium with quartz at 200°C.

5.8. Compute the value of the log Q for the alkali feldspar reaction for each of the waters in Table 5.1. Assuming a temperature of 200°C, which waters are in equilibrium with the potassium end member and which are in equilibrium with the sodium end member.

5.9. If a geothermal water at 150°C had an aqueous silica concentration of 0.006 molal, would it be in equilibrium with any silica polymorphs.

5.10. Compute the charge balance on the analysis for the Iceland geothermal fluid in Table 5.1 and decide whether this analysis is sufficiently accurate for geothermometric work.

REFERENCES

Aagaard, P. and Helgeson, H.C., 1977. Thermodynamic and kinetic conditions on the dissolution of feldspars. *Geological Society of America: Abstracts with Programs*, 9, 873.

Alvarez, J., Corti, H.R., Fernandez-Pirini, R., and Japas, M.L., 1994. Distribution of solutes between coexisting steam and water. *Geochimica et Cosmochimica Acta*, 58, 2789–2798.

Barton, P.B., Jr., 1984. High temperature calculations applied to ore deposits. In *Fluid-Mineral Equilibria in Hydrothermal Systems*, vol. 1. Reviews in Economic Geology, eds. R.W. Henley, A.H. Truesdell, and P.B. Barton, Jr. Chelsea, MI: Society of Economic Geologists, pp. 191–201.

Brook, C.A., Mariner, R.H., Mabey, D.R., Swanson, J.R., Guffanti, M., and Muffler, L.J.P., 1979. Hydrothermal convection systems with reservoir temperatures ≥90°C. In *Assessment of geothermal resources of the United States—1978: US Geological Survey Circular 790*, ed. L.J.P. Muffler. Washington, DC: US Department of the Interior, Geological Survey, pp. 18–85.

Carroll, S., Mroczek, E., Alai, M., and Ebert, M., 1998. Amorphous silica precipitation (60°C to 120°C): Comparison of laboratory and field rates. *Geochimica et Cosmochimica Acta*, 62, 1379–1396.

Chen, Y. and Brantley, S.L., 1997. Temperature- and pH-dependence of albite dissolution rate at acid pH. *Chemical Geology*, 135, 275–290.

Fernandez-Prini, R., Alvarez, J.L., and Harvey, A.H., 2004. Aqueous solubility of volatile non-electrolytes. In *Aqueous Systems at Elevated Temperatures and Pressures: Physical Chemistry in Water, Steam and Hydrothermal Solutions*, eds. D.A. Palmer, R. Fernandez-Prini, and A.H. Harvey. Boston, MA: Elsevier, pp. 73–98.

Fournier, R.O., 1992. Water geothermometers applied to geothermal energy. In *Applications of Geochemistry in Geothermal Reservoir Development*. Series of Technical Guides on the Use of Geothermal Energy, ed. F. D'Amore. Rome, Italy: UNITAR/UNDP Center on Small Energy Resources, pp. 37–69.

Giggenbach, W.F., 1988. Geothermal solute equilibria: Derivation of Na-K-Mg-Ca-geoindicators. *Geochimica et Cosmochimica Acta*, 52, 2749–2765.

Giggenbach, W.F., 1992. Chemical techniques in geothermal exploration. In *Application of Geochemistry in Geothermal Reservoir Development*. Series of Technical Guides on the use of Geothermal Energy, ed. F. D'Amore. Rome, Italy: UNITAR/UNDP Centre on Small Energy Resources, pp. 119–144.

Giggenbach, W.F., 1995. Geochemical exploration of a "difficult" geothermal system, Paraso, Vella Lavella, Solomon Islands. *Proceedings of the World Geothermal Congress*, Florence, Italy, pp. 995–1000.

Glassley, W.E., Nitao, J.J., Grant, C.W., Boulos, T.N., Gokoffski, M.O., Johnson, J.W., Kercher, J.R., Levatin, J.A., and Steefel, C.I., 2001. Performance prediction for large-scale nuclear waste repositories: Final Report. Lawrence Livermore National Laboratory UCRL-ID-142866, pp. 1–47.

Glassley, W.E., Nitao, J.J., Grant, C.W., Johnson, J.W., Steefel, C.I., and Kercher, J.R., 2003. Impact of climate change on vadose zone pore waters and its implications for long-term monitoring. *Computers & Geosciences*, 29, 399–411.

Harvey, A.H. and Prausnitz, J.M., 1989. Thermodynamics of high-pressure aqueous systems containing gases and salts. *AIChE Journal*, 35, 635–644.

Henley, R.W., Truesdell, A.H., Barton, P.B., and Whitney, J.A., 1984. *Fluid-Mineral Equilibria in Hydrothermal Systems*, vol. 1. Reviews in Economic Geology. Chelsea, MI: Society of Economic Geologists.

Johnson, J.W., Knauss, K.G., Glassley, W.E., DeLoach, L.D., and Tompson, A.F.B., 1998. Reactive transport modeling of plug-flow reactor experiments: Quartz and tuff dissolution at 240°C. *Journal of Hydrology*, 209, 81–111.

Kühn, M. and Gessner, K., 2006. Reactive transport model of silicification at the Mount Isa copper deposit, Australia. *Journal of Geochemical Exploration*, 89, 195–198.

Laidler, K.J., 1987. Chemical Kinetics. New York, NY: Harper & Row, 513 pp.

Lasaga, A.C., 1981. Transition state theory. In *Kinetics of Geochemical Processes*, vol. 8. Reviews in Mineralogy, eds. A.C. Lasaga and R.J. Kirkpatrick. Washington, DC: Mineralogical Society of America, pp. 135–169.

Lasaga, A.C., 1986. Metamorphic reaction rate laws and development of isograds. *Mineralogical Magazine*, 50, 359–373.

Lasaga, A.C., Soler, J.M., Ganor, J., Burch, T.E., and Nagy, K.L., 1994. Chemical weathering rate laws and global geochemical cycles. *Geochimica et Cosmochimica Acta*, 58, 2361–2386.

Meike, A. and Glassley, W.E., 1990. In-situ observation of the alpha/beta cristobalite transition using high voltage electron microscopy. *Materials Research Society Symposium Proceedings V*, 176, 631–639.

Palandri, J.L. and Reed, M.H., 2004. Geochemical models of metasomatism in ultramafic systems: Serpentinization, rodingitization, and sea floor carbonate chimney precipitation. *Geochimica et Cosmochimica Acta*, 68(5), 1115–1133.

Palmer, D.A., Fernandez-Prini, R., and A.H. Harvey, eds., 2004. *Aqueous Systems at Elevated Temperatures and Pressures: Physical Chemistry in Water, Steam and Hydrothermal Solutions*. Boston, MA: Elsevier.

Reed, M.H., 1997. Hydrothermal alteration and its relationship to ore field composition. In *Geochemistry of Hydrothermal Ore Deposits*, ed. H.L. Barnes. New York: John Wiley & Sons, pp. 303–366.

Reed, M.H., 1998. Calculation of simultaneous chemical equilibria in aqueous-mineral-gas systems and its application to modeling hydrothermal processes. In *Techniques in Hydrothermal Ore Deposits Geology*, Vol. 10. Reviews in Economic Geology. eds. J.P. Richards and P.B. Larson. Littleton, CO: Society of Economic Geologists, pp. 109–124.

Steefel, C.I. and Lasaga, A.C., 1994. A coupled model for transport of multiple chemical species and kinetic precipitation/dissolution reactions with application to reactive flow in single phase hydrothermal systems. *American Journal of Science*, 294, 529–592.

Steefel, C.I., DePaolo, D.J., and Lichtner, P., 2005. Reactive transport modeling: An essential tool and new research approach for the Earth sciences. *Earth and Planetary Science Letters*, 15, 539–558.

Steefel, C.I. and Yabusaki, S.B., 1996. OS3D/GIMRT: Software for modeling multicomponent-multidimensional reactive transport. User's Manual and Programmer's Guide, Version 1.0.

Velbel, M.A., 1989. Effect of chemical affinity on feldspar hydrolysis rates in two natural weathering systems. *Chemical Geology*, 78, 245–253.

Whittaker, E.J.W. and Muntus, R., 1970. Ionic radii for use in geochemistry. *Geochimica et Cosmochimica Acta*, 34, 945–966.

Wood, B.J. and Walther, J.V., 1983. Rates of hydrothermal reaction. *Science*, 222, 413–415.

Xu, T., Sonnenthal, E., Spycher, N., and Pruess, K., 2004. TOUGHREACT: A simulation program for non-isothermal multiphase reactive geochemical transport in variably saturated geologic media. Lawrence Berkeley National Laboratory Report, 38 pp.

SIDEBAR 5.1   Water Analyses

Most reported chemical analyses of water are reported in mg/kg of solution, which equals ppm by weight, or in mg/l of solution. However, most chemical calculations are carried out using moles/kg of solvent (which can be either steam or liquid water). Moles/kg of solvent is called molality, and the unit is molal, abbreviated m.

Using a variety of methods, it has been established that one mole of a substance contains 6.0221e23 atoms (or molecules), which is often called Avogadro's number. The gram formula weight of a substance, also known as the molecular weight, is the mass, in grams, of Avogadro's number of atoms of that substance. These masses have been tabulated for known elements and compounds and can be accessed either in reference materials or online. An excellent reference for molecular weights of compounds is the National Institute of Standards and Technology (NIST) (http://www.nist.gov and http://webbook.nist.gov/chemistry/).

Converting from mg/kg of solute species i to the molal concentration of i requires calculating the number of moles of i in the solution and multiplying that quantity by the amount of solvent in 1 kg of the solution:

$$\text{Molal concentration of } i = \left(\frac{\text{mg}/\text{kg}_{\text{of } i}}{1000 \times \text{mol. wt. of } i}\right) \times \frac{1000}{1000 - \Sigma(z)/1000} \tag{5S.1}$$

where $\Sigma(z)$ is the sum of the concentration of all dissolved species.

Conversions such as this, where the number of moles of a species is determined, is an important first step in establishing the quality of an analysis. A poor quality analysis can lead to miscalculations of reservoir temperature or other important resource estimates, which can become costly errors. An important test of the quality of an analysis is to determine the charge balance. Because all solutions are electrically neutral, the total negative charge attributable to the number of moles of anions must exactly match the total positive charge attributable to the number of moles of cations. For example, in a chemical analysis of a water that has only dissolved salt (i.e., halite = NaCl), the only solutes in the water will be sodium (Na^+) and chlorine (Cl^-) ions and a negligible amount of NaCl (aq). If the analysis is of high quality, the number of moles of Na^+ multiplied by the charge on the sodium atom (i.e., 1.0) computed from the reported concentrations should be within a few percent of the computed number of moles of Cl^- multiplied by the charge on the chlorine atom (i.e., −1.0). If the difference is greater than 10%, the analysis should be considered inadequate and not used. For more complex solutions, the total positive and negative charges (computed by multiplying the charge of the species by its molal concentration and summing all positive and negative charges separately) should be within 10% of being equal.

Chemical analyses of water almost always include an evaluation of the acidity of the water. The acidity is a function of the concentration of the hydrogen ion (H^+) in solution. The measure of acidity is pH, which is equal to $-\log aH^+$, that is, the negative log of the hydrogen ion activity. Neutral pH at 25°C is defined as 7.0, which is based on the reaction

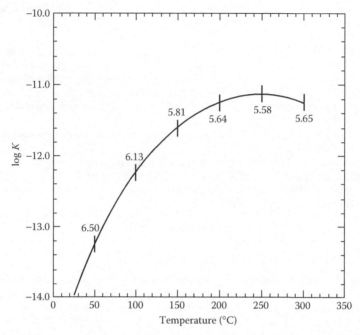

FIGURE 5S.1 Variation of the log of the equilibrium constant for the dissociation of H_2O, as a function of temperature. The vertical tick marks indicate neutral pH values at temperatures of 50°C, 100°C, 150°C, 200°C, 250°C, and 300°C.

$$H_2O \Leftrightarrow H^+ + OH^- \qquad \log K = -13.9951$$

Neutral pH, in other words, is conceptually based on the condition where the hydrogen ion and hydroxyl ion have equal activities in a solution. Acidic solutions have pH values less than 7 and basic or alkaline solutions have pH values greater than 7, at 25°C. At elevated temperatures the log K value for the water dissociation reaction becomes less negative, shifting neutral pH to lower values (Figure 5S.1). There is no absolute measure of a H^+ so it is based on measured differences in the activity of the hydrogen ion in fluids of different composition.

Most analyses of geothermal fluids can be considered to represent the concentration of the fluid in the reservoir, since reaction rates are slow for most minerals compared to the time between exiting the reservoir and sampling. However, in cases where gas separation has occurred, the pH measured in the laboratory is not likely to reflect the pH in the reservoir (note the reactions discussed in this section involving volatile species, and their partitioning between steam and liquid). Under some conditions, correction can be made for this situation (Fernandez-Prini et al. 2004).

6 Exploring for Geothermal Systems
Geology and Geochemistry

Geothermal systems that have the potential to provide economically useful heat are located throughout the world in diverse geological settings. Finding them can be as easy as noting the presence of a geyser, a boiling mud pot, or a steaming pool. Indeed, the early development of geothermal power installations commonly occurred in places that had obvious surface manifestations, such as Larderello in Italy and The Geysers in California. However, there are many resources that have little or no surface expression (so-called hidden resources). Over the course of the last few decades, increasingly sophisticated geochemical, geophysical, and statistical techniques have been developed to aid in the discovery, location, and evaluation of geothermal resources. These techniques assist with identifying regions that possess a thermal resource and establishing its magnitude. They are also needed to determine whether there is sufficient permeability to allow fluid flow and whether there is a high probability that sufficient fluid is present to transfer the heat to the surface. All of these attributes contribute to establishing the degree of risk in developing a resource, and the potential economic value and environmental impacts that developing a resource may have.

The general approach that is followed for exploration purposes is to first analyze available geological information (such as geological maps, drill records, and surface features) to identify environments likely to possess a resource. Key questions that are considered are whether there is evidence that hot geothermal fluids have been active in the region within the last million years or so and, if so, what are their characteristics? If fluids have been active, is there evidence that the fluids are hot enough to support power generation or some other application?

Once a target is identified, field studies involving such things as mapping, and geochemical and geophysical surveys are undertaken to determine the subsurface conditions. If the results of such studies are promising, then more expensive techniques are employed to further evaluate a potential target. These can include more extensive geophysical surveys involving aeromagnetic or resistivity studies, or remote sensing surveys using infrared and hyperspectral techniques. Finally, once a target is delineated, drilling programs are undertaken to fine-tune concepts and models. This chapter describes many of the chemically affected geological features, and analytical techniques that consider chemical interactions that are currently employed to address these issues, as they relate to conventional hydrothermal resources. In Chapter 7, the geophysical techniques useful for this purpose are described.

CLASSIFYING GEOTHERMAL ENVIRONMENTS

The most obvious first step for identifying likely sites that may possess geothermal resources is to focus attention on those geological environments that are known to host active geothermal resources. Referring back to the global map of geothermal sites where power production is currently supported by geothermal systems (Figure 2.6), it is clear that most such systems are in close proximity to regions that are volcanically active or have been volcanically active in the recent past. But there are also regions, such as the Imperial Valley in California, that do not fit this scheme.

111

This raises the question of what geological settings are worth considering when it comes to exploring for geothermal sites?

Several efforts have been made to categorize geothermal systems, based on their geological settings. In Chapter 2, a classification scheme based on very broad geological regimes was outlined. However, such schemes, although useful from the perspective of geological principles, do not assist in refining targets for further exploration. The criteria one chooses to use for exploration purposes depend upon the scale of the systems one is interested in considering and the specific purpose of the classification scheme. For example, an academic perspective in which one is primarily interested in relationships between fundamental driving forces and locations of geothermal resources may be more interested in global or regional scale criteria than a perspective that is solely interested in identifying a specific drilling target. Of course, getting to the point of identifying a drilling target will, at some point, most likely involve consideration of all relevant information at all scales. For that reason, the classification scheme outlined in Chapter 2 can be useful as a starting point for exploration. Refining that process to a more local scale must rely on schemes similar to that presented in Section "Classification of Geothermal Systems from a Regional Perspective."

CLASSIFICATION OF GEOTHERMAL SYSTEMS FROM A REGIONAL PERSPECTIVE

At the regional scale, specific geological attributes, rather than broad tectonic classifications, can be used to focus exploration activities. The following classification scheme is a synthesis of concepts discussed in Chapter 2 (Section "Classification of Geothermal Systems by Their Geological Context") and further developed by Paul Brophy (former President of the Geothermal Resources Council, pers. comm.). The attributes of these types of systems are summarized in Table 6.1. This classification scheme is different from, but complementary to, that described in Chapter 2, in that the scheme here focuses on physiographic features rather than structural features. It therefore does not consider plate tectonics per se and, as a consequence, results in different plate tectonic

TABLE 6.1

Geological Settings That Host Geothermal Systems

Type of System	Topography	Depth to Resource (m)	Surface Manifestations	Permeability
Isolated continental volcanic centers (A)	Mountainous	Moderate to deep (2000 to >4000)	Hot springs, pools	Low to moderate
Andesitic volcano (B)	Mountainous	Moderate to deep (2000 to >4000)	Restricted, depending upon depth and groundwater level	Low to moderate, but fracture permeability can be high
Caldera (C)	Rugged ring fractures, gentle floor	Shallow to moderate (1500–2500)	Springs, pools, geysers, mud pots common	High permeability in tuffaceous units, some fault permeability
Extensional sedimentary basin and spreading centers (D, E)	Rugged on horsts, gentle in grabens	Usually deep (>2500 m)	Normally along bounding faults	Mainly along bounding faults or transverse faults
Oceanic basaltic provinces, hot spots (F)	Rugged	Shallow (<2000 m)	Lava flows, hot springs common	High horizontal permeability along flow units/breccias, etc.; low vertical, mainly fractures

Source: Paul Brophy, pers. comm.

environments being considered within a single category, such as the features in *Type A Systems* which combine volcanic systems from hot spots and volcanic features in transitional settings. The advantage of this approach is that it is independent of the process underlying the feature of interest. This can be especially useful when the specific underlying process that caused the development of a volcanic system is unknown.

Isolated Continental Volcanic Centers (Brophy Type A)

The geological attributes of this category of a geothermal system are diverse, but they share the characteristic that they are active or recently active volcanic systems that are geographically separate from any other obvious volcanic systems. Examples of such volcanic environments are those in The Geysers in California and El Chichón in Chiapas, Mexico (Figure 6.1). These systems are composed of diverse volcanic rocks, commonly have hot springs and other obvious geothermal activity present, and exhibit low levels of seismic activity. The geothermal resource in such systems is commonly at depths of several kilometers. Permeability is low to moderate and mainly derived from fracture systems, since these settings are within continental crust that is mainly crystalline. Such systems are scattered around the globe, but are not common.

Andesite Volcanoes (Brophy Type B)

Andesite volcanoes have excellent potential for geothermal resource development. They are relatively abundant and tend to occur in close proximity to ocean basins. These systems are physiographically prominent because they form stratovolcanoes, which have a well-defined

FIGURE 6.1 **(See color insert.)** El Chichon volcanic center, Mexico. The most recent eruption of this volcano was in 1982. Pictured here is a steaming ground and an acidic lake in the volcano floor. (Photograph from the US Geological Survey, http://hvo.wr.usgs.gov/volcanowatch/uploads/image1-155.jpg.)

FIGURE 6.2 (**See color insert.**) Mt. Augustine, an andesitic stratovolcano, Aleutian Islands. (Photograph by US Geological Survey.)

conical form. They commonly have high fracture permeability that is localized, thus requiring careful mapping of structural features to delineate them. Geothermal reservoirs in such systems are located at a variety of depths, from moderate to deep. Access to the reservoirs is along the flanks and plains surrounding the volcano. Seismic activity associated with the local volcano is generally low, but because these systems are most commonly associated with subduction zones, major devastating earthquakes that originate at significant depths but which are not directly related to the volcanic centers can occur. In addition, hazards associated with major andesitic volcanoes, such as large ash eruptions and ash flows, as well as lahars and avalanches, can be problematic, thus requiring thoughtful evaluation of the local risks. Examples of andesitic volcanic systems include the Aleutian Islands (Figure 6.2), numerous volcanoes in Japan (e.g., Mt. Fuji), the Cascade Mountains volcanoes that extend from British Columbia, Canada, through Washington, Oregon, and Northern California (e.g., Mt. Baker, Mt. Hood, and Mt. Shasta) and San Jacinto in Nicaragua, to name just a few.

CALDERAS (BROPHY TYPE C)

Calderas, as discussed previously in Chapter 4, form when the magma reservoir of a large volcano partially or completely empties through a series of large eruptions. As the magma chamber empties, the overlying rock burden can no longer be supported and the volcanic edifice catastrophically collapses. The resulting physiography is that of a large central depression surrounded by a circular or oval ring of higher topography. This topographic characteristic has associated with it a series of ring faults that lie between the elevated region surrounding the collapse feature and the collapsed central basin. The central depression of these systems is often rejuvenated after the collapse, resulting in the formation of a central dome within the depression. Generally, the geothermal resource will be located at relatively shallow depths (within a few thousand meters). Surface manifestations are common, with bubbling hot springs, boiling mud pots, and steaming ground. Access to the

FIGURE 6.3   (See color insert.) Aniakchak Caldera, Alaska. The caldera is 10 km in diameter and between 500 and 1000 m deep. The caldera formed as the result of explosive eruption of more than 50 km^3 of magma approximately 3450 years ago. After caldera collapse, later small eruptions formed domes and cinder cones. (Photograph by M. Williams, National Park Service, 1977.)

geothermal reservoir will generally be in close proximity to the collapsed part of the system, on either side of the ring of elevated topography. Permeability is low to moderate, but can be quite high if there are porous rock units of coarse ash that were deposited during the eruption, and also within fracture and fault systems that form in association with caldera collapse. Examples of such systems, in addition to Long Valley, California, described previously in Chapter 4, are Yellowstone Caldera in Montana, the Valles Caldera in New Mexico, Los Humeros Caldera in Mexico, Ngorongoro Crater in Tanzania, and Aniakchak Caldera in Alaska (Figure 6.3).

FAULT-BOUNDED SEDIMENTARY BASINS (BROPHY TYPE D)

Geothermal resources can be present in sedimentary basins that are bounded by faults. The mitigating factor that determines if such settings host a geothermal reservoir is whether there has been recent igneous activity. If igneous activity is present, such systems have the potential to be exceptional geothermal resources. Their favorability results from the fact that such settings favor high permeability because of the presence of faults, both along the boundaries of the basin and traversing and cross-cutting the sediments of the basin. In addition, the sedimentary fill can have both high permeability sedimentary layers, such as porous sandstones, and fracture networks. Furthermore, the magmatic bodies that are the heat sources for such systems, often penetrate to relatively shallow levels, resulting in geothermal reservoirs that are within two thousand to three thousand meters of the surface. Surface manifestations can be well developed and prominent, including lava flows, steaming ground, and hot springs. The geothermal system at Cerro Prieto, Mexico, is an example of such a system. However, high sedimentation and subsidence rates can mask the presence of a geothermal system in such settings—the Imperial Valley/Salton Trough (Figure 6.4) geothermal complex in Southern California, which is an extension of the same system as in Cerro Prieto, has no surface manifestations, and was discovered accidentally during drilling for oil and gas.

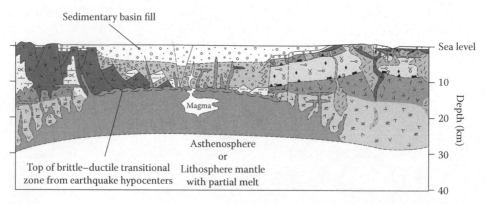

FIGURE 6.4 Cross section through the Salton Trough. (Modified from Fuis, G.S. and Mooney, W.D., 1990. Lithospheric structure and tectonics from seismic-refraction and other data. In *The San Andreas Fault System, California*, ed. R.E. Wallace. *U.S. Geological Survey Professional Paper*, 1515, 207–236.)

FAULT-BOUNDED EXTENSIONAL (HORST AND GRABEN) COMPLEXES (BROPHY TYPE E)

Complexes of horsts (ridges, which are often flat-topped) and grabens (flat-bottomed valleys) occur in regions where there has been extension and thinning of the crust (Figure 6.5). As the crust is pulled apart, it tends to fracture, forming steeply dipping faults that are perpendicular to the general direction of extension. Blocks of crust subside between these faults, forming the grabens, whereas the surrounding ground on the opposite side of the bounding faults remains elevated, forming horsts. The high-angle faults that bound horsts and grabens can extend to considerable depth. Such settings are places where magmas often rise in the crust, in response to the decrease in lithostatic pressure caused by the crustal thinning during extension. As a result of the presence of these heat sources, numerous geothermal reservoirs can be present. Geothermal resources in such settings tend to be relatively deep. Permeability is commonly restricted to fault-controlled zones in the vicinity of horst–graben boundaries (Figure 6.6). Surface manifestations of geothermal resources in these settings are often quite limited and mainly restricted to the presence of hot springs, although evidence of past volcanic activity is sometimes preserved along horst summits and faces. A classic example of this type of setting is the Great Basin of the western United States, particularly as exemplified in Nevada, but also extending into California, Idaho, Utah, and Arizona.

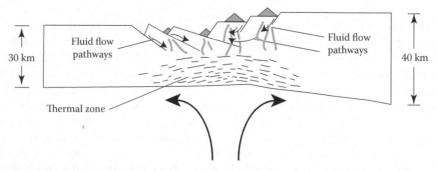

FIGURE 6.5 Schematic cross-sectional diagram of horst–graben structures in an extensional environment in continental crust. Heat rises into the faulted zone from the heated base of the continental crust. Magmatic feeders to volcanic centers cut through the crust. Fluids migrate along fault zones, descending into hot crust and, once heated, ascend through buoyancy along the same or other faults.

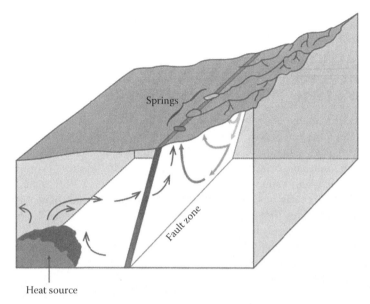

FIGURE 6.6 (**See color insert.**) Schematic block diagram of flow along a fault. Surface water flows from the mountainous region on the right into the fault zone at the base of the mountain range. Water then descends along the fault and becomes heated by the buried heat source in the basin. The heated water ascends, through buoyancy forces, along the fault to the surface where it emerges as springs. The relative temperature of the spring, compared to that of surrounding springs, provides a qualitative indication of the proximity of the spring to the heat source.

OCEANIC BASALTIC PROVINCES (BROPHY TYPE F)

Numerous oceanic island chains and archipelagos have the common characteristics of being relatively young volcanic systems that are predominately composed of basaltic lava and related magmatic products. Although hot springs are occasionally present, the primary manifestation of the existence of geothermal resources is the pervasive, massive, recent outpourings of basaltic lava. These systems have significant potential for the development of geothermal applications. The geothermal reservoirs are generally shallow (~1000–2000 m) and commonly have extensive lateral permeability resulting from the formation of breccias, lava tubes, and other high porosity and high permeability zones associated with the formation of lava. Vertical permeability is more restricted and forms primarily as a result of joint development within individual flows. Seismicity generally is low in these settings. Examples of these types of systems, in addition to those listed in the "hot spot" classification above, include Ascension Island in the Atlantic Ocean and Reunion Island in the Indian Ocean.

Classification schemes, such as this, provide a conceptual framework within which hydrothermal geothermal systems can be viewed. Particular features and characteristics that have commonly been observed in a specific type of system can be expected to be found in other examples of that same type of system when new exploration is undertaken. For example, caldera systems usually have ring faults and resurgent central domes, both of which can be good exploration sites for geothermal fluids. Knowing that such features are likely to be present can focus exploration efforts.

Once a geothermal system has been observed and its "type" identified, more detailed and focused exploration efforts can be initiated. The remainder of this chapter considers the geological, geochemical, and geophysical studies that help establish the extent, magnitude, and accessibility of a geothermal reservoir. But, before addressing exploration methods, it is important to address the issue of the origin of geothermal fluids and their characteristics. This background provides a qualitative conceptual framework within which more rigorous, quantitative geochemical approaches can be applied with an improved ability to understand the significance of particular observations and results.

ORIGIN OF GEOTHERMAL FLUIDS: SIGNIFICANCE FOR RESOURCE EXPLORATION AND ASSESSMENT

The Brophy classifications described throughout the Section "Classification of Geothermal Systems from a Regional Perspective" are based primarily on physiographic features. They involve mountain ranges, volcanoes, valleys, and ridges. Obviously, to be classified as a geothermal feature, a heat source in the subsurface must also be present. Many of these geothermal features also have a natural hydrological system with circulating water associated with them. The origin of that water and its history often can be deduced from chemical and isotopic characteristics of the water. Understanding the processes that affect these signatures allows inferences to be made regarding the characteristics of a resource and its history. The remainder of this section discusses the origins and characteristics of geothermal waters and how those characteristics can be used to learn more about the properties of a geothermal resource. Important in this, too, is understanding the history of a geothermal site. This history can be useful in defining whether a particular system is juvenile and just beginning to evolve, or if it has had an extensive lifetime and is in the process of declining.

Figures 6.6 and 6.7 show the various water sources and their interactions that occur in two of the most common geothermal settings in the Brophy classification scheme—volcanic edifices and ridge–valley pairs. The greatest volume of surface water in these settings is derived from rainfall and snowmelt. This is, by definition, meteoric water—meaning, it is water that is generated as a result of meteorological processes. This water feeds streams, and recharges subsurface aquifers. Meteoric water is obviously present in streams and lakes. It can also occur as the sole component of most springs. In such cases, the local water table intersects the ground surface, and meteoric water escapes to the surface. Such waters are generally dilute. In the United States, drinking water standards stipulate that the load of dissolved constituents, also known as total dissolved solids or TDS, must be less than 500 mg/l, or 500 ppm by weight. Many natural surface waters fall within

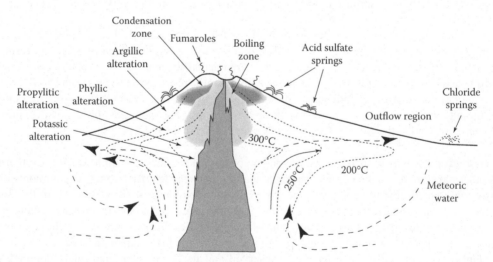

FIGURE 6.7   (See color insert.) A schematic diagram of fluid flow regimes in volcanic systems. The heat source (both magma and solidified igneous rocks) is shown as the orangish body. Surrounding it at shallow levels is a steam zone (light blue) and a condensation zone (dark blue). Surface manifestations such as fumaroles and springs are indicated. The isotherms are shown as finely dashed, labeled lines. Flow paths of meteoric water are shown by the heavy, dashed lines. Approximate regions where argillic, phyllic, propylitic, and potassic alteration would occur are also shown.

this range. However, water interacting with rock will dissolve the minerals in the rock, as discussed in Chapters 5 and 6, and will pick up a variety of constituents, or deposit them if saturation levels become high enough. The fraction of meteoric water that circulates to depths in excess of hundreds of meters to kilometers, depending upon the interconnected permeability of the rock units making up the local geology, will inevitably take on chemical characteristics that are influenced by the rocks through which the water has flowed, the duration of contact, and the temperature history. Geothermal waters fall in this latter category.

Deep circulation of meteoric water is favored by zones of high permeability, such as fault zones that occur in calderas or along range fronts, as depicted in Figure 6.6. This deep circulation can result in the generation of geothermal water if there is a heat source, such as a magma body, or if there is high heat flow because the crust is thin, or if there is a granitic body rich in K, U, Th, or other radioactive elements. As the water heats up, its density will decrease and it will tend to form a circulation pattern that is typical of a convection cell. These circulation systems can extend for many kilometers, allowing the heated water to chemically interact with rocks along the flow path. When these fluids emerge at the surface, they will form springs of varying temperatures (depending upon the temperature of heat source they have interacted with, the distance from the heat source, and the circulation path). The chemical composition of these springs will tend to be relatively dilute, neutral pH waters with some carbonate and chloride present due to the presence of dissolved atmospheric carbon dioxide (CO_2) and the presence of the common incompatible element chloride, which is invariably present in the parts per million range in all natural waters.

Much more chemically complex waters can form, however, if the circulating water interacts with fluids that escape from a crystallizing magma (Figure 6.7). Molten rock contains some small quantity of gasses and water dissolved in the melt. As the magma cools and begins to crystallize, gasses, water, and certain elements that will not fit into the crystal lattice of the minerals that are crystallizing (also called *incompatible* elements) will concentrate in the remaining melt as discussed in Chapter 5. Hydrogen sulfide (H_2S), HCl, CO_2, and He are examples of gasses that will be concentrated in the melt. Incompatible elements, such as B, Ba, Rb, Sr, and Cs, to name a few, will also concentrate in the remaining melt. As the concentration of these incompatible gasses and elements reaches saturation values, which are specific to the magma and chemistry of the system, the individual gases will begin to separate and take with them some proportion of the other incompatible elements. These *exsolved* fluids then migrate out of the melt and into the country rock, where they have the potential to mix with circulating meteoric waters. This mixing is profoundly important because of the chemical signatures it imposes on the circulating water.

One such water, that is, the product of mixing fluids, is *bicarbonate water*. CO_2-rich waters form bicarbonate solutions, as implied by the equilibria described in Chapter 5 in the Section "Gases in Geothermal Fluids." These bicarbonate waters can form when circulating meteoric water encounters and mixes with CO_2-rich steam or water that has condensed from such steam. These mixed fluids can occur in the upflow zones above geothermal reservoirs. They are also common at the distal regions that are peripheral to geothermal systems and form by dissolving atmospheric CO_2 in the geothermal fluid. Both of these scenarios are indicated in Figure 6.7. These bicarbonate waters are usually moderately dilute and commonly form travertine terraces, such as those depicted in Figure 6.8 at their outflow sites.

Another water encountered in geothermal systems is *acid-sulfate water*. In this instance, the chemical composition reflects the interaction of meteoric waters with fluids that originated as H_2S-rich solutions that formed via degassing of magmas and interaction with local rocks. H_2S is a strongly chemically reduced compound. In the presence of even the low partial pressure of oxygen normally encountered in the subsurface and dissolved in meteoric water, H_2S oxidizes to hydrogen sulfate (H_2SO_4) which dissociates to hydrogen ions and sulfate anions, as described in Chapter 5. H_2SO_4 is also known as sulfuric acid. Outflow of these acid-sulfate waters generally results in low-pH hot springs, with the obvious odor of sulfur. In many instances in close proximity to volcanic centers, *fumaloric* activity is evident where native sulfur will be deposited at the vents of fumaroles (Figure 6.9) as the reduced H_2S is oxidized to native sulfur.

FIGURE 6.8 Travertine deposits at Yellowstone National Park. (Photograph by David Monniaux http://commons.wikimedia.org/wiki/Yellowstone_National_Park.)

FIGURE 6.9 (**See color insert.**) Photograph of a fumarole with sulfur deposits at Kilauea Volcano, Hawaii. (Photograph by R.L. Christiansen, US Geological Survey.)

Research into the characteristics and origins of geothermal waters has been extensive. Figure 6.10a is one diagram that is often used for identifying the basic properties of geothermal waters. It relies on the fact that the anions that counter-balance the electrical charge of the cations in solution are generally limited to Cl^-, SO_4^{2-}, and HCO_3^-. These anions, as described above, dissolve in the circulating waters as a result of distinct flow paths. This allows a preliminary evaluation of the water in hot springs and in the subsurface and also provides an opportunity for mapping flow paths, as will be discussed below.

This classification scheme is accomplished by summing the dissolved masses of each of the three anions, as mg of the anion per kg of solution, and then dividing that sum into the mass of each of the anions. This result, when multiplied by 100, gives the percent of each anion in this three component, or *ternary* system. When plotted in a ternary diagram, the solution can be classified as indicated in Figure 6.10. Points that fall within the Cl^--rich portion of the figure are classified as chloride waters and are usually near neutral pH and somewhat alkaline. Waters that plot near the HCO_3^- apex are classified as carbonate or bicarbonate waters, whereas those that plot toward the SO_4^{2-} apex are classified as acid-sulfate waters, because they usually have low pH values.

Once a water has been characterized, general, conceptual models can be developed that provide a framework for testing various hypotheses about the nature of the geothermal resource being considered. For example, a spring with acid sulfate waters might indicate a relatively shallow magmatic resource that is generated from a relatively young igneous body. A spring with bicarbonate waters may indicate the presence of a geothermal resource that is located at some distance from the spring. However, such conceptualizations must be treated as little more than speculative guides that can help in the development of strategies for further research. As shown in Figure 6.10b, the range of temperatures a particular water type can have spans the entire observed range of temperatures. Thus, more research is needed to refine the conceptual model of the chemical signature a water sample develops.

Furthermore, the discussion thus far has treated geothermal fluids as modified meteoric water. Volcanic systems or other heat sources in close proximity to a coastline may also interact with seawater that infiltrates through the subsurface. Such systems, which are epitomized by Iceland, can have fluid-mixing systems in which the chemical signatures of the water are the result of complex interactions between the heat source, meteoric water and magmatic gases, and seawater.

Additionally, sedimentary basins can have very old waters present that have been preserved within the local geological framework for millions of years. Such waters are called *connate* waters. These waters remain out of contact with the atmosphere and undergo extensive water–rock interaction, usually becoming highly concentrated brines over hundreds of thousands to millions of years. These waters, when interacting with a geothermal heat source, can result in exceptionally complex chemical systems.

When using water classification schemes as a guide toward developing further understanding of a resource, it thus becomes important to bear in mind the potential complexities that can result from the local geological setting. With that awareness in mind, rigorous observations and thoughtful research questions can be developed that will improve the conceptual model that is being applied to a geothermal system.

SURFACE MANIFESTATIONS

SPRINGS

Geothermal systems often reveal themselves by the presence of various surface features. Careful observation of these, by mapping their location and noting their properties, can provide valuable information for characterizing a resource.

Water flowing from underground is relatively common. It can manifest itself as seeps that emerge as a trickle from a hill slope or cascades of water that rapidly flow to the surface. These springs can be cold or hot, or perhaps even boiling. All of these attest to the presence of permeable zones

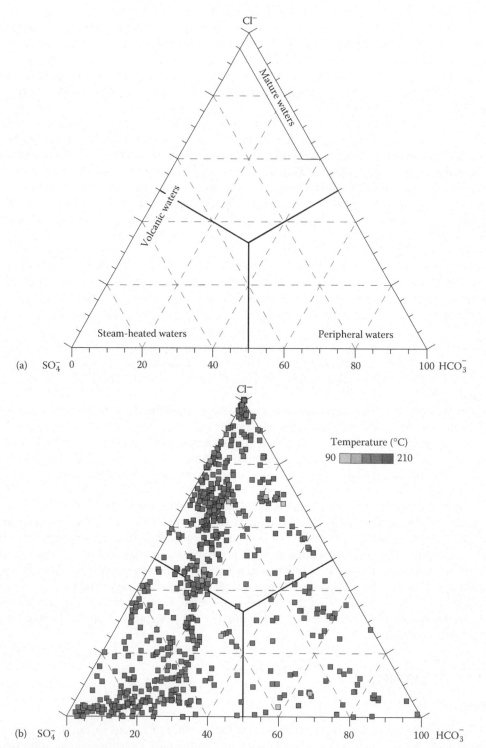

FIGURE 6.10 **(See color insert.)** Ternary diagram of the Cl—SO₄—HCO₃ system. (a) The approximate fields of various water types are delineated and labeled. (b) The ternary compositions of waters in the Argonne National Laboratory Geochemical database. (Modified from Clark, C.E. et al., *Water Use in the Development and Operation of Geothermal Power Plants*, United States Department of Energy, Argonne National Laboratory, ANL/EVS/R-10/5, Washington, DC, 2010.)

through which subsurface water flows. Determining whether or not they are related to the presence of a geothermal system requires a systematic mapping effort. The discussion above regarding water types also makes it clear that sampling of the waters during the mapping effort for chemical analysis is a critical part of an exploration program.

Hot water pools, bubbling mud pots, steaming streams, and geysers are obvious expressions of the presence of an active geothermal system in which water is heated at depth and then escapes to the surface. If the springs occur in a well-defined zone that extends for some distance, they are likely to indicate the presence of a fault-controlled flow system. Figure 6.11 is an example of such a system that occurs in Surprise Valley of northeastern California (Eggers et al. 2010). Note that both cold springs and hot springs are present over distances of tens of kilometers. Note, too, that hot springs are present along the faulted valley margin, as well as in the central part of the sedimentary valley bottom, in locations with and without observable surface faulting. This pattern suggests that there is an extensive geothermal resource at depth and that the flow of hot geothermal fluid in this region is not confined to the faulted margins of the valley, but also extends into the valley, implying that there are likely to be recently buried faults that provide high-permeability fluid flow pathways within the valley bottom.

Springs often have deposits of various kinds associated with them. This reflects the fact that hot waters that circulate in the subsurface are good solvents, dissolving minerals with which they are undersaturated, as discussed in Chapter 5. As these fluids approach the surface, the solubility of minerals often decreases as the temperature drops. When they reach the surface, cooling is rapid and the dissolved load often precipitates, forming mineral deposits around the spring. These deposits can be of several types, reflecting the type of geothermal resource and the associated water chemistry and temperature, as discussed above. These surface deposits generally fall into three types and correspond to water chemistries discussed above, shown in Table 5.1 and summarized in Table 6.2.

As previously noted, the three main types of geothermal waters are acid-sulfate waters, carbonate/bicarbonate waters, and chloride waters. Acid-sulfate waters are most commonly found in the vicinity of active volcanic systems. Their characteristics, which are low pH and moderate to high temperature, develop when meteoric groundwater interacts with the highly reduced, H_2S-rich fluids that may outgas from cooling magmas. Their often moderate temperatures are not necessarily an indication that a high temperature geothermal resource is not present. Rather, it often signals a fluid mixing system in which large volumes of cool, meteoric water have mixed with small volumes of magmatic gases and fluids. These systems can have a variety of mineral deposits at their outflow zones. Sulfur deposits, which are one example, commonly form around fumarolic vents, as shown in Figure 6.9. Bubbling mud pots are also often present. These can form when the acidic fluids chemically interact with the local host rock. The low pH solutions weaken the rock and chemically alter it.

The principle chemical reactions that occur in these acidic settings can result in the formation of clay minerals and sulfates via reactions such as the following (at temperatures less than about 200°C; see discussion of alteration types at the end of this Chapter):

$$3\,NaAlSi_3O_8 + 2\,SO_4^{2-} + 6\,H^+ \Leftrightarrow NaAl_3(SO_4)_2(OH)_6 + 9\ SiO_2 + 2\ K^+$$

Alkali feldspar   Na–Alunite   Quartz

and

$$CaAl_2Si_2O_8 + H_2O + 2H^+ \Leftrightarrow Al_2Si_2O_5(OH)_4 + Ca^{++}$$

Plagioclase feldspar   Kaolinite

Feldspars are among the most common rock-forming minerals. These reactions show how feldspars can convert to sulfates and clays when they interact with sulfur-bearing acidic solutions at temperatures that are usually less than 200°C. This suite of minerals—kaolinite, quartz, and alunite—is characteristic of what is called *argillic* alteration. Also present in argillic alteration are clays such as montmorillonite and illite.

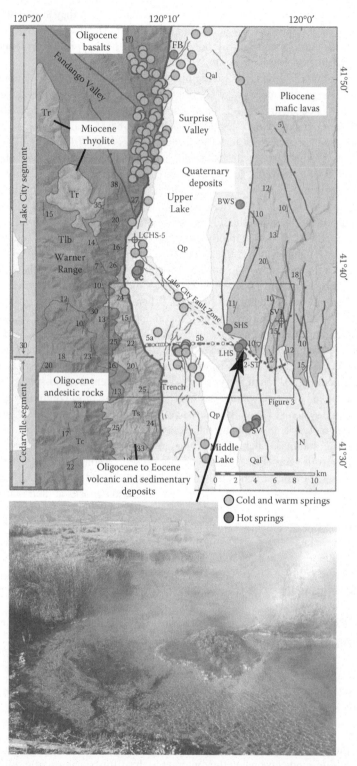

FIGURE 6.11 **(See color insert.)** Geologic map of the Surprise Valley, California, region. Blue dots show the locations of cold and warm springs; red dots show the locations of hot springs. The photograph at the bottom shows one of the hot springs. (Modified from Eggers, A.E. et al., *Tectonophysics*, 488, 150–161, 2010; photograph by the author.)

TABLE 6.2

Geothermal Water Classifications and Properties

Water Type	Cl (mg/kg)	pH	Temperature (°C)	Surface Features
Acid sulfate	<10	<3.0	Usually <100	Mud pools; collapsed pits
Carbonate/bicarbonate	<100	5.0–6.0	~100–180	Warm or hot springs; minor carbonate deposits
Chloride	>400	6.0–7.5	>200	Silica sinter; geysers

Source: Henley, R.W. and Hedenquist, J.W., 1986. Introduction to the geochemistry of active and fossil geothermal systems. In *Guide to the Active Epithermal Systems and Precious Metal Deposits of New Zealand.* Monograph Series on Mineral Deposits, eds. R.W. Henley, J.W. Hedenquist, and P.J. Roberts. Berlin, Germany: Gebrüder Borntraeger, pp. 129–145.

The interaction of bicarbonate waters with host rocks along their flow paths results in a suite of minerals that is distinct from that involved with argillic alteration. Bicarbonate-rich waters invariably precipitate carbonate minerals, such as calcite and dolomite, as they degas CO_2 at shallow levels. In addition, as they cool they often precipitate a SiO_2 phase (such as quartz or cristobalite), along with the carbonates. A variety of clay minerals commonly occur as well, because bicarbonate solutions can be slightly acidic and will interact with the rock in a manner similar to that depicted by the kaolinite-forming reaction given above. A distinctive mineral phase associated with these phases in high temperature bicarbonate systems is adularia, which is a potassium feldspar ($KAlSi_3O_8$) that forms at the expense of clays through the reaction

$$KAl_3Si_3O_{10}(OH)_2 + 6SiO_2 + 2HCO_3^- + 2K^+ \Leftrightarrow 3KAlSi_3O_8 + 2CO_2 + 2H_2O$$

Illite   Quartz   Adularia

When such waters flow from springs at the surface, they precipitate some subset of this suite of minerals as they cool, forming terrace deposits called *travertine* (Figure 6.8). If the outflow is underwater, tufa towers may form, as shown in Figure 6.12.

Near-neutral pH, chloride waters can form a variety of deposits, depending upon the rock type within which they are hosted. Such fluids are commonly associated with higher temperature flow histories resulting in, among other things, a dissolved load that is SiO_2-rich. If the fluid has reached temperatures in excess of about 210°C, they commonly form siliceous *sinter* deposits at the pool outlet of the spring. This type of deposit, sometimes called *geyserite*, is formed by the precipitation of opal or amorphous silica in porous encrustations around outflow pools or geysers.

Springs can be ephemeral or persistent. In areas where chemically aggressive waters exist, springs and other surface features may change their outflow characteristics repeatedly in short time periods (weeks to a few years). Fluid circulation paths will also be modified as hot waters move through permeable zones. As water is heated, it will tend to dissolve minerals along the flow path, as discussed in Chapter 5, thus increasing the permeability. However, as the fluid migrates away from the heat source and begins to cool, precipitation of minerals can occur. Consider, for example, the silica solubility diagram (Figure 5S.1). A geothermal fluid in equilibrium with quartz at 300°C will become strongly supersaturated in all silica polymorphs as it cools. Such a fluid ascending along a fault toward the surface would likely precipitate a silica polymorph, most likely cristobalite, opal, or chalcedony, thus decreasing permeability and eventually sealing the fluid flow path. When that happens, the pressure that drives the fluid flow will result in the formation of a new flow path, with the emergence at the surface of a new hot spring. If the flow system is fault controlled, it is likely the new spring will develop within relatively close proximity to the trace of a fault or other high-permeability zone on the ground surface. Such evolving patterns of springs over time can occur with any type of geothermal water.

FIGURE 6.12   Exposed tufa towers. (Photograph by Doug Dolde.)

Surface Deposits Lacking Active Springs

Many geothermal systems, however, do not have springs or other dramatic surface manifestations. In many instances, this absence of surface activity is due to the stage of geological evolution the site has reached. However, certain geological features can provide evidence of past geothermal activity that would potentially justify additional study.

Silica sinter and travertine deposits will persist long after a hot spring has ceased activity. Silica sinter, as noted above, is composed of fine-grained, banded mixtures of opal, cristobalite, amorphous silica, and/or quartz. These will cement together any detrital clasts that have settled onto the surface of the deposit. It is common for these deposits to develop a complex surface that can be laminated or rounded on a variety of scales and is called *botryoidal*. Under microscopic examination, these surface deposits often contain microspheres, as well as surfaces that are coated with very small crystals of quartz. The later characteristic is termed *druzy quartz*. Bacteria that thrive in hot water commonly colonize these surface deposits and leave behind casts of their form on the silica sinter. These casts of *thermophilic bacteria* can sometimes be useful interpreting the history of a region where the only evidence of past hot spring activity are silica-rich horizons and zones.

Travertine deposits that occur in the absence of flowing springs may also have microscopic evidence useful for interpreting their past history. Bacterial casts can occur, as well as other minerals. Tufa towers that form in the bottoms of lakes or other water bodies are also good indicators of the recent presence of geothermal fluids. Rapid accumulation of these deposits causes the formation of hollow tubes and mounds that can grow to considerable heights. If later geological processes cause the lake to drain, these towers become exposed. In geothermal regions, it is relatively common to find travertine deposits and tufa towers located along linear trends that usually mark the fault system along which the fluids ascended. Mapping and dating such features can be useful for locating potential geothermal resources.

Often associated with such surface deposits are other minerals that can be good indicators of geothermal activity. Borate, sulfate, and chloride minerals, in particular, can be suggestive that geothermal activity may have been persistent in a region. Borates are especially significant because geothermal fluids tend to have elevated boron contents, relative to meteoric groundwaters (Coolbaugh et al. 2006).

This compositional characteristic reflects the fact that boron is preferentially partitioned into the fluid phase, relative to most minerals because of its small size and high charge (3^+). The most common boron-bearing minerals associated with geothermal fluids that evaporate at the surface are borax ($Na_2B_4O_7 \cdot 10H_2O$) and tincalconite ($Na_2B_4O_7 \cdot 5H_2O$). These minerals are often associated with halite (NaCl) and other salts common in evaporite settings.

Using surface mineralogy as a means to undertake reconnaissance exploration efforts is a cost-effective way to identify geothermal resource targets. However, this approach is necessarily limited by the local climate patterns. Areas of high rainfall can prevent deposition of noticeable volumes of minerals that require high evaporation rates. Conditions that favor formation of such deposits are readily achieved in the arid western United States, the East African Rift Valley, and central Australia, for example.

FLUID GEOCHEMISTRY AS AN EXPLORATION TOOL

FLUID COMPOSITION AND GEOTHERMOMETRY

As noted at the beginning of this chapter, and discussed in Chapter 5, geothermal fluids exhibit a broad compositional range. However, with the exception of boron, their compositions completely overlap those of many groundwaters that have interacted with a broad range of geological environments. Although boron is often elevated in geothermal systems, it alone is insufficient to be an indicator of a potential geothermal prospect. The concentrations of individual solutes are rarely sufficient to provide good evidence that a potential geothermal resource is available at depth.

However, as previously discussed, thermodynamic and kinetic relationships determine how fluids interact with the rock matrix through which they flow. As a consequence, heated water contains chemical signatures reflecting rock–water interaction along its flow path. Although part of this signature is represented by the bulk water characteristics that allow classification schemes such as that shown in Figure 6.10 to be developed, there is additional information contained in the proportions of specific elements in solution. Provided fluids migrate to the surface at a rate sufficiently high to prevent extensive re-equilibration with the surrounding rocks, a water that has interacted with a geothermal reservoir will possess a chemical record of that high temperature interaction. It is on the basis of this conceptual model that geothermometers have been developed. Their use in geothermal exploration is now a common practice.

Consider, for example, a hypothetical geothermal reservoir that is composed of the minerals α-cristobalite, alkali feldspar, and calcite. Three of the many possible chemical reactions that can be written for this system are

$$SiO_2(\alpha\text{-cristobalite}) \Leftrightarrow SiO_2(aq)$$

$$NaAlSi_3O_8 + K^+ \Leftrightarrow KAlSi_3O_8 + Na^+$$

$$CaCO_3 + H^+ \Leftrightarrow Ca^{++} + HCO_3^-$$

where SiO_2 (aq) indicates dissolved silica and all of the charged species are part of the solute load. The log K values for these reactions as a function of temperature are tabulated in Table 6.3.

For water that has interacted with these minerals and achieved chemical equilibrium, it is a thermodynamic requirement that the activities of the solute components simultaneously satisfy the individual log K expressions for each reaction. In principle, a plot of log (Q/K) (Chapter 5, Equation 5.6) for each reaction, as a function of temperature, will exhibit a series of lines (one for each reaction) that intersect at a value of 0.0 at the temperature at which the fluid last equilibrated with the geothermal reservoir.

Table 6.4 is a hypothetical geothermal water that equilibrated with the assemblage mentioned above at a temperature of 100°C. Plotted in Figure 6.13 is the variation with temperature of log (Q/K) for this water. Although log (Q/K) for each reaction varies independently with temperature, the ratio Q/K will necessarily acquire a value of 1.0 at the temperature of equilibrium. Consequently, as shown in Figure 6.13, a plot of log (Q/K) versus temperature for each reaction will exhibit a common intersection point of the curves at a value for log (Q/K) equal to 0 at the temperature at which equilibrium was achieved.

TABLE 6.3

Variation in the Log of the Equilibrium, as a Function of Temperature, for the Hydrolysis Reaction for α-Cristobalite, the Na–K Exchange Reaction for Alkali Feldspar, and Hydrolysis Reaction for Calcite

Temperature (°C)	25°C	60°C	100°C	150°C	200°C	250°C	300°C
α-Cristobalite	−3.45	−2.99	−2.66	−2.36	−2.13	−1.94	−1.78
Feldspar exchange	1.84	1.54	1.28	1.06	0.90	0.79	0.71
Calcite	1.85	1.33	0.77	0.09	−0.58	−1.33	−2.21

Source: Wolery, T.J., 1992. EQ3/6, A Software Package for Geochemical Modeling of Aqueous Systems: Package Overview and Installation Guide (Version 7.0). UCRL-MA-110662-PT-I. Lawrence Livermore National Laboratory.

Note: Data are from the data0 file for the EQ3/6 computer code.

TABLE 6.4

Hypothetical Composition of a Dilute Geothermal Water at 100°C

Temperature (°C)	pH	Na	K	Ca	SiO$_2$ (aq)	HCO$_3^-$
100	6.7	0.43	0.026	2.49	2.50	0.44

Note: All concentrations are in millimolal.

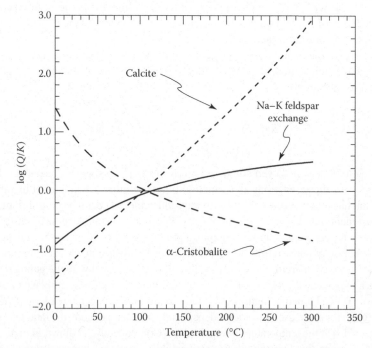

FIGURE 6.13 Variation of log (Q/K) for the water composition in Table 6.2 and the reactions listed in the text.

Theoretically, hydrolysis reactions similar to those given above can be written for every mineral phase in a geothermal reservoir. If those mineral phases are known, and the equilibrium constants for reactions involving the minerals are well established as a function of temperature, graphs like Figure 6.13 can be used with the water analysis for a geothermal site to establish the temperature of the reservoir. This technique can provide a crucial data point in a geothermal resource exploration program.

However, a number of constraints limit the application of this approach. One constraint is the absence of high-quality thermodynamic data for many of the mineral phases that are often present in geothermal reservoirs. Without that data, the equilibrium constants are not known and accurate Q/K values cannot be obtained. An assumption that also must be made is that there has been no chemical exchange between the rock and water after the water has left the reservoir. Because the water flow path after it exits the geothermal reservoir may be long and circuitous, this assumption may be problematic. Another constraint is lack of knowledge of the mineral phases that are actually present in a reservoir. This is especially true during exploration when little or no drilling information is available. Additionally, water analyses are often incomplete or of insufficient quality. Although SiO_2 (aq), Na, K, and Ca are routinely done with high accuracy and precision, Al and Mg are often present at low abundances and are difficult to analyze with high accuracy and precision (Pang and Reed 1998). Also, fluids migrating to the surface exsolve a certain proportion of their dissolved gases, including CO_2, as discussed in Chapter 5. As a result, the carbonate–bicarbonate equilibria that can have a strong influence on pH are likely to be representative of conditions during fluid ascent, not of bicarbonate and pH values in the reservoir. Finally, if steam separation has taken place, partitioning of gases such as CO_2 into the steam phase will influence pH. Silica, too, is also somewhat partitioned into the steam phase. This can result in low apparent silica concentrations in the collected water (Truesdell 1984).

Geochemical thermometers, or geothermometers, can nevertheless be constructed from empirical data that are conceptually grounded in the approach delineated above. As an example, if there is a silica polymorph in the reservoir with which the fluid has equilibrated, the silica concentration will be directly controlled by that polymorph. That, in turn, will exert influence on the equilibration of other minerals with the fluid. The end result is a cascading coupled relationship of interdependencies that will be expressed as a temperature-dependent evolution of the concentration ratios of various solutes.

Since 1960s, formulation of a variety of geothermometers has progressively developed a suite of applicable relationships useful for exploring for potential geothermal reservoirs. Currently, more than 35 different formulations of geothermometers have been published (see Verma et al. 2008 for a recent compilation). Each geothermometer has been formulated by numerically fitting a functional form to datasets that have known or well-constrained temperatures and compositional data. Thus far, equations have been published for concentrations as a function of temperature for the following systems: SiO_2, Na–K, Na–Ca–K, K–Mg, Na–K–Mg, Na–K–Ca–Mg, Na–Li, and Mg–Li. A subset of those that have been published is provided in Table 6.5.

TABLE 6.5
Selection of Published Geothermometers for Geothermal Waters

System	Equation, Temperature (°C)	Applicable Temperature Range	Source
SiO_2 (quartz) $\Leftrightarrow$ SiO_2 (aq)	$T = -42.2 + (0.28831 \times SiO_2) - [0.00036686 \times (SiO_2)^2]$ $+ [(3.1665 \times 10^{-7}) \times (SiO_2)^3] + [77.034 \times \log(SiO_2)]$	25–400	Fournier and Potter (1982)
SiO_2 (chalcedony) $\Leftrightarrow SiO_2$ (aq)	$T = \{1032/[4.69 - \log(SiO_2)]\} - 273.15$	0–250	Fournier (1977)
Na–K (feldspars)	$T = 733.6 - [770.551 \times (Na/K)] + [378.189 \times (Na/K)^2]$ $- [95.753 \times (Na/K)^3] + [9.544 \times (Na/K)^4]$	0–350	Arnórsson (2000b)

To appreciate the difficulty in developing rigorous and quantitatively accurate geothermometers, it is useful to examine in detail the differences in geothermometer results. To illustrate these effects, we will consider the SiO_2 system and the $Na^+ \Leftrightarrow K^+$ exchange geothermometer.

Shown in Figures 6.14 and 6.15 are examples of computed results from different formulations for the SiO_2 geothermometers using five natural geothermal systems and one experimental study.

In Figure 6.14, agreement between observed and computed temperatures varies significantly among the different formulations for the SiO_2 geothermometer. For the experimental study, quartz was the controlling silica polymorph (Pang and Reed 1998), and the computed temperatures for the quartz geothermometers are all well within ±20°C of the actual value. The chalcedony geothermometer significantly underestimates the temperature, as would be expected because of the higher solubility of chalcedony (Chapter 5). At temperatures below 150°C, the quartz geothermometer significantly overestimates the temperature, as does the chalcedony geothermometer, although the latter does so with a smaller error. At higher temperatures, the empirical geothermometers show a broad array of results, differing from the measured values by as much as 120°C and as little as 5°C.

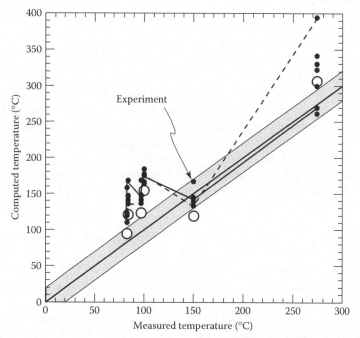

FIGURE 6.14 Comparison of measured water temperatures and computed reservoir temperatures using eight different formulations of the silica geothermometer for five geothermal waters and one experimental water. The solid circles are computed temperatures assuming quartz is the controlling SiO_2 polymorph, whereas the open circle assumes the controlling polymorph is chalcedony. The broad line indicates where points would fall if there was perfect correlation between measurements and computed temperatures, and the shaded band encloses the region within 20°C of that line. To show how the results from individual formulations of the geothermometer vary, the thinner solid line and the dashed line connect points that were computed using two different forms of the geothermometer. (Modified from Fournier, R.O., *Geothermics*, 5, 31–40, 1977; Fournier, R.O., Application of water geochemistry to geothermal exploration and reservoir engineering. In *Geothermal Systems: Principles and Case Histories,* eds. L. Ryback and L.J.P. Muffler. Wiley, New York, 109–143, 1981; Fournier, R.O. and Potter II, R.W., *Geothermal Resources Council Bulletin,* 11, 3–12, 1982; Verma, S.P. and Santoyo, E., *Journal of Volcanology and Geothermal Research,* 79, 9–23, 1997; Verma, S.P., *Proceedings of the 2000 World Geothermal Congress,* Kyushu-Tohoku, Japan, 1927–1932, 2000; Arnórsson, S., *Isotopic and Chemical Techniques in Geothermal Exploration, Development and Use: Sampling Methods, Data Handling, Interpretation.* International Atomic Energy Agency, Vienna, Austria, 351, 2000a; Pang, Z.-H. and Reed, M., 1998. Theoretical chemical thermometry on geothermal waters: Problems and methods. *Geochimica et Cosmochimica Acta,* 62, 1083–1091, 1998.)

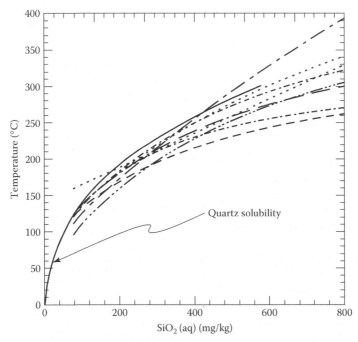

FIGURE 6.15 Comparison of the functional form of the empirical curves for eight different formulations of the silica geothermometer used to compute temperatures in Figure 6.14. The solid curve, for reference, is that for quartz solubility. (Modified from Fournier, R.O., *Geothermics*, 5, 31–40, 1977; Fournier, R.O., Application of water geochemistry to geothermal exploration and reservoir engineering. In *Geothermal Systems: Principles and Case Histories*, eds. L. Ryback and L.J.P. Muffler. Wiley, New York, 109–143, 1981; Fournier, R.O. and Potter II, R.W., *Geothermal Resources Council Bulletin*, 11, 3–12, 1982; Verma, S.P. and Santoyo, E., *Journal of Volcanology and Geothermal Research*, 79, 9–23, 1997; Arnórsson, S., *Isotopic and Chemical Techniques in Geothermal Exploration, Development and Use: Sampling Methods, Data Handling, Interpretation.* International Atomic Energy Agency, Vienna, Austria, 351, 2000a; Verma, S.P., *Proceedings of the 2000 World Geothermal Congress*, Kyushu-Tohoku, Japan, 1927–1932, 2000.)

The discrepancies between measured and computed temperatures provide insight into the caveats associated with empirical geothermometers, some of which are similar to the caveats associated with the Q/K approach described above. Empirical geothermometers represent a strategy for overcoming the inherent limitations of laboratory experimental systems. Geothermal waters are compositionally very diverse. Chloride-, bicarbonate-, and acid-sulfate-dominated solutions characterize water types that have been reported. These natural waters are invariably compositionally complex, requiring detailed and accurate representation of activity–composition relationships in the mass action expressions for the solubility equations, in order to accurately compute a temperature from a silica concentration. Because activity–composition relationships are not generally available in a simple computational scheme but, rather, require accurate and thorough compositional data for use in sophisticated computer models (discussed in Chapters 5 and 6), the strategy has been developed to fit curves of an appropriate form to observed data to allow an *approximate* temperature to be computed. The majority of waters (but by no means all) used to fit curves for geothermometers are of the sodium chloride—type at near-neutral pH. As a result, the fitted curves that are generated will have varying degrees of validity for waters for which the compositions differ from those used in the fitting database. As well, because different authors use different datasets for fitting, each calibration will result in a different form to the geothermometer curve. This is emphasized by the points connected by the dashed and solid lines. For each line, a specific geothermometer was used. Note that the formulation that was used for the points along the dashed line gives the closest results at the low temperature end, but the most deviant results at the high temperature end of the group, but just the opposite is true for the points connected by the solid line. This suggests that each formulation has its own range of "best fit," which is a common problem for empirical geothermometers. In Figure 6.15 are plotted the forms of the solubility curves used to compute the points in Figure 6.14. Clearly, the different formulations will result in widely different results, especially at temperatures greater than about 200°C.

An additional compositional effect comes from the pH of the solution. As pointed out by Fournier (1992), the silica geothermometer works best in conditions where the pH is between about 5 and 7. At alkaline conditions (pH > 8) and acidic conditions (pH < 3), the speciation of silica in water affects the kinetics of reactions involving silica precipitation and dissolution. This can impact equilibrium processes and thus the solute load in the solution.

The discrepancies may also reflect an inherent problem with assumptions regarding the controlling mineral or minerals for the compositional parameters being considered. Giggenbach (1992) and Williams et al. (2008) note the potentially important role of chalcedony in controlling SiO_2 (aq) at temperatures below about 180°C. Whether or not it is, in fact the controlling phase, or even if a silica polymorph exists in contact with the fluid at the conditions of the geothermal reservoir, is unknown until sufficient drilling has been done to elucidate the reservoir mineralogy.

An additional assumption that is problematic but which must always be invoked is that the ascending waters have not changed their compositional characteristics during ascent to the point of sampling. Several processes can occur that will make such an assumption inappropriate. Waters that have flashed to steam or have had gasses exsolve from them during ascent will invariably have some degree of compositional modification through the partitioning of elements between vapor and liquid. Silica, in particular, can be significantly impacted by its partitioning into the vapor phase. The extent to which this process will occur is not directly evident, however, because exsolution of CO_2 or its partitioning into the steam phase will change the fluid pH, which will also indirectly impact silica solubility.

Another aspect of the assumption regarding invariant composition during ascent is that no exchange of solutes occurs between the fluid and the surrounding rock within which it is in contact. Barton (1984) has shown that for exchange of components in feldspars, reaction times vary between hours and years, depending upon the temperature. As previously noted, reaction times are also affected by exposed surface area. Fluids that flow sluggishly on their ascent to the point of sampling are likely to react to some degree with the host rock. The extent to which that happens remains a matter of some debate.

Finally, the fluid will invariably cool as it comes from depth. Obviously, were that not true, the need for geothermometers to estimate the reservoir temperature would be eliminated. Cooling will tend to drive solutions toward saturation or supersaturation in potentially dissolved silicate mineral species. The extent to which minerals precipitate from solution during cooling cannot be derived through any a priori means. As a result, it is unknown to what extent a fluid may have been affected by such a process.

Given these caveats, the extent to which geothermometers provide useful information is quite impressive. Figure 6.16 shows temperatures computed using the Na–K geothermometer and the same dataset as that used for Figure 6.14. The fact that these computations are based on a completely different controlling mineral suite and yet similar temperatures are obtained (Figure 6.17) provides a reasonable indication that these geochemical geothermometers provide an ability to identify possible geothermal resources, as well as a reasonable approximation for reservoir temperatures. Williams et al. (2008) have also concluded that the K–Mg thermometer provides good results for geothermal assessments and recommend its use when high-quality K–Mg data are available.

Mapping features on the ground surface and sampling surface waters are critical tools for establishing the presence of a geothermal resource and understanding some of its characteristics. Additional information is gained from thorough examination of other properties of sampled fluids. One such analytical dataset that provides insight into the properties of a geothermal system are isotope systematics.

ISOTOPES

Stable isotopes provide a powerful way to explore the origin and evolution of geothermal waters that complements information gained from standard water analyses. Isotopic analyses require much greater care in sample collection and sample treatment and are more expensive and time-consuming

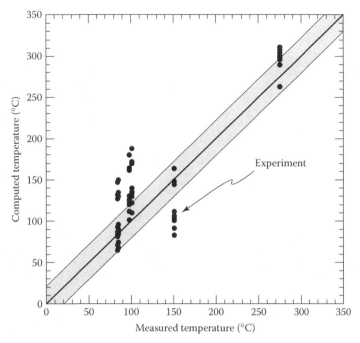

FIGURE 6.16 Comparison of measured water temperatures and computed reservoir temperatures using nine different formulations of the Na–K geothermometer for five geothermal waters and one experimental water. (Modified from Fournier, R.O. and Truesdell, A.H., *Geochimica et Cosmochimica Acta*, 37, 1255–1275, 1973; Truesdell, A.H., *Proceedings of the 2nd U.N. Symposium on the Development and Use of Geothermal Resources 1975*. San Francisco, CA, 831–836, 1976; Fournier, R.O., *Geothermal Resources Council Transactions*, 3, 221–224, 1979; Tonani, F., *Proceedings of the 2nd Symposium on Advances in European Geothermal Research*, Strasbourg, France, 428–443, 1980; Arnórsson, S. et al., *Geochimica et Cosmochimica Acta*, 47, 567–577, 1983; Giggenbach, W.F., *Geochimica et Cosmochimica Acta*, 52, 2749–2765, 1988; Díaz-González, L. et al., *Revista Mexicana de Ciencias Geológicas*, 25, 465–482, 2008; Pang, Z.-H. and Reed, M., *Geochimica et Cosmochimica Acta*, 62, 1083–1091, 1998.)

to conduct than bulk water chemical analyses. However, they often can provide results that resolve the ambiguity commonly associated with standard geochemical surveys. The most frequently used isotopes are those of oxygen, hydrogen, helium, and sulfur.

Table 6.6 tabulates the masses and abundances of these stable isotopes. Note that in each case, there is one dominant isotope for each element and that the mass differences for the isotopes decrease percentage wise as the mass of the element increases. For hydrogen and deuterium, for example, the mass difference is a factor of 2, whereas for ^{16}O and ^{18}O, and ^{32}S and ^{35}S, the difference is approximately 10%. Given that the vibrational frequency of a particle is inversely proportional to its mass, and vibrational frequency influences chemical behavior, it is evident that the isotopes for a given element will behave differently under any given set of conditions. This results in *fractionation*, whereby one isotope of an element will naturally be more readily accommodated in a crystal structure or vapor phase than will the other isotope of the element, at a given set of pressure and temperature conditions. Generally, the lighter isotope will be favored by the phase in which occurs the higher vibrational energy, at a given temperature. Hence, water vapor will naturally have lower $D/^{1}H$ and $^{18}O/^{16}O$ abundance ratios than coexisting liquid water.

Fractionation between coexisting phases is represented as a *fractionation factor*. Fractionation factors are computed by comparing the proportions of the element or isotope of interest in phases that exist at equilibrium with each other. For example, oxygen isotope fractionation between coexisting quartz and calcite can be represented as

$$Si^{18}O_2 + CaC^{16}O_3 \Leftrightarrow Si^{16}O_2 + CaC^{18}O_3$$

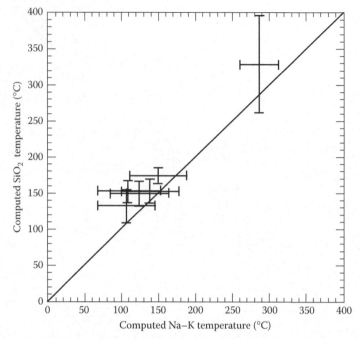

FIGURE 6.17 Comparison of SiO_2 (aq) and Na–K computed temperatures. The capped bars define the range of each geothermometers values for each sample, and the crossing point of the bars is the median value for all computed temperatures for each sample.

TABLE 6.6
Stable Isotopes and Their Abundances

Isotope	Mass	Abundance (%)
^1H	1.007824	99.985
^2H (D)	2.014101	0.015
^3He	3.01603	0.00014
^4He	4.00260	99.99986
^{16}O	15.9949	99.762
^{17}O	16.9913	0.038
^{18}O	17.9991	0.200
^{32}S	31.972	95.02
^{33}S	32.971	0.75
^{34}S	33.968	4.21
^{36}S	35.967	0.02

Source: Walker, F.W. et al., *Chart of the Nuclides*, General Electric Co., Nuclear Engineering Operations, San Jose, CA, 59, 1984.

with an equilibrium constant (K) for this exchange reaction being

$$K = \frac{(Si^{16}O_2 \times CaC^{18}O_3)}{(Si^{18}O_2 \times CaC^{16}O_3)} = \frac{(^{18}O/^{16}O)_{calcite}}{(^{18}O/^{16}O)_{quartz}} = \alpha_{calcite-quartz} \qquad (6.1)$$

In this expression, $\alpha_{calcite-quartz}$ is the isotope fraction factor and is equivalent to the equilibrium constant for this exchange reaction.

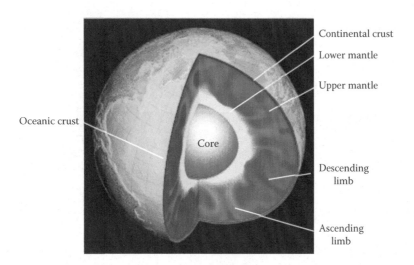

FIGURE 2.1 Interior of the earth, shown in a cutaway that depicts the outer edge of the liquid core (reflecting orange sphere), the lower mantle (yellow), the upper mantle (pink and purple), and the crust. Ascending limbs of convection cells are shown as the orange-tinted plumes extending from the lower mantle through the upper mantle to the base of the crust. Descending limbs of convection cells are shown as the darker purple features extending into the mantle from the base of the crust. (From United States Geological Survey, http://geomag .usgs.gov/about.php.)

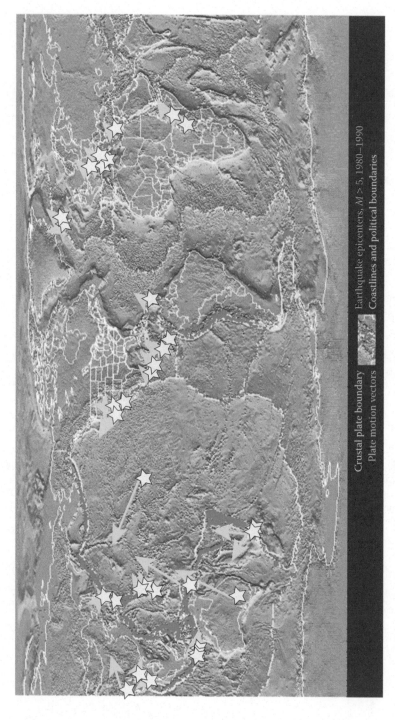

FIGURE 2.6 Global map showing the locations of earthquakes (red dots) that indicate plate boundaries (yellow lines), political boundaries (in white), and the locations of the world's geothermal power plants (stars). The directions of some plate motions are shown by the green arrows, with the length of the arrow corresponding to relative velocity of plate motion. Note the strong correlation between power plant sites and plate boundaries. There are many more power plants than stars because many sites have several power plants. The global map, earthquake data, and boundaries are from the National Oceanic and Atmospheric Administration Plates and Topography Disc and the power plant sites from the International Geothermal Association website (http://iga.igg.cnr.it/geo/geoenergy.php).

Crustal plate boundary   Earthquake epicenters, $M > 5$, 1980–1990
Plate motion vectors   Coastlines and political boundaries

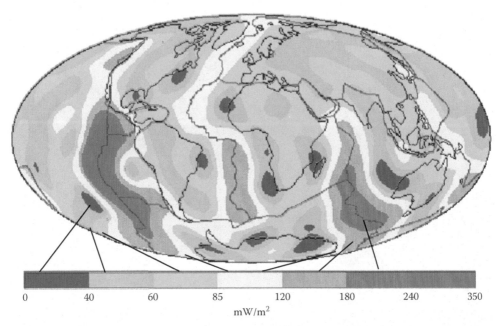

FIGURE 2.7 Low resolution global map showing the distribution of heat flow at the surface. Compare this figure with that in Figure 2.6 to see the relationship between plate boundaries, geothermal power plants, and heat flow. (From International Heat Flow Commission, http://www.geophysik.rwth-aachen.de/IHFC/heatflow.html.)

FIGURE 2.9 Heat flow map of North America 2004. (From Geothermal Laboratory, Southern Methodist University. http://smu.edu/geothermal/2004NAMap/Geothermal_MapNA_7x10in.gif.)

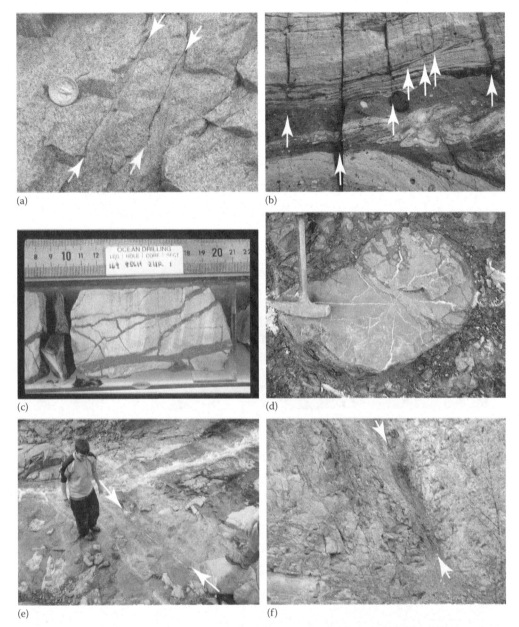

FIGURE 4.4 Examples of the diversity of fracture forms in various rock types. (a) Two parallel, planar fractures (indicated by arrows) cutting granite. Note the change in color and texture along the fractures, compared to the bulk rock. Note, too, that the fracture has some open space. This change indicates fluid movement and chemical alteration took place along the fractures. (Coin for scale. Photograph by the author; Sangre de Cristo Mountains, New Mexico.) (b) Planar, parallel fractures in marble, indicated by arrows. Note that some fractures cut through all of the different compositional layers, whereas some fractures terminate at the boundary between one compositional layer and another. (Lens cap for scale. Photograph by the author; West Nordre Strømfjord, West Greenland.) (c) Irregular, bifurcating, and cross-cutting filled fractures in hydrothermally altered turbidite. Turbidite is a sedimentary rock, but in this instance, it has been slightly metamorphosed by hot fluids circulating through it. Recovered from a borehole drilled by the Ocean Drilling Project in the northeast Pacific Ocean. (Photograph courtesy of Robert Zierenberg; Middle Valley, Northeast Pacific Ocean.) (d) Radial, filled fractures in a pillow basalt. (Photograph courtesy of Robert Zierenberg; Clear Lake, California.) (e) Fracture with alteration halo from an exhumed ~5–6 million-year-old geothermal system in Geitafell, Iceland. The arrows indicate the location of the planar, linear fracture. Note the green alteration halo that extends for more than a meter around the fracture. (Photograph courtesy of Peter Schiffman.) (f) Small fault zone (arrows) with fault gouge. (Width of image is 5 m. Photograph by the author; Sangre de Cristo Mountains, New Mexico.)

FIGURE 6.1 El Chichon volcanic center, Mexico. The most recent eruption of this volcano was in 1982. Pictured here is a steaming ground and an acidic lake in the volcano floor. (Photograph from the US Geological Survey, http://hvo.wr.usgs.gov/volcanowatch/uploads/image1-155.jpg.)

FIGURE 6.2 Mt. Augustine, an andesitic stratovolcano, Aleutian Islands. (Photograph by US Geological Survey.)

FIGURE 6.3 Aniakchak Caldera, Alaska. The caldera is 10 km in diameter and between 500 and 1000 m deep. The caldera formed as the result of explosive eruption of more than 50 km³ of magma approximately 3450 years ago. After caldera collapse, later small eruptions formed domes and cinder cones. (Photograph by M. Williams, National Park Service, 1977.)

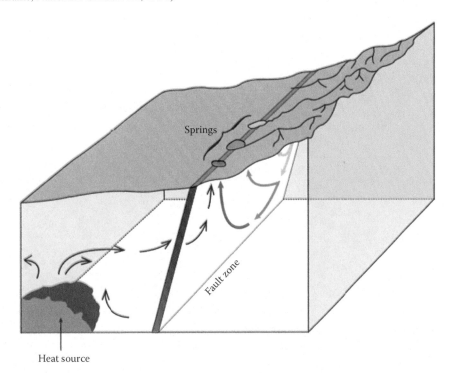

FIGURE 6.6 Schematic block diagram of flow along a fault. Surface water flows from the mountainous region on the right into the fault zone at the base of the mountain range. Water then descends along the fault and becomes heated by the buried heat source in the basin. The heated water ascends, through buoyancy forces, along the fault to the surface where it emerges as springs. The relative temperature of the spring, compared to that of surrounding springs, provides a qualitative indication of the proximity of the spring to the heat source.

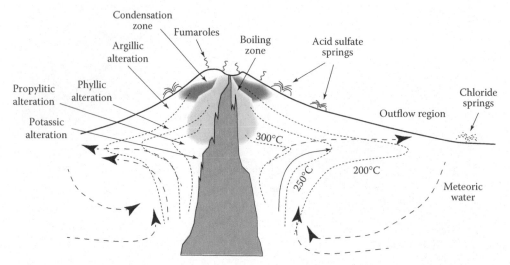

FIGURE 6.7 A schematic diagram of fluid flow regimes in volcanic systems. The heat source (both magma and solidified igneous rocks) is shown as the orangish body. Surrounding it at shallow levels is a steam zone (light blue) and a condensation zone (dark blue). Surface manifestations such as fumaroles and springs are indicated. The isotherms are shown as finely dashed, labeled lines. Flow paths of meteoric water are shown by the heavy, dashed lines. Approximate regions where argillic, phyllic, propylitic, and potassic alteration would occur are also shown.

FIGURE 6.9 Photograph of a fumarole with sulfur deposits at Kilauea Volcano, Hawaii. (Photograph by R.L. Christiansen, US Geological Survey.)

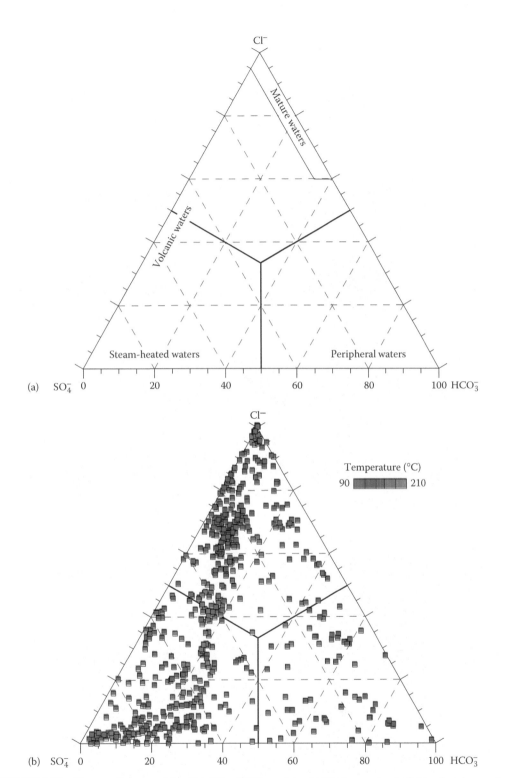

FIGURE 6.10 Ternary diagram of the Cl—SO₄—HCO₃ system. (a) The approximate fields of various water types are delineated and labeled. (b) The ternary compositions of waters in the Argonne National Laboratory Geochemical database. (Modified from Clark, C.E. et al., *Water Use in the Development and Operation of Geothermal Power Plants*, United States Department of Energy, Argonne National Laboratory, ANL/EVS/R-10/5, Washington, DC, 2010.)

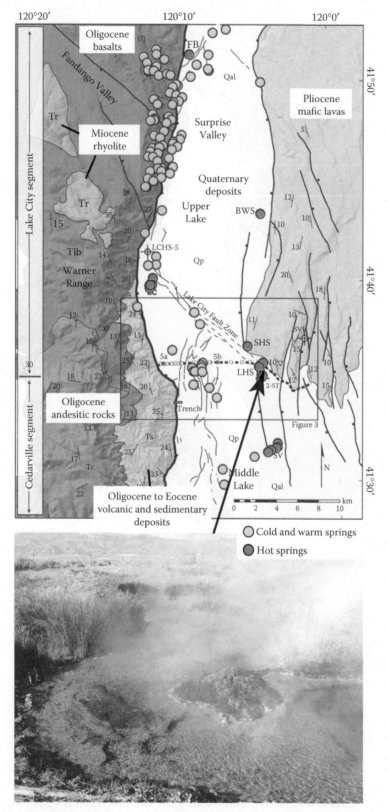

FIGURE 6.11 Geologic map of the Surprise Valley, California, region. Blue dots show the locations of cold and warm springs; red dots show the locations of hot springs. The photograph at the bottom shows one of the hot springs. (Modified from Eggers, A.E. et al., *Tectonophysics*, 488, 150–161, 2010; photograph by the author.)

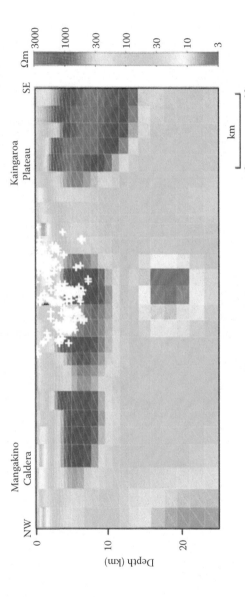

FIGURE 7.3 A model based on a MT survey conducted in the Wairakei Valley, New Zealand. The low resistivity region at a depth of about 20 km is interpreted to be a magma body. The white crosses indicate sites of microseismic events. Possible fluid flow pathways are suggested by the green and pale-blue bands at depths less than 10 km, as well as in the very shallow levels where low resistivity (<10 Ωm) occur. (Modified from Heise, W. et al., *Geophysical Research Letters,* 34, 6, 2007.)

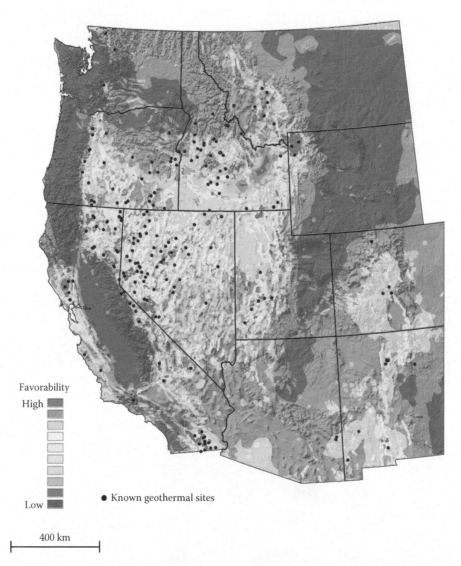

FIGURE 8.7 Western United States showing the locations of individual geothermal sites (black dots). Also shown in color are those regions for which geological evidence suggests the possibility that geothermal reservoirs may be located in the vicinity. The likelihood that a geothermal resource may exist in the region is ranked from low to high "favorability," based on the nature of the geological evidence. (Modified from Williams, C.F. et al., Assessment of moderate- and high-temperature geothermal resources of the United States; US Geological Survey Fact Sheet 2008-3082, 4, 2008b.)

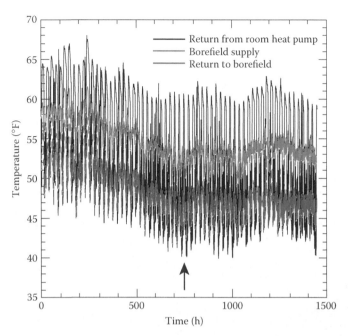

FIGURE 11.14 Temperature variation of the working fluid coming from the borefield (red line labeled "Borefield supply"), returning to the borefield (blue line labeled "Return to borefield"), and exiting the heat pump from a specific room (black line labeled "Return from room heat pump"). The time period covered is from November 4, 2005, to January 4, 2006. The arrow points to the period represented in Figure 11.15.

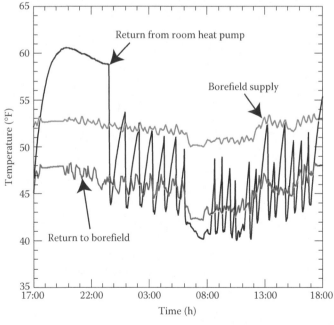

FIGURE 11.15 Detailed daily behavior of the temperature of the fluids. The time period covered in the figure is from Monday, December 5, 2005, 17:00, to Tuesday, December 6, 2005, 18:00. The curves are colored as in Figure 11.14.

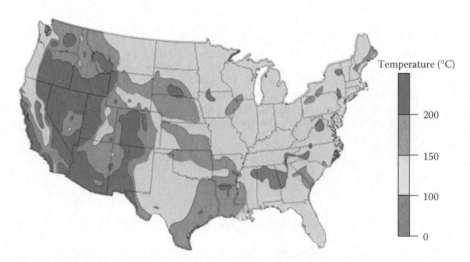

FIGURE 13.1 Subsurface temperatures at a depth of 6 km in the continental United States. (From US Department of Energy, Energy Efficiency and Renewable Energy Office, http://www1.eere.energy.gov/geothermal/geomap.html.)

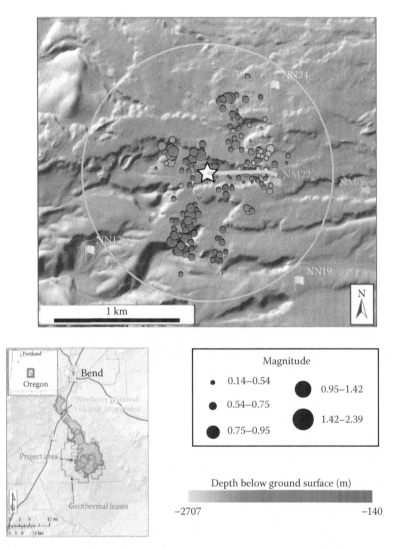

FIGURE 13.12 The location of the Newberry EGS demonstration project in Oregon (lower left). The upper figure is a topographic map that shows the location of the wellhead (star), the trajectory of the well (light blue), and the locations of microseismic events (depth and magnitude are indicated by the color scale and magnitude chart below the figure). (Modified from Cladouhos, T.T. et al., *Geothermal Resources Council Transactions*, 37, 133–140, 2013.)

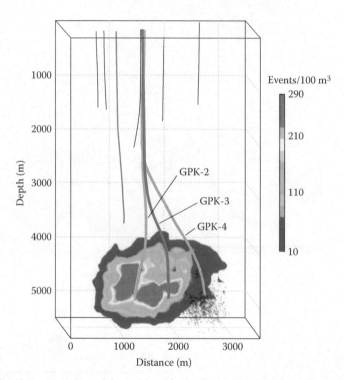

FIGURE 15.6 Density contour map of the number of seismic events associated with stimulation of the crystalline rock reservoir at the European Hot Dry Rock (EGS) project in Soultz-sous-Forêts, France. The contour map reflects the results from two periods of hydro-fracturing carried out on the wells labeled GPK-2 and GPK-3. Also shown are individual events (dots) associated with hydro-fracturing from GPK-4. The other wells shown in the figure were used for seismic monitoring. (Modified from material in Baria, R. et al. Creation of an HDR reservoir at 5000 m depth at the European HDR project, *Proceedings of the 31st Workshop on Geothermal Reservoir Engineering*, Stanford University, Stanford, CA, SGP-TR-179, 8, 2006.)

The values for these ratios can be measured by mass spectrometry, but the absolute values for each isotope are not routinely measured. Generally, a standard with an assigned or measured isotopic ratio is used as the basis for evaluating other samples of interest. For oxygen and hydrogen isotope measurements, the standard has generally been Standard Mean Ocean Water (SMOW), and the isotopic composition of a sample is expressed as the difference from SMOW in parts per thousand:

$$\delta^{18}O = \left\{ \frac{\left[(^{18}O/^{16}O)_{sample} - (^{18}O/^{16}O)_{SMOW} \right]}{(^{18}O/^{16}O)_{SMOW}} \right\} \times 1000 \qquad (6.2)$$

$$\delta D = \left\{ \frac{\left[(D/H)_{sample} - (D/H)_{SMOW} \right]}{(D/H)_{SMOW}} \right\} \times 1000 \qquad (6.3)$$

Substances that are isotopically lighter than SMOW will consequently have a negative $\delta^{18}O$ and δD. Other standards have also been used, resulting in different $\delta^{18}O$ and δD values. It is thus important to know what standard was used in an analysis in order to allow quantitative comparison of different samples.

Meteoric water has a range between $\delta^{18}O$ and δD values that depends upon the latitude and elevation at which precipitation occurs. This reflects the significant impact isotopic fractionation has during evaporation and condensation at different temperatures. However, the range of values is constrained to a linear array by the fact that the primary source for meteoric water is the ocean. Fractionation of oxygen and hydrogen differs by the ratio of their respective fractionation factors, which necessitates that they will be linearly related. Figure 6.18 shows the line that delineates the

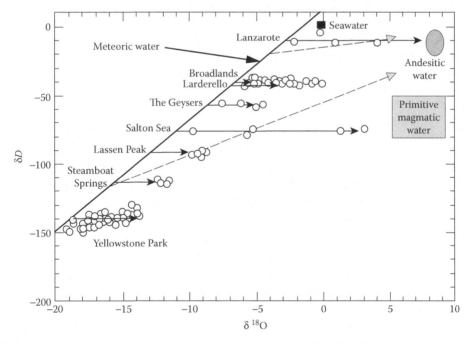

FIGURE 6.18 δD and $\delta^{18}O$ values for waters from geothermal areas. (Modified from Craig, H., The isotopic geochemistry of water and carbon in geothermal areas. In *Nuclear Geology on Geothermal Areas: Spoleto, 1963*, ed. E. Tongiorgi, Consiglio Nazionale delle Ricerche, Laboratorio di Geologia Nucleare, Pisa, Italy, 17–53, 1963; Craig, H., Science, 154, 1544–1548, 1966; Truesdell, A.H. and Hulston, J.R., Isotopic evidence on environments of geothermal systems. In *Handbook of Environmental Isotope Geochemistry*, vol. I. The Terrestrial Environment, eds. P. Fritz and J.Ch. Fontes. Elsevier, Amsterdam, the Netherlands, 179–226, 1980)

$\delta^{18}O$ and δD meteoric water linear relationship. Most surface waters and groundwaters fall on this line. Deviations from this line indicate that the measured water has a history more complex than single-stage evaporation and precipitation and can be used to identify potential geothermal targets.

Also shown in Figure 6.18 is the distribution of points from a few selected geothermal sites. The geothermal sites delineated by open circles exhibit a trend that is commonly observed in geothermal systems. For these sites, the local meteoric waters and the geothermal waters define a trend that origi-nates on the meteoric water line (MWL), with the geothermal waters displaced horizontally to higher $\delta^{18}O$ with little or no corresponding change in δD. Craig (1963) realized that the most likely explana-tion for this behavior was that the geothermal waters must have originated as meteoric water that had circulated to deeper, hotter regions where the oxygen in the H_2O molecule exchanged with oxygen in the minerals at elevated temperatures. Because the $\delta^{18}O$ of minerals is generally in the range of $+1$–$+10$, $\delta^{18}O$ of meteoric water interacting with these minerals will increase. Because the amount of oxygen in rocks far exceeds the hydrogen content, little shift in the δD values would be expected.

Giggenbach (1992), however, recognized that not all geothermal waters follow this horizontal trend toward higher $\delta^{18}O$ values with little or no change in δD. From an extensive suite of analyses, he showed that geothermal waters associated with volcanic systems that occur at subduction zones commonly have a trend toward more positive δD as well as the commonly observed enrichment in ^{18}O. He attributed that trend to mixing of water associated with the magmatic degassing of sub-duction zone volcanic rocks and the resulting interaction with local meteoric fluids.

These results document that the variation in oxygen and hydrogen isotopes observed in geo-thermal systems can reflect competing processes. But, whichever process is dominant at any given site, these stable isotopes can provide evidence of the presence of a geothermal reservoir at depth, if the isotopic analyses show significant displacement away from the local meteoric water values. In the absence of such a displacement, the presence of a significant geothermal resource connected to the surface via circulating water is problematic.

Helium isotopic studies are another means for identifying potential geothermal resources. In the case of this isotope system, however, the underlying processes controlling isotopic anomalies are funda-mentally different and thus provide a different insight into the likely presence of a geothermal resource.

The primary reservoirs for helium isotopes are the original ^3He that the earth inherited when it formed, and ^4He that accumulates as U and Th decay via emission of alpha particles (which are ^4He nuclei). Because the crust has the bulk of the earth's U and Th, ^4He accumulates in the continental crust and its abundance increases over time, progressively enriching the crust and the atmosphere in this isotope. The mantle, however, holds the earth's main store of ^3He. Any volatile emissions from the mantle that escape into the crust will increase the local ratio of ^3He–^4He and result in a helium isotope anomaly. Similarly, melts that are generated in the mantle will contain some small amount of ^3He and virtually no ^4He. When such magmas rise from the mantle and enter the crust, the magma will cool and slowly degas. As the degassed volatiles rise through the crust, they will interact with any local meteoric waters, imposing on them a ^3He signature. A survey of helium isotopes conducted on well or spring waters where such interactions have occurred will detect an anomalously high ratio of ^3He:^4He if a crystallizing magma is degassing in the nearby subsurface or if mantle volatiles are escaping through the crust at that location. The current atmospheric ^3He/^4He ratio (designated Ra) is 1.2×10^{-6}, and crustal fluids lacking any mantle influence have values of approximately 0.02 Ra. Mantle-derived fluids have values of 6–35 Ra (Kennedy and van Soest 2007), which are readily distinguished from crustal values.

Figure 6.19 depicts the results of a survey of helium isotopes across the western United States (Kennedy and van Soest 2007), extending from the Cascade Mountains in Oregon to Colorado. This region encompasses the volcanic mountain system of the Sierras and Cascades that are or were part of the tectonic regime known as the eastern Pacific subduction zone, as well as the Basin and Range extensional zone that includes Nevada, Arizona, New Mexico, Utah, and Colorado. The well-defined peaks that fall at Rc/Ra values greater than 1.0 demark well-defined geographi-cal zones where mantle-derived helium is reaching the surface. This requires either that recent

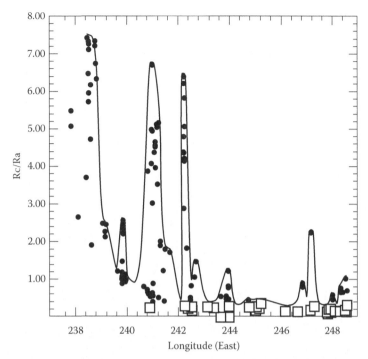

FIGURE 6.19 Variation in He isotope ratios across the western United States. The ratio Rc/Ra is the ^{3}He/^{4}He ratio, corrected for crustal values. The open squares are samples that have a purely crustal signature, whereas the filled circles have some component of mantle-derived He. The curve that encloses the points is drawn simply as an envelope to delineate the pattern. (Data from Kennedy, B.M. and van Soest, M.C., *Science*, 318, 1433–1436 2007.)

igneous intrusions reside below the surface or that permeability exists in the deep crust sufficient for volatiles to migrate in sufficient and persistent quantity to allow sampling at the surface. Such spikes in helium isotopes identify good targets for further exploration for geothermal resources. Importantly, helium isotope surveys such as this can provide data useful for identifying sites for Enhanced Geothermal Systems (EGS), an important topic that is explored in detail in Chapter 13.

Sulfur isotopes are an additional isotopic system that provides information regarding geothermal systems that complements that derived from other isotope systems as well as geochemical analyses. Sulfur differs from the systems described above because there are four stable isotopes of importance, ^{32}S, ^{33}S, ^{34}S, and ^{36}S, and it can exist in five different valence states of +6, +4, 0, −1, and −2. As a result, it can occur in nature in minerals and solution as sulfides, sulfites, sulfates, and native sulfur. Examples of likely mineral hosts are given in Table 6.7, which also summarizes some of the chemical properties important for this system.

Fractionation of the sulfur isotopes among these various molecular hosts is measured and represented similarly to the approach used for hydrogen and oxygen isotopes, namely, the *del* notation. For sulfur isotopes, the most common measurement is made using ^{32}S as the reference and considering the relative enrichment or depletion of ^{34}S, through the relationship:

$$\delta^{34}S = \left\{ \frac{\left[\left(^{34}S / ^{32}S \right)_{sample} - \left(^{34}S / ^{32}S \right)_{VCDT} \right]}{\left(^{34}S / ^{32}S \right)_{VCDT}} \right\} \times 1000 \tag{6.4}$$

where VCDT is the Vienna Canyon Diablo Troilite, which, by definition, $\delta^{34}S = 0.0$.

TABLE 6.7

Sulfur Valence States, Speciation, and Common Mineral Phases

Valence	Gas	Aqueous	Minerals	Molecular Weight (g/mole)
+6	–	SO_4^{2-}	Anhydrite—$CaSO_4$	96.06
			Barite—$BaSO_4$	–
			Alunite—$KAl_3(SO_4)_2(OH)_6$	–
+4	SO_2	–	–	64.066
0	–	–	Native sulfur—S	32.066
–1	–	–	Pyrite, marcasite—FeS_2	–
–2	H_2S	HS^-	Pyrrhotite—FeS	H_2S—32.081
			Sphalerite—ZnS	HS^-—33.07

Source: Zierenberg, R., Sulfur isotopes, *Recent Advances in Geothermal Geochemistry Workshop*, Geothermal Resources Council, Davis, CA, 2012.

However, because VCDT is no longer available, convention has established the IAEA S-1 silver sulfide as a standard, with a defined value of −0.3 per mil, that is, −0.3‰.

The chemical complexity of the sulfur system, with its variety of masses and oxidation states, results in the possibility of *mass-dependent isotope fractionation*. Because of the differences in masses of the chemical species sulfur isotopes can partition among, it is now well-established that the heavier sulfur isotopes are preferentially taken up by the heavier sulfur species. As a result, the higher oxidation state sulfur species are enriched in the heavier sulfur isotopes. Hence, the sequence $SO_4^{2-} > SO_3^{2-} > S^o > S^{2-}$ also defines the relative enrichment in the heavy sulfur isotopes relative to the light isotopes.

The fractionation factor (α) for a system involving two sulfur-bearing minerals, A and B, can be represented as

$$\alpha_{A-B} = (1000 + \delta_A)/(1000 + \delta_B) \tag{6.5}$$

An alternative method, which provides similar results, is the delta convention, which is defined as

$$\Delta_{A-B} = \delta_A - \delta_B \tag{6.6}$$

As an example of how these notations are used, consider the mineral pair galena (PbS) and sphalerite (ZnS). If we treat ZnS as mineral A and PbS as mineral B, the 300°C value for $\alpha_{Sph-Gal}$ is measured as 1.0022 (Seal 2006). This means that ZnS is enriched in ^{34}S by 2.2 per mil (i.e., 2.2 ‰) relative to PbS. Although this is a small number, it is readily measurable and thus of analytical significance.

The fractionation of sulfur isotopes is temperature sensitive, providing a potential means to evaluate the thermal characteristics and history of a system. Figure 6.20 summarizes data on sulfur isotope fractionation between coexisting sulfate and sulfide minerals in hydrothermal systems. The temperature signature for these systems is clear. Similar results could potentially be developed for fluid species, if relevant coexisting aqueous species and mineral species can be identified and measured. However, rates of isotopic equilibration between coexisting mineral phases, and as mineral reactions take place, can vary between days to decades, depending upon temperature and the local chemical conditions of the solution, particularly pH (Seal 2006).

Because of the diversity of chemical species and minerals that sulfur can form in hydrothermal systems, sulfur isotope systematics hold great potential to characterize active geothermal systems, as well as delineate the history of such systems. However, this potential has not yet been exploited sufficiently to be used as a standard tool for geothermal exploration. With further research, however, this isotope system is likely to become a very powerful geochemical tool for geothermal exploration.

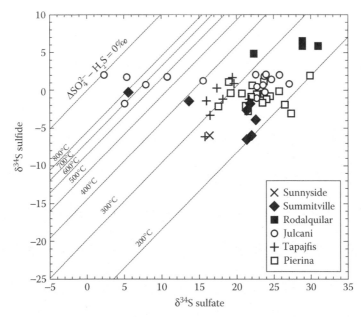

FIGURE 6.20 Sulfur isotopic variation from a suite of geothermally altered systems in which sulfide and sulfate $\delta^{34}S$ were measured. The isotherms indicate the equilibrium $\delta^{34}S$ values for the indicated temperatures. (Modified from Seal II, R.R., *Reviews in Mineralogy and Geochemistry*, 61, 633–677, 2006.)

FLUID INCLUSIONS

A unique source of information regarding geothermal systems is fluid inclusions. As minerals grow, they can entrap fluids with which they are in contact. Since the trapped fluids are captured at a specific pressure and temperature, their phase state will be fixed at the time of entrapment—if the fluid is trapped at P and T above the critical point of water, the inclusion will consist of a single supercritical fluid phase, whereas if it is trapped at a set of conditions below the triple point, the fluid will either be a liquid or a vapor.

As the rock containing the mineral with fluid inclusions cools, the fluid phase will separate into a vapor bubble with coexisting liquid, if it was originally a supercritical fluid. The two-phase inclusion will consist of a small bubble volumetrically dominated by the liquid, if the P–T conditions at the time of entrapment were at significantly higher pressures than the critical point, or it will be a large gas bubble surrounded by a volumetrically minor liquid phase if the supercritical fluid was trapped at temperatures well beyond the critical point. An example of two-phase fluid inclusions from a geothermal system is shown in Figure 6.21, which is of a core sample collected from a drill hole on the Reykjanes Peninsula of Iceland.

Fluid inclusions are small pressure vessels which, if they have remained intact and unruptured, preserve the pressure at the time of entrapment. Assuming the fluid is predominately H_2O, the temperature at which the fluid was trapped can be determined by heating the inclusions up to the point at which the vapor bubble and liquid recombine into a single phase, or *homogenize*. That homogenization temperature uniquely determines the temperature of entrapment. Additionally, it is well known that geothermal waters are not pure H_2O, but rather have a dissolved load that makes them complex solutions, or brines, of varying salinities. The salinity of the water affects the thermodynamic properties of water, including its heat of vaporization and freezing point. The salinity of a fluid inclusion can also be determined by measuring its freezing temperature. Such measurements are conducted using heating–freezing stages that are mounted on petrographic microscopes. These combined data points establish how saline the fluid is and the temperature at which it was trapped.

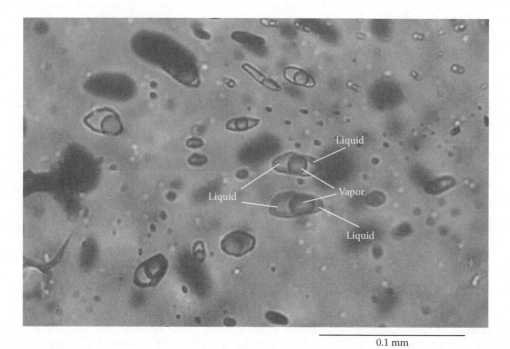

0.1 mm

FIGURE 6.21   Two-phase fluid inclusions from the Icelandic geothermal system. The vapor bubble and enclosing liquid are labeled. (Courtesy of Andrew Fowler.)

Figure 6.22 shows the results obtained from the core sample depicted in Figure 6.21 when such measurements are carried out (Andrew Fowler, pers. comm.). The upper figure is of the rock slab on which some of the measurements were made. The sample consists of a volcanic rock (right side of upper figure) that is cut by an epidote vein (flat gray material between 'Vein center' and mottled rock). The location of the vein center is depicted by the dashed line. The circles and squares indicate where some of the heating–freezing measurements were made. The lower figure shows the salinities and homogenization temperatures of fluid inclusions in various parts of the vein.

If it is assumed that the vein minerals grew from the host rock outward toward the vein center, the fluid inclusion data suggest that the first growth period of the vein epidote occurred in the presence of a low-salinity water at temperatures between 380°C and 400°C. This was followed by a drop in fluid temperature to between 340°C and 380°C that occurred as the salinity increased to values exceeding that of seawater. These data suggest that a change in the main reservoirs that recharged the flow regime in this geothermal system also coincided with a change in the heat content of the circulating fluids.

Such information regarding the history of a geothermal system can form the basis for models of the system, including the magnitude of the thermal reservoir, the locations of recharge areas and possible flow pathways, the role of fluid mixing in the hydrological regime, and whether the system is heating up or cooling down. Since the drill hole from which this sample was obtained occurs within less than 2 km of the Atlantic coast of the Reykjanes Peninsula, one possible model that could be developed would involve a shift in the physical framework of this geothermal system such that seawater obtained access to the hydrological regime (Andrew Fowler, pers. comm.). The value of such models is that they can focus research on specific tests to invalidate a particular concept, thus improving the understanding of the system under consideration. Improved understanding inevitably results in reduced risk for resource development and thus is of great importance.

The geochemical methods described above allow investigators to develop a description of the thermal and chemical properties of possible geothermal reservoirs. They also support the development of geological and hydrological models that allow conceptualization of the physical framework within which a geothermal reservoir exists. Taken together the shape, flow field, stage of

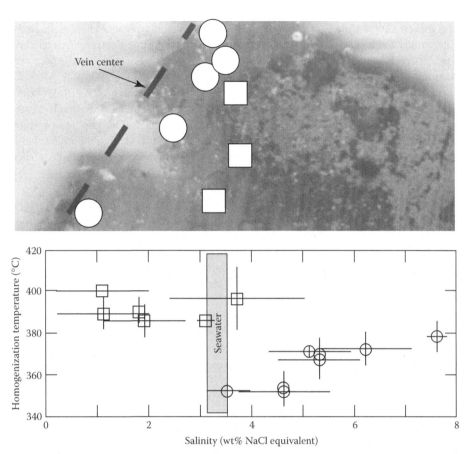

FIGURE 6.22   Upper figure is a photograph of an epidote vein (vein center indicated by dashed line) cutting a basaltic rock. The locations of heating–freezing stage measurements on fluid inclusions are shown by the circles and squares. The lower figure shows the homogenization temperatures as a function of salinity. Some locations have multiple measurements, hence the difference in number of circles and squares between the upper and lower figures. (Courtesy of Andrew Fowler.)

development, location of high-temperature zones, and possible drilling targets can be approximately delineated. However, refining these models to a higher spatial resolution is required before serious consideration is given to whether a resource is a good candidate for further development. To obtain this higher resolution, geophysical methods are employed to better define the resource.

ALTERATION AND EXPLORATION

Studies of mineralogical and geochemical characteristics associated with geothermal systems have led to the recognition of certain patterns that have been persistently observed. These, in turn, have led to a variety of ways of classifying the type of alteration that can be present in geothermal systems. In general, the purpose of these schemes is to expedite the exploration process by providing a summary of properties and characteristics of a system, based on a key set of critical mineral and chemical indicators. These schemes provide conceptual models that can be useful in guiding geothermal exploration. However, it must be borne in mind that these generalizations are developed from a limited number of systems and using diverse classification criteria. Given the diverse, complex, and heterogeneous nature of geological systems, these classification schemes are not universally applicable—each natural geothermal system will have its own set of characteristics that may or may not directly correspond to those that are the focus of a given classification scheme. Therefore, the various nomenclatures, although very useful, must also be considered approximate. Described below are some alteration types that have become common in the literature and that have proven to be helpful in exploration efforts. Their approximate relationship to geological conditions and alteration is highlighted in Figure 6.7. For more detailed description, see Hedenquist and Houghton (1987), White and Hedenquist (1990), Damian (2003), and Moore (2012).

Argillic—It is characterized by the presence of mixed-layered clays (11 Å–14 Å, illite–smectite clays), possibly with sericite, sulfides, zeolites, quartz, and calcite. This alteration forms as a result of chemically aggressive fluids interacting with feldspars and mafic minerals. Temperatures of formation are generally below 200°C. If the solutions are particularly acidic, the mineral assemblage may be composed of kaolinite and alunite, with some silica polymorph, for example, chalcedony or cristobalite, or possibly quartz.

Phyllic—It is characterized by the presence of illitic clays, without smectite. Sericite is consistently present, along with quartz. Pyrite is common. This alteration forms as a result of chemically aggressive fluids interacting with feldspars and mafic minerals, as with argillic alteration, but at higher temperatures, generally between 200°C and 250°C.

Propylitic—It is characterized by the general absence of clay minerals and the presence of epidote, actinolite, albite, and chlorite. Carbonate minerals may also be present. This alteration also forms as a result of chemically aggressive fluids interacting with the host rock, but at temperatures that are generally above ~300°C.

Potassic—It is characterized by the presence of biotite, potassium feldspar, and quartz, possibly with epidote, chlorite, and muscovite. Such alteration generally occurs with potassium feldspar replacing the plagioclase in the original rock. This alteration also forms as a result of potassium-rich fluids interacting with the host rock at elevated temperatures, generally at or above 320°C.

SYNOPSIS

Exploration for geothermal resources requires employing numerous techniques in order to identify a resource. The initial phase must always rely on available geological information, from which refinement of an exploration strategy can proceed. Important approaches include using geochemical evidence to establish possible reservoir temperatures through the use of geothermometers and isotopic analyses to identify anomalies that may indicate interaction with a thermal source. A variety of isotope systems can be employed to gain information about temperature, fluid sources, and fluid mixing. Examination of the mineralogy of rocks in an exploration program is a critical method for establishing the type of system that has operated in the region in the past. Specific alteration types can give information regarding fluid chemistry as well as temperature history.

PROBLEMS

6.1. List three surface features that are important indicators of near surface geothermal resources and discuss what they indicate.

6.2. Why might a warm spring occur at the surface but there would be no geothermal resource in the subsurface directly below it?

6.3. Listed below are three geothermometers (Henlet et al., 1984)—all concentrations are in mg/kg:

SiO_2: T (°C) = $[1032/(4.69 - \log SiO_2)] - 273.15$
Na/K: T (°C) = $[855.6/(\log (Na/K) + 0.8573)] - 273.15$
Na-K-Ca: T (°C) = $[1647/(\log (Na/K) + \{0.33*[\log(sqrt(Ca)/Na) + 2.06]\} + 2.47)] - 273.15$

Compute the temperatures for the waters listed in Table 5.1. Which samples have the best agreement among the geothermometers?

6.4. Provide an explanation for the disagreement among the samples. Is there anything systematic about the results?

6.5. Assuming that the water at The Geysers is mixing with *Primitive Magmatic Water*, what proportion of magmatic water is in The Geysers sample with the highest oxygen isotope ratio?

6.6. Using the quartz and chalcedony geothermometers in Table 6.5, compute the temperatures for the waters in Table 5.1. What would explain the discrepancies? Provide an argument, in each case, for which geothermometer you think is the most accurate.

6.7. What might explain the large range of sulfate sulfur isotopes for the Julcani samples in Figure 6.20?

REFERENCES

Arnórsson, S., 2000a. *Isotopic and Chemical Techniques in Geothermal Exploration, Development and Use: Sampling Methods, Data Handling, Interpretation.* Vienna, Austria: International Atomic Energy Agency, p. 351.

Arnórsson, S., 2000b. The quartz- and Na/K geothermometers. I. New thermodynamic calibration. *Proceedings of the World Geothermal Congress*, Kyushu-Tohoku, Japan, pp. 929–934, May 28–June 10.

Arnórsson, S., Gunnlaugsson, E., and Svavarsson, H., 1983. The chemistry of geothermal waters in Iceland. III. Chemical geothermometry in geothermal investigations. *Geochimica et Cosmochimica Acta*, 47, 567–577.

Barton, P.B. Jr., 1984. High temperature calculations applied to ore deposits. In *Fluid-Mineral Equilibria in Hydrothermal Systems*, vol. 1. Reviews in Economic Geology, eds. R.W. Henley, A.H. Truesdell, and P.B. Barton, Jr. Chelsea, MI: Society of Economic Geologists, pp. 191–201.

Clark, C.E., Harto, C.B., Sullivan, J.L., and M.Q. Wang., 2010. *Water Use in the Development and Operation of Geothermal Power Plants*. Washington, DC: United States Department of Energy, Argonne National Laboratory, ANL/EVS/R-10/5.

Coolbaugh, M.F., Kraft, C., Sladek, C., Zehner, R.E., and Shevenell, L., 2006. Quaternary borate deposits as a geothermal exploration tool in the Great Basin. *Geothermal Resources Council Transactions*, 30, 393–398.

Craig, H., 1963. The isotopic geochemistry of water and carbon in geothermal areas. In *Nuclear Geology in Geothermal Areas: Spoleto, 1963*, ed. E. Tongiorgi. Pisa, Italy: Consiglio Nazionale delle Ricerche, Laboratorio di Geologia Nucleare, pp. 17–53.

Craig, H., 1966. Isotopic composition and origin of the Red Sea and Salton Sea geothermal brines. *Science*, 154, 1544–1548.

Damian, F., 2003. The mineralogical characteristics and the zoning of the hydrothermal types alteration from Nistru ore deposit, Baia Mare metallogenetic district. *Geologia*, XLVIII(1), 101–112.

Díaz-González, L., Santoyo, E., and Reyes-Reyes, J., 2008. Tres nuevos geotermómetros mejorados de Na/K usando herramientas computacionales y geoquimiométricas: Aplicación a la predicción de temperaturas de sistemas geotérmicos. *Revista Mexicana de Ciencias Geológicas*, 25, 465–482.

Eggers, A.E., Glen, J.M.G. and Ponce, D.A., 2010. The northwestern margin of the Basin and Range province Part 2: Structural setting of a developing basin from seismic and potential field data. *Tectonophysics*, 488, 150–161.

Fournier, R.O., 1977. Chemical geothermometers and mixing models for geothermal systems. *Geothermics*, 5, 31–40.

Fournier, R.O., 1979. A revised equation for the Na/K geothermometer. *Geothermal Resources Council Transactions*, 3, 221–224.

Fournier, R.O., 1992. Water geothermometers applied to geothermal energy. In *Applications of Geochemistry in Geothermal Reservoir Development–Series of Technical Guides on the Use of Geothermal Energy*, Chapter 2. coordinator, Franco D'Amore. Rome, Italy: UNITAR/UNDP Center on Small Energy Resources, pp. 37–69.

Fournier, R.O., 1981. Application of water geochemistry to geothermal exploration and reservoir engineering. In *Geothermal Systems: Principles and Case Histories*, eds. L. Ryback and L.J.P. Muffler. New York: Wiley, pp. 109–143.

Fournier, R.O. and Potter II, R.W., 1982. A revised and expanded silica (quartz) geothermometer. *Geothermal Resources Council Bulletin*, 11, 3–12.

Fournier, R.O. and Truesdell, A.H., 1973. An empirical Na-K-Ca geothermometer for natural waters. *Geochimica et Cosmochimica Acta*, 37, 1255–1275.

Fuis, G.S. and Mooney, W.D., 1990. Lithospheric structure and tectonics from seismic-refraction and other data. In *The San Andreas Fault System, California*, ed. R.E. Wallace. *U.S. Geological Survey Professional Paper* 1515, 207–236.

Giggenbach, W.F., 1988. Geothermal solute equilibria. Derivation of Na-K-Mg-Ca geoindicators. *Geochimica et Cosmochimica Acta*, 52, 2749–2765.

Giggenbach, W.F., 1992. Chemical techniques in geothermal exploration. In *Application of Geochemistry in Geothermal Reservoir Development*, ed. F. D'Amore. Rome, Italy: UNITAR/UNDP Centre on Small Energy Resources, pp. 119–144.

Hedenquist, J.W. and Houghton, B.F., 1987. Epithermal gold mineralization and its volcanic environments. Earth Resources Foundation Conference, The University of Sydney, Australia, p. 422, November 15–21.

Henley, R.W. and Hedenquist, J.W., 1986. Introduction to the geochemistry of active and fossil geothermal systems. In *Guide to the Active Epithermal Systems and Precious Metal Deposits of New Zealand*. Monograph Series on Mineral Deposits, eds. R.W. Henley, J.W. Hedenquist, and P.J. Roberts. Berlin, Germany: Gebrüder Borntraeger, pp. 129–145.

Kennedy, B.M. and van Soest, M.C., 2007. Flow of mantle fluids through the ductile lower crust: Helium isotope trends. *Science*, 318, 1433–1436.

Moore, J., 2012. The hydrothermal framework of geothermal systems. *Recent Advances in Geothermal Geochemistry Workshop*, Geothermal Resources Council, Davis, CA, June 19–20.

Pang, Z.-H. and Reed, M., 1998. Theoretical chemical thermometry on geothermal waters: Problems and methods. *Geochimica et Cosmochimica Acta*, 62, 1083–1091.

Seal II, R.R., 2006. Sulfur isotope geochemistry of sulfide minerals. *Reviews in Mineralogy and Geochemistry*, 61, 633–677.

Tonani, F., 1980. Some remarks on the application of geochemical techniques in geothermal exploration. *Proceedings of the 2nd Symposium on Advances in European Geothermal Research*, Strasbourg, France, pp. 428–443, March 4–6.

Truesdell, A.H., 1976. GEOTHERM, a geothermometric computer program for hot spring systems. *Proceedings of the 2nd U.N. Symposium on the Development and Use of Geothermal Resources 1975*. San Francisco, CA, pp. 831–836, May 20–29.

Truesdell, A.H., 1984. Chemical geothermometers for geothermal exploration. In *Fluid-Mineral Equilibria in Hydrothermal Systems*, vol. 1. Reviews in Economic Geology, eds. R.W. Henley, A.H. Truesdell, and P.B. Barton, Jr. Chelsea, MI: Society of Economic Geologists, pp. 31–43.

Truesdell, A.H. and Hulston, J.R., 1980. Isotopic evidence on environments of geothermal systems. In *Handbook of Environmental Isotope Geochemistry*, vol. I. The Terrestrial Environment, eds. P. Fritz and J.Ch. Fontes. Amsterdam, The Netherlands: Elsevier, pp. 179–226.

Verma, S.P., 2000. Revised quartz solubility temperature dependence equation along the water-vapor saturation curve. *Proceedings of the 2000 World Geothermal Congress*, Kyushu-Tohoku, Japan, pp. 1927–1932, May 28–June 10.

Verma, S.P., Pandarinath, K., and Santoyo, E., 2008. SolGeo: A new computer program for solute geothermometers and its application to Mexican geothermal fields. *Geothermics*, 37, 597–621.

Verma, S.P. and Santoyo, E., 1997. New improved equations for Na/K, Na/Li and SiO2 geothermometers by outlier detection and rejection. *Journal of Volcanology and Geothermal Research*, 79, 9–23.

Walker, F.W., Miller, D.G., and Feiner, F., 1984. *Chart of the Nuclides*. San Jose, CA: General Electric Co., Nuclear Engineering Operations, p. 59.

White, N.C. and Hedenquist, J.W., 1990. Epithermal environments and styles of mineralization: Variations and their causes, and guidelines for exploration. *Journal of Exploration Geochemistry*, 36, 445–474.

Williams, C.F., Reed, M.J., and Mariner, R.H., 2008. A review of methods by the US geological survey in the assessment of identified geothermal resources. US Geological Survey Open File Report 2008-1296, 27 pp.

Wolery, T.J., 1992. EQ3/6, A software package for geochemical modeling of aqueous systems: Package overview and installation guide (Version 7.0). UCRL-MA-110662-PT-I. Lawrence Livermore National Laboratory.

Zierenberg, R., 2012. Sulfur isotopes, *Recent Advances in Geothermal Geochemistry Workshop*, Geothermal Resources Council, Davis, CA.

ADDITIONAL INFORMATION SOURCES

Abrahamian, Y., Martirossyan, R., Gasparyan, F., and Kocharyan, K., 2003. *Methods and Materials for Remote Sensing: Infrared Photo-Detectors, Radiometers and Arrays, Proceedings of SPIE*, v. 6678.

Krauskopf, K.B. and Bird, D.K., 2003. *Introduction to Geochemistry*, 3rd Edition. New York: McGraw-Hill.

Reynolds, J.M., 1997. *An Introduction to Applied and Environmental Geophysics*. New York: Wiley.

These textbooks provide good introductions to geochemistry and geophysics that are required for understanding and using many of the methods discussed in this chapter and Chapter 7.

Strojnik-Scholl, M. ed., 2007. *Infrared Spaceborne Remote Sensing and Instrumentation XV, Proceedings of Society of Photo-Optical Instrumentation Engineers*.

These and other books on remote sensing infrared technology and methods provide a good introduction to the basics of this technique.

7 Exploring for Geothermal Systems
Geophysical Methods

The geochemical methods described in Chapter 6 allow investigators to develop a description of the thermal and chemical properties of possible geothermal reservoirs. They also support the development of geological and hydrological models that allow conceptualization of the physical framework within which a geothermal reservoir exists. Taken together, the shape, flow field, stage of development, location of high-temperature zones, and possible drilling targets can be approximately delineated. However, refining these models to a higher spatial resolution is required before serious consideration is given to whether a resource is a good candidate for further development. To obtain this higher resolution, geophysical methods are employed to better define the resource.

GEOPHYSICS AS AN EXPLORATION TOOL: AEROMAGNETIC SURVEYS

Although the dominant magnetic field that is sensed by a handheld compass is the earth's magnetic dipole field (the geodynamic field), which establishes the magnetic north and south poles, there exist other magnetic fields that can also be detected and measured. One of these fields is that which is imposed on the earth by the solar wind, which interacts with the geodynamic field. Less significant, but of use for geophysical surveys, is the local magnetic field resulting from the magnetic characteristics of the local geology.

Rocks are magnetic to varying degrees, depending upon their mineralogy, history, and temperature and pressure conditions. A few iron-bearing minerals, particularly magnetite (Fe_3O_4) and maghemite (Fe_2O_3), are themselves magnetic, whereas a variety of other minerals are responsive to magnetic fields by developing an induced magnetic field of their own when placed within an external magnetic field (such as the earth's magnetic field), a phenomenon called *magnetic susceptibility*. Rocks vary in their content of magnetic minerals and their magnetic susceptibility and thus possess weak magnetic fields of their own.

If one were to carry a sensitive magnetometer across the land surface, it would be possible to detect the local magnetic field, which would be a combination of the intrinsic field due to the rocks in the region plus that of the solar and geodynamic magnetic fields. By knowing the magnitude and direction of the nonlocal magnetic fields, their effects can be subtracted from measurements made at a particular location, and the resulting magnetic field (direction and magnitude) can be determined. The challenge then becomes how to interpret the magnetic pattern that one has obtained.

Magnetic surveys are now routinely done using airplane- or helicopter-mounted magnetometers, allowing large areas to be surveyed and magnetically mapped within a matter of days. The resolution of the survey results will depend on the spacing of the flight lines that are flown, their elevation, and the sampling rate of the magnetometer, relative to the speed of the aircraft. High-resolution surveys allow surface features as small as a few meters in diameter to be resolved. However, most features of interest in geothermal surveys are many meters in the subsurface, and the resolution obtainable for such magnetic features will be significantly poorer than that.

Magnetic anomalies of interest for geothermal exploration result from the fact that magnetic properties of rock are sensitive to hydrothermal alteration. When a hot, flowing aqueous fluid migrates through a rock, alteration of the original mineralogy will occur. This usually includes transformation of magnetic minerals such as magnetite and maghemite to hydrous oxide minerals that are not magnetic and have a low magnetic susceptibility, thus reducing the overall magnetic susceptibility of the rock. Clay minerals often end up replacing other minerals that have significant magnetic susceptibilities, as well, the net consequence being that low magnetic anomalies can indicate hydrothermal alteration that can be associated with geothermal systems.

To rigorously evaluate the significance of the mapped anomaly patterns, it is important to know the magnetic susceptibility of the rocks in the region. Accomplishing that requires conducting a field sampling effort in which representative samples are collected and their magnetic susceptibilities are measured in the laboratory. Using the surface geology as a constraint for how one distributes the known rock types in the subsurface, models are then constructed of how the rocks might be distributed underground, in an attempt to reproduce the observed magnetic field pattern. This approach cannot provide a unique answer to what is in the subsurface, because there will be a large number of ways to distribute the known rock types and generate the observed anomaly pattern. However, when combined with the local known geology and geological history, along with geochemical data and other geophysical measurements, only a few plausible configurations are likely to emerge.

Shown in Figure 7.1 is an example of data that can be collected on a flight line and a model that was constructed to fit the data (Hunt et al. 2009). The unit used for magnetic surveys is the tesla (T)

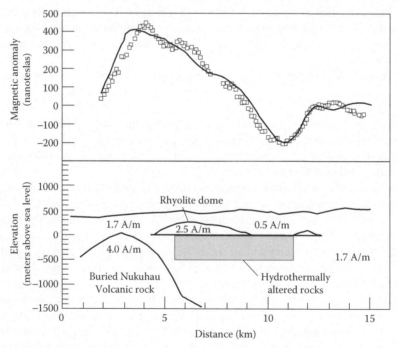

FIGURE 7.1 Aeromagnetic measurements and computed model for a transect near the Wairakei power station, New Zealand. The open squares in the upper figure indicate corrected values of the measured magnetic field, and the solid line is the magnetic field that would be measured for the model shown in the lower part of the figure. A/m is the magnetic intensity of the rock in the various regions of the model. (Modified from Hunt, T.M. et al., *Geothermics*, 38, 85–97, 2009.)

or nanotesla (nT), which is a measure of the magnetic flux density. The units in the lower figure are amperes/meter (A/m) and are referenced to the tesla, but represent intensity of magnetization. The relationship between these units is 1×10^{-4} T and is equivalent to $(1 \times 10^3)/4\pi$ A/m.

Several features of this model are worth noting. One point that is important to observe is that the low point in the magnetic pattern does not fall directly over the hydrothermally altered zone in the model. This results from the combination of rocks and their respective intensity of magnetization. The buried volcanic rock with the very high intensity of magnetization (4.0 A/m) essentially masks the low magnetization of the hydrothermally altered zone. To quantitatively account for these effects, numerical simulations, such as the one used to generate the lower figure, are required.

The presence of a low magnetic intensity region in the subsurface at a depth of 500–1000 m below the ground surface is a potentially attractive target. It may indicate the presence of rocks that have interacted with hot fluids, resulting in the formation of clays and the alteration of magnetic minerals. However, given the nonuniqueness of models such as this, it would be important to develop additional data that could allow the model to be tested.

RESISTIVITY AND MAGNETOTELLURIC SURVEYS

Complementary to aeromagnetic surveys are studies that evaluate the electrical and induced magnetic responses of rocks. Electrical resistance is a function of material properties. Geological materials are generally poor electrical conductors and therefore have a high resistivity, which is measured in units of ohm-meters. The presence of fluids in pores and fractures, especially fluids that have an elevated concentration of dissolved species that are electrically charged, increases the conductivity of the rock substantially and correspondingly reduces the resistance. These basic concepts have been known for decades, but they were not applied to geothermal exploration until the mid-1960s and early 1970s, when they were used to explore for geothermal resources in New Zealand (Hatherton et al. 1966; Macdonald 1967; Risk 1983). Those early studies, and many since then, have shown the usefulness of such measurements.

Resistivity measurements are made using a series of probes distributed tens to hundreds of meters apart, to detect the electrical response of the earth to injection of electrical impulses. By conducting a series of such experiments using different probe distributions, it is possible to reconstruct the distribution of electrical resistance in the rocks several hundred meters below the ground surface. Because flowing geothermal waters can be detected as zones of low resistance, it is possible to map geothermal resources using such a technique. Generally, flow zones are delineated by steep gradients that bound low resistivity areas. Zones of high resistivity can be interpreted as either dry rock or rock with low permeability and high resistance mineralogy, such as tight clay or alteration zones. However, care must be exercised when interpreting resistivity zones because they may also be caused by changes in rock type, complex intermingling of a variety of rock types, and temperature.

An example of a resistivity survey is shown in Figure 7.2, which was conducted in the Boku geothermal region in the Ethiopian portion of the East African rift valley (Abiye and Haile 2008). The results of the survey indicate the presence of two regions of very low resistivity at depths greater than 1500 m above sea level (asl), separated by a zone of higher resistivity at a distance of 0.9–1.0 km along the profile. Capping both areas of low resistivity are very shallow, very high resistivity regions, one of which corresponds to the Boku thermal area, where there occur warm pools, warm ground, and steam emissions. The thermal area is characterized by rocks that have been altered to kaolinite clay and within which fractures are sealed by calcite and quartz. It is this tightly sealed, clay lithology that gives rise to the measured high resistivity values. It is inferred that the low resistivity region maps the area where a high-temperature aquifer exists that feeds the surface features. The shallow nature of this aquifer

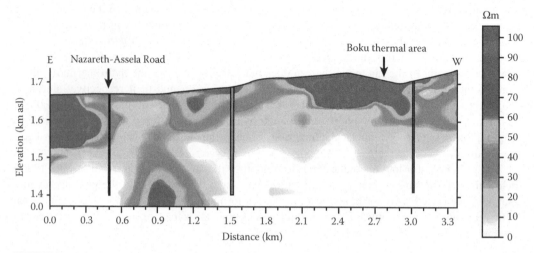

FIGURE 7.2 A cross-sectional figure of the results of a resistivity survey conducted in the Boku geothermal field, Ethiopia. The vertical axis is in km above sea level. The dark regions are areas of high resistivity, while the light areas are low resistivity. Favored locations for wells are indicated by the vertical lines. (Modified from Abiye, T.A. and Haile, T., *Geothermics*, 37, 586–596, 2008.)

makes it an attractive potential geothermal resource. The vertical lines in Figure 7.2 indicate where the researchers who conducted this survey propose drill sites for a small-scale geothermal project.

The limitation of such resistivity studies is their relatively shallow depth. Most significant geothermal resources that could support geothermal power generation at the MW level occur at depths greater than a kilometer or more. Such systems are not easily resolved using electrical resistivity studies. More recently, *magnetotelluric* (MT) surveys have become an effective means to image deeper structures. MTs takes advantage of the fact that the earth's magnetic field varies in intensity and orientation throughout the day, thus inducing small but detectable electrical currents in the earth's crust. The frequency of those currents spans a broad range, thus allowing a multispectral analysis of the variation in the local electromagnetic field. Such surveys are carried out over hours or days, with sampling rates varying from continuous to seconds-long intervals. Deploying sensors across the landscape allows a tomographic reconstruction of geology, because the frequency and magnitude of the currents that are detected are determined by the underlying response of the different rocks in the subsurface to the changing magnetic field. The result is that an unusually clear picture of the distribution of electromagnetic properties in the subsurface can be obtained. The result of such a survey carried out in New Zealand is shown in Figure 7.3. This survey was conducted across a portion of the Wairakei Valley, which is one of the world's great geothermal regions on the North Island. A MT survey can delineate features to depths in excess of 20 km, thus providing a means to image such things as the location of the actual source of heat for a geothermal resource region—in this case, it is interpreted to be ~5 km by 5 km magma body at a depth of about 15–20 km. Such surveys can also indicate the location of low resistivity zones that may be up flow regions at shallow levels. In this case, three separate regions of low resistivity were mapped and interpreted as possible zones of fluid flow.

Although MT surveys provide a powerful means for accessing information about the deep subsurface, heat sources, and regional fluid flow pathways, shallow-level resistivity surveys are needed in order to identify specific drilling targets. In this way, they are admirable complements of an exploration program.

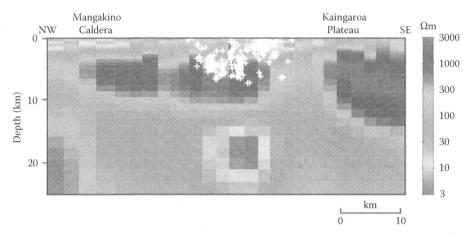

FIGURE 7.3 (**See color insert.**) A model based on a MT survey conducted in the Wairakei Valley, New Zealand. The low resistivity region at a depth of about 20 km is interpreted to be a magma body. The white crosses indicate sites of microseismic events. Possible fluid flow pathways are suggested by the green and pale-blue bands at depths less than 10 km, as well as in the very shallow levels where low resistivity (<10 Ωm) occur. (Modified from Heise, W. et al., *Geophysical Research Letters*, 34, 6, 2007.)

GRAVITY SURVEYS

An additional measurement technique often employed to determine the distribution of rock types in the subsurface is gravimetery. This technique uses highly sensitive instruments to detect and quantify slight differences in the strength of the gravitational field at a point on the ground surface. By conducting gravity surveys in which many tens to hundreds of individual measurements are made, or by using continuous measurements as one flies or sails over the land and ocean surface, detailed maps can be made of the local variation in gravity.

If the earth were a homogeneous perfect sphere, the resulting gravitational acceleration measurements would be the same everywhere. The equation describing this phenomenon is the classic Newtonian representation of gravitational force:

$$F = \frac{G \times m_1 \times m_2}{r^2} \tag{7.1}$$

where:
The constant G equals 6.67×10^{-11} Nm^2/kg^2
m_1 and m_2 are the masses of the interacting objects, respectively
r is the distance between their respective centers of mass

At the surface of the earth, the resulting nominal value for the acceleration due to gravity is 9.80665 m/s^2. But, topography, the oblate spheroidal form of the planet and the complex geological structures that make up the earth result in a more complex gravitational field than a simple spherically homogeneous body would provide. Two factors contribute to this situation. One factor reflects the density stratification and plastic nature of the deep, hot earth. Because the crust of the earth, whether it be oceanic crust or continental crust, literally floats on the mantle, and, to a first approximation, the mantle and core are homogeneous in density, they can be treated as a constant mass that underlies and supports the more heterogeneous, floating crust. As a result, measurements of the value of the gravitational acceleration at the earth's surface can be interpreted in terms of differences in the density and thickness of the various rock units in the crust that exist below the point at which a gravity measurement is made.

The second factor results from the effect of the r^2 term in the gravitational force equation. A small variation in rock density near the surface of the earth will have a much larger effect on the measured value of the acceleration of gravity than would the same-sized density contrast at a greater distance. Hence, the elevation at which a gravity measurement is made can influence the measured value of the acceleration of gravity, as will differences in the depth at which a specific rock unit may occur.

These variations in the gravitational field are called gravity anomalies and can be either negative or positive, depending upon whether they result in a local gravity measurement that is, respectively, less than or greater than a locally derived baseline value. Sensitive gravimeters, capable of measuring variations in the earth's gravitational field at a level of precision exceeding a fraction of a part per million, have become routine instruments to employ in field surveys. As a result, it is possible to map variations in rock density in the subsurface in remarkable detail.

Consider, for example, what would be detected using such an instrument if the subsurface structure described previously in the New Zealand aeromagnetic survey (see Section "Geophysics as an Exploration Tool: Aeromagnetic Surveys" and the model that was derived from it (Figure 7.1) were subjected to a gravity survey. The density of the Nukuhau volcanic rock will be approximately 10%–30% greater than that of the hydrothermally altered rocks. Because they are both surrounded, for the most part, by the same sedimentary and volcanic rocks, one would expect to detect a stronger gravitational field over the Nukuhau location than over the area above the hydrothermally altered rock. Given that they are at approximately the same depth but differ in thickness by at least a factor of 2, one would predict that the gravitational signal would vary by a few parts per hundred across this region, a variation that is well within the sensitivity of most high-precision gravimeters.

However, interpreting the results of gravity surveys can be difficult. Rock density can vary by 50% in the near surface, and the thickness of individual rock units, such as lava flows or sedimentary layers that occur in a place like Wairakei, for example, can vary by hundreds of percent over short distances. Given that these kinds of rocks can often overlap each other, there are many possible ways in which rock sequences can be combined to produce models that result in identical gravity anomaly patterns.

Consider, too, that the processes that result in the development of a geothermal field can change rock densities through fluid–rock interaction, as well as through affects on the density of the fluid that fills fractures and rock pores. Rock densities can be increased by replacement of clay minerals and carbonates by silica, or they can be decreased by replacement of feldspars by clays and zeolites. In addition, fracturing will decrease rock density, as will replacement of pore fluids by steam. Hence, the ability to identify a potential geothermal site using gravimetry depends to a great extent on additional information about the processes that have been operating at the site of interest. Consider, for example, the contrasting signatures of the geothermal sites at Heber, California, and The Geysers, California. The Heber site, which was originally explored as a potential oil and gas field, was identified, in part, by the coincidence of a positive gravity anomaly with a thermal anomaly (Salveson and Cooper 1979). The positive gravity anomaly was related to an increase in the local rock density due to geochemical processes that altered the country rock, resulting in the replacement of porosity and lower-density minerals with higher-density minerals. The Geysers, however, fall within a regional gravity low, related both to the local geology, and the fact that the geothermal field is charged with low-density dry steam (Stanley and Blakely 1995).

Even so, gravity surveys, when combined with other geophysical techniques can greatly diminish the ambiguity associated with proposed subsurface models, since the models must satisfy the constraints of each technique. It is precisely for this reason that a combination of techniques is always employed when exploring for geothermal resources. In addition, recent advances in signal processing technology have improved the ability to extract information and develop model structures, enhancing the usefulness of combined surveys.

SEISMICITY AND REFLECTION SEISMOLOGY

The rocks of the earth are very good transmitters of low-frequency energy. Such energy can be propagated for thousands of miles and detected at remote locations. Earthquakes are the classic example of generators of low-frequency energy—seismic waves generated by large earthquakes can travel around the earth several times and are readily detected by sensitive seismometers. Techniques that utilize this behavior for exploring and characterizing the subsurface have been developed for the oil and gas industry and are being refined and adapted to the needs of the geothermal community.

Seismic energy is transmitted in several different modes. *Body waves* propagate through the interior of a body and are of two types. Pressure waves or P-waves compress and expand materials in the direction in which the wave is traveling. These waves are the fastest seismic waves, traveling at velocities between approximately 1 and 8 km/s when in the crust of the earth. Shear waves or S-waves propagate via a shearing motion perpendicular to the direction of travel. S-waves have velocities that are approximately 60% of that of a P-wave in the same material.

The velocities of body waves strongly depend on the material through which the seismic waves are traveling. Very dense materials have the highest velocities (and therefore are said to have low impedance), and low-density materials have much lower velocities and much higher impedance. When two materials of different impedance are in contact, reflection of seismic waves can occur, and it is this phenomenon that is particularly important for exploration purposes and has led to the development of *reflection seismology*. This behavior is qualitatively similar to the reflection and refraction of light at the interface between two materials with different refractive indices.

Shown in Figure 7.4 is a schematic representation of the basic principles of seismic reflection. The "source" represents an impulse of seismic energy that is transmitted to the underlying rock mass, usually by detonating a small explosive charge or by using large vibrating trucks, also known as thumper trucks. The arrowed lines, R_1 and R_2, are two of an infinite number of ray traces that could be drawn to show the path seismic waves would follow as they propagate in the subsurface. These arrows are perpendicular to wave fronts that expand outward from the "source," much like

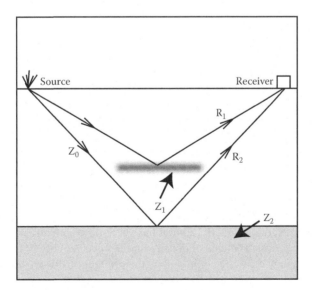

FIGURE 7.4 Schematic representation of a reflection seismology experiment. The source is the location of an energy source (impact or explosion) and the receiver is a mechanism for detecting seismic energy, usually a seismometer. The Z's are materials that have different impedances and, hence, velocities for seismic waves. R_1 and R_2 represent ray paths for seismic waves following different trajectories, one of which reflects off material imbedded within the Z_0 unit.

waves radiate outward on the surface of a pond when an object is dropped into the pond. Three geological materials are represented in the figure by their impedances, Z_0, Z_1, and Z_2. The rays will reflect off of materials that have an impedance that is different from that of the material the ray is traversing. The strength of a reflection from a body will depend upon how great the difference is in the impedance of each material, the angle of incidence of the incoming seismic wave, how sharp the boundary is between the materials, and the extent to which the local pore spaces are filled with fluid.

Factors that influence impedance can be important for geothermal exploration. High fracture densities, which can dramatically improve fluid flow rates, can greatly increase impedance, especially if the void space is filled with steam rather than liquid. Being able to identify such zones from seismic reflection studies is one important contribution seismic studies can make to an exploration program. Impermeable cap rocks that trap geothermal fluids are another potential target for seismic reflection studies. In this instance, such rocks are likely to be cemented or welded rock units, in which case they would be expected to possess low impedance and would potentially be strong reflectors.

Identifying such features in the subsurface is crucial for delineating drilling targets. But equally important is knowing how deep such a target may be. By recording the time between impulse release and receiving a seismic signal, the "two-way travel time" can be determined. That travel time provides the means to determine the depth to a reflector, using the relationship

$$t = 2 \times (D/V) \tag{7.2}$$

where:
  t is the two-way travel time
  D is the distance traveled
  V is the velocity of the seismic wave in the medium

For a geological system more complex than that depicted in Figure 7.4, additional terms would be required to account for different velocities of different rock units traversed by the ray, and the effects they would have on the ray path.

Recent advances in computer-assisted signal processing techniques have allowed much more information to be extracted from received seismic waves in seismic reflection studies than was previously possible. Estimates can now be made of the density and elastic properties of a reflector, and in some cases, the porosity of the material can be inferred (Abramovitz 2011).

An example of a relatively recently completed seismic study in Larderello, Italy, provides an example of what can be achieved (Cappetti et al. 2005). Shown in Figure 7.5 is the result of sophisticated signal processing techniques applied in that seismic reflection study. The region marked as the "Geothermal target" has high reflectivity, in part due to the fact that it is a zone of high fracture density, as confirmed by the drilling (dashed vertical line) that intersected that horizon about 400 m to the left of the target. The high fracture density resulted in a higher impedance contrast and, hence, greater reflectivity.

Mapping subsurface porosity anomalies has become a relatively common endeavor in the oil and gas industry (e.g., Abramovitz 2011), but it has yet to be systematically employed in geothermal exploration. Using measurements of density and P-wave velocity in a borehole that has been logged, acoustic impedance can be computed as the product of P-wave velocity and density and has the units of kg/m^3-s or g/cm^3-ms. By also measuring total porosity of core samples, a quantitative relationship between porosity and acoustic impedance can be derived that applies to the region being studied. Such an approach has the potential to allow identification of high porosity zones in geothermal reservoirs. Of course, whether such regions have adequate permeability to support sufficient fluid flow for power generation would require additional studies, especially flow testing.

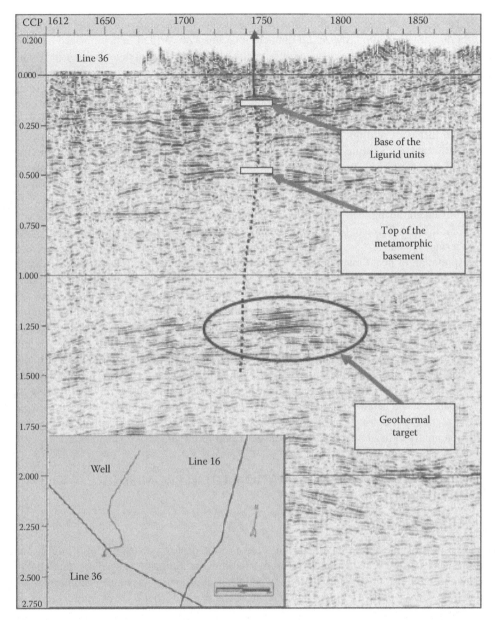

FIGURE 7.5 Results of signal processing techniques applied to a seismic reflection study conducted in Larderello, Italy. (From Cappetti, G. et al., A new deep exploration program and preliminary results of a 3D seismic survey in the Larderello-Travale geothermal field [Italy], *Proceedings of the World Geothermal Congress*, Antalya, Turkey, April 3, 2005.)

TEMPERATURE MEASUREMENTS

Once geological and geophysical techniques have identified potential geothermal resources, it is imperative that measurements be made in the subsurface of the temperature gradient. Such measurements are fundamentally important because it is only through them that the presence of a thermal resource can be established. However, drilling programs that are used to acquire these data are expensive and hence undertaken only after there is good evidence from other sources that a geothermal resource is available at depth.

The principle means for accomplishing this is to identify a target region and then conduct a drilling program in which numerous small diameter boreholes are drilled to depth. Because drilling is one of the most expensive components of an exploration and development project, it has become routine to use slim hole drilling technology rather than large diameter rotary drilling methods. Slim holes (diameters less than about 15 cm) are drilled with drilling rigs that are smaller than large diameter rotary drilling equipment, require fewer materials for drilling, and are easier to complete. Slim holes have been drilled to depths of 2000 m, thus making them suitable for exploration programs in a variety of settings.

Heat flow and temperature gradient measurements can be made using downhole equipment that is lowered on a wire line. The equipment commonly is small diameter ($\sim$6.25 cm) and allows relatively rapid measurement of temperatures at many levels in a borehole. Such data allow construction of temperature profiles (such as those shown in Figure 4.11, page 77). Once samples are collected from cores taken during drilling, and the thermal conductivity of the material is measured in the laboratory, the heat flow can be calculated using Equation 2.3. Such data also allow the calculation of geothermal gradients.

Interpretation of heat flow values and geothermal gradients requires knowledge of the region's baseline, or average, heat flow in order to reasonably identify shallow-level thermal anomalies. Rarely are such data available in sufficient coverage of a region to allow detailed thermal anomaly maps to be drawn. Generally, though, heat flow values on the order of 100 mW/m^2 are reasonable indicators that elevated temperatures can be reached at modest depths.

However, regions in which geothermal reservoirs exist are notorious for possessing complex thermal structure in the subsurface. The Long Valley Caldera case study in Chapter 4 demonstrated linear geothermal gradients that are often inadequate to describe the subsurface thermal regime. Indeed, because a geothermal heat source is expected to drive convective flow of groundwater, the ideal signature that a geothermal resource had been found would be a rapid increase in temperature near the surface and constant temperature over a considerable depth below that. Such a geometry would potentially indicate that a zone of convective upwelling had been encountered.

REMOTE SENSING AS AN EXPLORATION TOOL: PROMISING NEW TECHNIQUE

MULTISPECTRAL STUDIES

Advances in instrumentation and electronics have resulted in the development of highly sensitive measurement tools that can be used in aircraft, satellites, and drones. These allow rapid mapping, at moderate to high resolution, of virtually any geophysical property relevant to geothermal resource exploration. This section briefly describes some of the methods that have been utilized to conduct such remote sensing surveys.

The fact that minerals can be distinguished by their color is apparent to even the most untrained observer. The fact that color is the result of complex interactions between electromagnetic radiation, mineral composition, and mineral structure is less apparent, but nevertheless, important. The electromagnetic spectrum covers wavelengths from about 10^{-6} microns (gamma rays) to more than 10^8 microns (TV and radio waves). The visible portion of the spectrum occupies the region from about 0.4 microns (blue light) to about 0.7 microns (red light). Just beyond the longer wavelength portion of the visible spectrum is the infrared portion of the spectrum. The infrared includes the near infrared, at 0.7–1.2 microns, the solar reflected infrared at 1.2–3.2 microns, the mid-infrared at 3.2–15 microns, and the far infrared at >15 microns. Laboratory measurements of the reflected visible and infrared light from samples of pure minerals allow the characteristic reflectance spectra of minerals to be determined and used as a standard against which field measurements can be compared. It is this observation that holds the promise of providing the ability to sense from a distance the mineralogical make-up of soils and rocks.

Airborne and satellite surveys using *optical* wavelengths have become relatively straightforward. Generally, such surveys use an approach in which the visible spectrum obtained from a single measurement of a selected region, or measured along a flight line, is subjected to spectral analysis, whereby the relative or absolute intensity of specific wavelengths over a range of wavelengths is measured along a flight path. Using laboratory measurements that have established the spectral reflectance and adsorption "signatures" of specific minerals, the mineralogy on the ground surface can be approximately determined by comparing the measured spectra with the laboratory-determined signatures. Such information can be used to remotely map rock types and rock structure, but they currently do not provide sufficient discrimination to map detailed mineralogy. However, combining spectral analysis at infrared and visible wavelengths provides greater discriminatory power, allowing the identification of many surface minerals useful for identifying regions where hydrothermal fluids have affected the surface mineralogy, either through deposition during evaporation of geothermal fluids or through alteration of local rocks. Significant in these studies is the ability to identify borate minerals (Crowley 1993; Stearns et al. 1999; Crowley et al. 2000; Khalalili and Safaei 2002; Kratt et al. 2006, 2009; Coolbaugh et al. 2006, 2007). As discussed in this section, boron is commonly elevated in geothermal waters. Evaporite deposits that contain borate minerals may thus provide an indication of a previously active geothermal spring or other water source that may not currently be active. In other words, detection of such deposits may be a means for identifying hidden resources.

Field measurements of the reflectance spectra of soils and rocks can be made using handheld instruments. This approach, coupled with spectra obtained remotely using aircraft or satellites, allows ground truthing of remote measurements. Such efforts can provide accurate analysis of the mineralogy of potential hidden resource regions.

An example of this capability has recently been documented by Kratt et al. (2006, 2009). Using satellite data from the Advanced Spaceborne Thermal and Emitted Reflectance Radiometer (ASTER), airborne instruments, and ground-based measurements, these researchers were able to identify tincalconite deposits in a playa that represented localized borate concentrations (Figure 7.6). The form and location of the concentrations are suggestive of a localized source of fluid discharge. This type of occurrence could be a useful indicator of geothermal fluids flowing into the surface or near-surface.

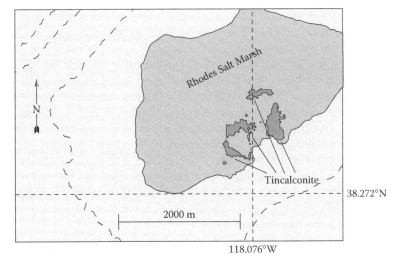

FIGURE 7.6 Map of Rhodes Salt Marsh in which the distribution of tincalconite is delineated. These surface deposits are identified on the basis of infrared spectra obtained from satellite data. (Modified from Kratt, C. et al., *Geothermal Resources Council Transactions*, 30, 435–439, 2006.)

Although use of these techniques currently is suitable only for regions where there is low rainfall and sufficient development of evaporite deposits to support such an analysis, it is likely that additional strategies will be developed for use of such remotely collected data. The advantages of such an approach are the specificity it allows for diagnostic mineral identification and the ability to survey large areas relatively quickly.

SYNOPSIS

In addition to the geochemical techniques employed for exploration for geothermal resources, numerous geophysical techniques are available to identify a resource. Geophysical tools can be very useful in identifying target areas, depth to a resource, and other parameters. Important geophysical methods include electrical resistivity, MT, gravity, seismic reflection and refraction, interferometric synthetic aperture radar (InSAR), and temperature gradient measurements. However, because each technique often can be interpreted in a variety of ways, ambiguous results may result, which requires judicious analysis. It is wise to develop multiple working hypotheses and test them repeatedly as new data are acquired. Thermal gradient wells are particularly important for resolving uncertainties, but they generally follow after all other analytical approaches have been used, since they are expensive. Remote sensing capabilities are rapidly evolving and may provide the ability, in the near future, to quickly identify hidden resources, potentially avoiding months of labor-intensive field-work.

CASE STUDY: FALLON, NEVADA

Exploration for geothermal resources, as discussed above, involves evaluation and integration of a variety of data. A study conducted by Combs et al. (1995) in western Nevada provides a good example of such an effort. It includes a review of earlier geological studies, an analysis of the seismic activity (and anomalies associated with that), geochemistry, geophysics, and exploratory drilling. In the end, a sound basis was developed for concluding that a significant resource was available, was accessible, and could be economically developed. The approach used could be considered a template for exploration strategies in other geological settings within which extensional tectonics are associated with volcanic activity.

The area studied was in the volcanically active Basin and Range Province. The city of Fallon is located in the Carson Desert, a valley of the Basin and Range Province where episodes of volcanic activity have occurred between 43 million years ago and 20,000 years ago (Stewart and Carlson 1978). Numerous hot springs occur throughout the area.

Previous geological studies had documented that the valley is enclosed by ranges that are bounded by normal faults, which is typical of the Basin and Range Province. The valley floor and subsurface geology are composed mainly of relatively recent sandstones, siltstones, clays, and other sedimentary rocks. This combination of a fault-bounded valley filled with porous sediments is a favorable situation for runoff from the topographic highs to recharge aquifers by infiltration along faults. Many of the faults are likely to extend to considerable depth, providing access of infiltrating water to the deeper hydrothermal system, if it is present. The presence of basin fill composed of porous sediments could provide sufficient permeability to accommodate flow rates sufficient to allow fluid extraction for power generation. Such a geological framework is hypothetically ideal for siting a geothermal power facility. The caveat, of course, is whether or not a source of heat of sufficient magnitude exists at accessible depths.

The record of recent volcanic activity and the location of the Carson Desert within the active portion of the Basin and Range would suggest that some seismicity would be expected. However, there is little in the way of recorded seismic events. Generally, as Combs et al. (1995) point out, geothermal sites have microseismic activity (earthquake magnitudes in the range of −2 to 3.0).

In this instance, they argue that the absence of recorded seismicity is more likely a reflection of the absence of appropriately placed seismometers than a real absence of seismicity, because the magnitude of these earthquakes makes it unlikely that they could be detected at any significant distance from the center of rupture. This approach to data analysis will be discussed further at the end of this section.

In 1980, an analysis was conducted on a local hot spring that had 70°C water flowing from depth (Bruce 1980). A version of the Na–K–Ca geothermometer gave a temperature for the reservoir of 204°C. On the basis of that work, several temperature gradient holes were drilled to determine the subsurface temperature distribution. The measured temperature gradients ranged between 97°C/km and 237°C/km. Fluids collected from the drilling effort were chemically analyzed and geothermometer calculations using the Na–K–Ca model also gave 204°C as the source region temperature, which matched the results for the artesian well (Combs et al. 1995). The geothermal gradients are high for continental crust, even within the Basin and Range Province, thus suggesting there exists at depth a significant thermal source. The geothermometer results are consistent with this conclusion.

Soil samples were collected and analyzed for mercury (Hg) (Katzenstein and Danti 1982). As noted in Chapters 3 and 12, Hg is a common trace metal present in geothermal fluids. The results of this soil survey showed elevated Hg concentrations in the vicinity of the temperature gradient holes (Figure 7.7). It was suggested that these elevated concentrations were the result of geothermal fluids escaping from the geothermal reservoir along faults.

Gravity and magnetic surveys were conducted in the area. A magnetic high was determined to be present approximately in the area where the Hg concentrations were elevated. A gravity anomaly was observed, but did not correlate with these other anomalies, and remains enigmatic.

In 1986, a temperature gradient hole was drilled to a depth of 1367 m. It was found that hydrothermal alteration had affected the rocks, resulting in the development of pyrite and other alteration of primary minerals. Hydrogen sulfide was also detected. The bottom hole temperature for this hole was found to be 155°C, and the temperature gradient was 104°C. However, there was no fluid flowing into the well.

On the basis of these results, a larger, deep test well (Figure 7.7) was drilled in the Hg anomaly and near the previous well that was drilled to 1367 m. This larger well terminated at 2119 m. A temperature of 191°C was measured during evaluation of the well, providing clear evidence that a substantial geothermal resource was present and accessible.

In order to induce flow, the hole was stimulated with liquid nitrogen to fracture the rock through thermal stress and pressure. This effort was successful, resulting in a flow rate of up to 40 kg/s of brine.

This effort to find a resource was successful and provides several important lessons for exploration efforts. These include the following:

- Previously collected data were used to delineate a target area that would be explored in detail. Note that the geological setting is one of the classical "extensional settings," which is also classified as a "fault-bounded extensional (horst and graben) complex" (Brophy type E).
- Geothermometers were used whenever suitable samples were available for analysis. This can provide a good basis for developing working hypotheses.
- The absence of recorded seismic activity could be an indication that the region was tectonically inactive and, hence, not likely to possess a suitable resource. However, in this instance, geological experience and expertise on the part of the investigators allowed them to consider alternative interpretations. The bulk of the other evidence was strongly suggestive of the presence of a resource, thus making it wise to develop multiple working hypotheses.

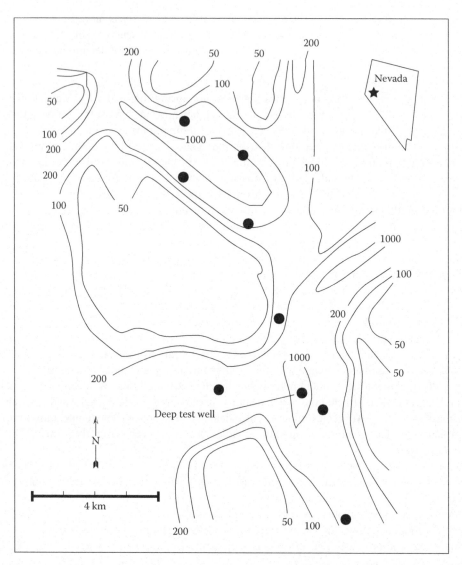

FIGURE 7.7   Contour map of soil mercury concentrations (in parts per billion) in the Fallon, Nevada region. Temperature gradient wells are indicated by the solid dots. The deep test well discussed in the text is indicated. (Modified from Combs, J. et al., Geothermal exploration, drilling, and reservoir assessment for a 30 MW power project at the Naval Air Station, Fallon, Nevada, USA. *Proceedings of the World Geothermal Conference*, 2, 1371–1378, 1995.)

- The use of soil analyses for trace metals provided an important and useful guide to where a potential resource might be located. Drilling of the deep test well, in fact, was centered in one such Hg anomaly.
- Other geophysical data (magnetic and gravity) provided some indication of a geological anomaly, but data were not sufficient to delineate a target. This is often the case with such data. They can be useful, but need additional techniques to provide a rigorous subsurface model.
- The absence of flowing water in a well is not necessarily an indication of an absence of water in the system. In this case, formation or enhancement of fracture permeability successfully allowed fluid flow to occur at a reasonable rate.

PROBLEMS

7.1. From the data in Figure 7.3, what would be the most likely circulation pattern of water above the *hot spot* located at a depth of about 18 km and in the middle of the figure?

7.2. What might be the explanation for the lack of correspondence between the magnetic and gravity surveys that was observed in the Fallon, Nevada study?

7.3. In the Boku cross section in Figure 7.2, where might other wells be drilled that would have a probability of accessing geothermal fluids? Explain your reasoning.

7.4. Brophy Type E systems provide opportunities for fluid to move through fault systems. Consider how fluids would move through Figure 6.6 and describe where multispectral analysis might provide indications of geothermal systems.

7.5. Given your answer to question 7.4, what would be the multispectral differences one might observe for the various deposits?

REFERENCES

Abiye, T.A. and Haile, T., 2008. Geophysical exploration of the Boku geothermal area, Central Ethiopian Rift. *Geothermics*, 37, 586–596.

Abramovitz, T., 2011. Mapping porosity anomalies in deep Jurassic sandstones—An example from the Svane-1A area, Danish Central Graben. *Geological Survey of Denmark and Greenland Bulletin*, 23, 13–16.

Bruce, J.L., 1980. Fallon geothermal exploration project, Naval Air Station, Fallon, Nevada, Interim Report. Naval Weapons Center Technical Report NWC-TP-6194, Naval Weapons Center, China Lake, CA.

Cappetti, G., Fiordelisi, A., Casini, M., Ciuffi, S., and Mazzotti, A., 2005. A new deep exploration program and preliminary results of a 3D seismic survey in the Larderello-Travale geothermal field (Italy). *Proceedings of the World Geothermal Congress*, Antalya, Turkey, April, 3 pp.

Combs, J., Monastero, F.C., Bonin, Sr., K.R., and Meade, D.M., 1995. Geothermal exploration, drilling, and reservoir assessment for a 30 MW power project at the Naval Air Station, Fallon, Nevada, USA. *Proceedings of the World Geothermal Conference*, 2, 1371–1378.

Coolbaugh, M.F., Kratt, C., Fallacaro, A., Calvin, W.M., and Taranik, J.V., 2007. Detection of geothermal anomalies using Advanced Spaceborne Thermal Emission and Reflection (ASTER) thermal infrared images at Brady hot springs, Nevada, USA. *Remote Sensing of Environment*, 106, 350–359.

Coolbaugh, M.F., Kraft, C., Sladek, C., Zehner, R.E., and Shevenell, L., 2006. Quaternary borate deposits as a geothermal exploration tool in the Great Basin. *Geothermal Resources Council Transactions*, 30, 393–398.

Crowley, J.K., 1993. Mapping playa evaporite minerals with AVIRIS data: A first report from Death Valley, California. *Remote Sensing of Environment*, 44, 337–356.

Crowley, J.K., Mars, J.C., and Hook, S.J., 2000. Mapping evaporite minerals in the Death Valley salt pan using MODIS/ASTER airborne simulator (MASTER) data. *Proceedings of the 14th International Conference on Applied Geologic Remote Sensing*, Las Vegas, NV, November, 6–8.

Hatherton, T., Macdonald, W.J.P., and Thompson, G.E.K., 1966. Geophysical methods in geothermal prospecting in New Zealand. *Bulletin Volcanologique*, 29, 484–498.

Heise, W., Bibby, H.M., Caldwell, T.G., Bannister, S.C., Ogawa, T., Takakura, S., and Uchida, T., 2007. Melt distribution beneath a young continental rift: The Taupo volcanic zone, New Zealand. *Geophysical Research Letters*, 34(L14313), 6.

Hunt, T.M., Bromley, C.J., Risk, G.F, Sherburn, S., and Soengkono, S., 2009. Geophysical investigations of the Wairakei Field. *Geothermics*, 38, 85–97.

Katzenstein, A.M. and Danti, K.J., 1982. Evaluation of geothermal potential of the Naval Air Weapons Training Complex, Fallon, Nevada. Naval Weapons Center Technical Report NWC-TP-6359, Naval Weapons Center, China Lake, CA.

Khalalili, M. and Safaei, H., 2002. Identification of clastic-evaporite units in Abar-Kuh playa (Central Iran) by processing satellite digital data. *Carbonates and Evaporites*, 17, 17–24.

Kratt, C., Coolbaugh, M., and Calvin, W., 2006. Remote detection of Quaternary borate deposits with ASTER satellite imagery as a geothermal exploration tool. *Geothermal Resources Council Transactions*, 30, 435–439.

Kratt, C., Coolbaugh, M., Peppin, B., and Sladek, C., 2009. Identification of a new blind geothermal prospect with hyperspectral remote sensing and shallow temperature measurements at Columbus Salt Marsh, Esmeralda County, Nevada. *Geothermal Resources Council Transactions*, 33, 481–485.

Macdonald, W.J.P., 1967. *A Resistivity Survey of the Taupo-Waiotapu Area at Fixed Spacing (1800 ft.)*. Wellington, New Zealand: Geophysics Division, Department of Scientific and Industrial Research, 10 pp.

Risk, G.F., 1983. Delineation of geothermal fields in New Zealand using electrical resistivity prospecting. *Proceedings of the 3rd Biennial Conference of the Australian Society of Exploration Geophysicists*, Brisbane, Australia, pp. 147–149, August 12–14.

Salveson, J.O. and Cooper, A.M., 1979. Exploration and development of the Heber geothermal field, Imperial Valley, California. *Geothermal Resources Council Transactions*, 3, 605–608.

Stanley, W.D. and Blakely, R.J., 1995. The Geysers—Clear Lake geothermal area, California: An updated geophysical perspective of heat sources. *Geothermics*, 24, 187–221.

Stearns, S.V., van der Horst, E., and Swihart, G., 1999. Hyperspectral mapping of borate minerals in Death Valley, California. *Proceedings of the 13th International Conference on Applied Geologic Remote Sensing*, Vancouver, BC, March 1–3.

Stewart, J.H. and Carlson, J.E., 1978. Geologic Map of Nevada. Nevada Bureau of Mines and Geology, Reno, NV. http://www.nbmg.unr.edu/Pubs/Misc/Stewart&Carlson500K.pdf.

ADDITIONAL INFORMATION SOURCES

Abrahamian, Y., Martirossyan, R., Gasparyan, F., and Kocharyan, K., 2003. *Methods and Materials for Remote Sensing: Infrared Photo-Detectors, Radiometers and Arrays*, Springer.

Krauskopf, K.B. and Bird, D.K., 2003 *Introduction to Geochemistry*, 3rd Edition. New York: McGraw-Hill.

Reynolds, J.M., 1997. *An Introduction to Applied and Environmental Geophysics*. New York: Wiley.

    These textbooks provide good introductions to geochemistry and geophysics that are required for understanding and using many of the methods discussed in this chapter.

Strojnik-Scholl, M., ed., 2007. *Infrared Spaceborne Remote Sensing and Instrumentation XV, Proceedings of Society of Photo-Optical Instrumentation Engineers*, Proceedings of SPIE—the International Society for Optical Engineering, v. 6678.

    These and other books on remote sensing infrared technology and methods provide a good introduction to the basics of this technique.

8 Resource Assessments

Assessing a resource is a fundamental activity that is common to all energy use enterprises in which the amount of energy must be established for economic, management, or scientific purposes. With respect to geothermal resources, the scale at which this effort is carried out can be quite local, such as assessing the long-term energy availability from a single hot spring, or it can be extremely broad, such as a national resource assessment. Despite these vastly different scales, the methodologies employed are surprisingly similar because, in all cases, what is sought is a measure of the amount of heat that is accessible, the extent to which it can be economically extracted using available technology, and what its lifetime is. Beyond this similarity, however, important differences exist in how the necessary information is collected and how it is processed. This chapter will first consider how a local resource is evaluated and then consider the ensemble of information that is necessary to conduct national and international assessments and how those data are processed.

ASSESSING A GEOTHERMAL RESOURCE

We will consider here the issues associated with establishing the available heat for applications that will either directly use the heat, so-called direct-use applications (discussed in detail in Chapter 12), or use the heat to generate power (Chapter 10). We will not consider resource assessments for ground source heat pump applications (Chapter 11) because the heat available for their deployment exists everywhere—the most significant challenge for their economic viability is the local geothermal gradient, drilling costs, and competition with other energy sources and is not directly a function of the availability of heat.

Geothermal resource assessments have been an integral part of developing a geothermal resource ever since the industry began producing power. Such an effort is the first step in deciding whether investment in a facility is an economically viable undertaking. However, to better understand the magnitude of the geothermal resource base and the ability of this energy source to contribute to local, regional, or national energy markets, it is necessary to develop a more systematic analysis than is usually pursued when considering the development of a single facility.

In 1979, the US Geological Survey published its second systematic assessment of the geothermal resource in the United States (Muffler 1979), which followed the first national assessment carried out in 1975 (White and Williams 1975). The preliminary results of a new assessment are currently available through the US Geological Survey (Williams et al. 2008b).

Since publication of the 1979 report, several other assessments have been carried out on a variety of scales, most of them regional or state-wide (Petty et al. 1992; Lovekin 2004; Gawell 2006; Western Governor's Association 2006). As summarized by Gawell (2006), different approaches and underlying methodologies have given rise to a broad range of results that are not directly comparable. As an example, the results of various resource assessments for the state of California are shown in Figure 8.1. This diversity of results emphasizes the important fact that resource assessments provide results that are sensitive to the methodology and assumptions employed in the analysis. As a result, it is entirely possible that different studies will produce widely different estimates of a resource, and yet they each may be correct. In the approach we will take here, we will follow the methodology that was developed by Nathenson (1975), White and Williams (1975), and Muffler and Cataldi (1978), employed by Muffler in his 1979 assessment, and modified by Williams et al. (2008a) in their current assessment.

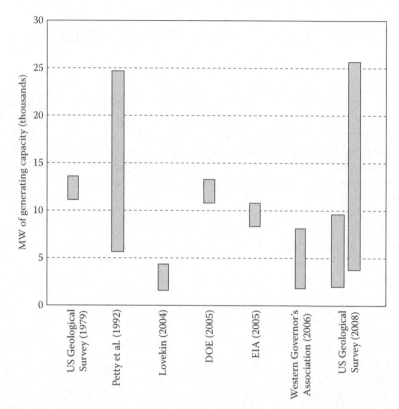

FIGURE 8.1   Range of resource assessments for the state of California. Assessment values are arranged chronologically, from oldest at the left to the most recent at the right. (From Muffler, L.P.J., Assessment of geothermal resources of the United States—1978, US Geological Survey Circular 790, 163, 1979; Petty, S. et al., Supply of geothermal power from hydrothermal sources: A study of the cost of power in 20 and 40 years, Sandia National Laboratory Contract Report SAND92-7302, 1992; Lovekin, J., *Geothermal Resources Council Transactions*, 33(6), 242–244, 2004; US Department of Energy 2005, as quoted in Gawell, 2006; Western Governor's Association, Geothermal task force report, Clean and Diversified Energy Initiative Report, 66, 2006; Williams et al., 2008b.)

The underlying concept is that a geothermal resource assessment is designed to establish the amount of heat that is available from a reservoir to produce electricity or some other form of work. To determine this value requires solving the relationship:

$$Q_R = \rho \times C \times V \times (T_R - T_0) \tag{8.1}$$

where:

Q_R is the heat available in the reservoir (J)

ρ is the density of the rock making up the reservoir (g/cm³)

C is the heat capacity of the reservoir rock (J/g-K)

V is the volume of the reservoir (cm³)

$(T_R - T_0)$ is the difference between the reservoir temperature T_R and the end-state temperature T_0

Consideration of the various terms of Equation 8.1 reveals several significant obstacles in establishing Q_R. We will consider these individually in the remainder sections of this chapter. However, in order to avoid confusion about what a resource assessment actually provides, it is important to define exactly what it is we are seeking when we undertake a resource assessment.

RESOURCE BASE AND RESERVES

When we undertake an evaluation of how much of a resource exists, there are usually very specific reasons why this effort is being made. From a purely scientific or academic perspective, it may be important to know how much of a material or substance exists in order to conduct global mass balance calculations. From a regulatory perspective, it may be important to establish the extent of a resource that is potentially extractable in order to determine how best to manage development of the resource in a way that is also environmentally intelligent. From an economic perspective, it may be that an estimate is needed to determine how much of a resource can be extracted using a specific kind of technology in order to establish the economic potential of a development project.

These different needs have given rise to a methodology that allows quantification of the different components that make-up the totality of a resource. Figure 8.2 graphically delineates the categories, nomenclature, and variables used in this approach.

For virtually any energy resource, some fraction of that resource has already been identified, mapped, and is readily accessible using existing technology. The suite of such occurrences, whether a population of local sites or the total ensemble of accessible locations on a national basis, is called the resource *reserve*. The reserve amounts to that portion of the resource that can be economically developed and brought online with little or no additional exploration effort. These are occurrences about which there would be relatively little uncertainty regarding the magnitude of what is available.

The reserve provides an important insight into how we might develop a strategy for estimating how much additional resource might exist that would be economically viable to pursue, but has not yet been discovered. Estimating the amount of a geothermal resource that has not yet been discovered is based on the reality that the existing geothermal resource reserves occur in specific types of geological environments. For example, The Geysers geothermal area in California and the Iceland geothermal systems exist in close proximity to recently active volcanic centers. For each of

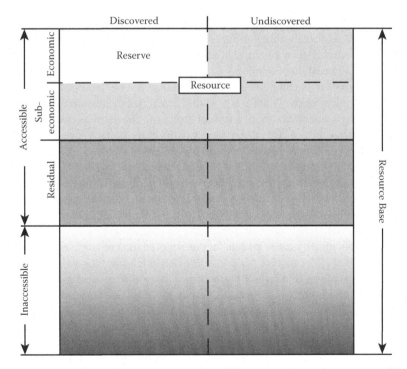

FIGURE 8.2 Nomenclature and parameters used to delineate geothermal resources. (Modified from Williams, C.F. et al., A review of methods by the US geological survey in the assessment of identified geothermal resources, US Geological Survey Open File Report 2008-1296, 27, 2008a.)

those types of known and characterized environments, experience gained through exploration and development provides a means to estimate the fraction of that type of environment that may possess a geothermal reservoir of a given quality. It is through this experience base that we can make an estimate of the undiscovered geothermal reservoirs that are potentially economically suitable for development. This undiscovered but potentially economically viable heat source, along with the known reserves, make up what is formally termed the *resource*.

UNCERTAINTY

Uncertainty is introduced into the process of identifying the formal resource by the fact that establishing the economic viability remains a contentious issue. In some instances, a site may be suitable for development using available technology but cannot be economically brought online because it is too remote from a transmission system. If a transmission line were brought in, for whatever reason, the site would then be economically viable. In other cases, the reservoir may be exploitable using a particular technology that is difficult to access, and therefore, the resource is not likely to be brought online in the near future. In some instances, some reservoirs may not be economically viable under existing tax and incentive structures but would be with a small change in either regulatory framework. In all of these cases, there is nothing about their physical attributes that make these reservoirs unsuitable for use. Rather, the immediate set of economic circumstances makes them economically marginal or unsuitable for immediate development. Because circumstances can change relatively quickly, these types of occurrences are added to the overall formal resource.

A so-called *residual* exists that represents occurrences that are theoretically accessible, in the sense that they could be drilled or otherwise accessed, but which pose significant challenges for development and are not seen to be economically sustainable. Finally, there is that portion of the total geothermal energy budget that simply is not accessible using any existing or foreseeable technology. The combined total of all of these sources of heat is called the *resource base*.

In summary, the resource base is made up of all heat, both accessible and inaccessible, that we wish to consider. The resource is that portion of the resource base that is technologically accessible, whether or not it is currently economically suitable for development under current market conditions. The reserve is that portion of the resource base that is accessible and can be economically developed.

It is important to realize that all of these values are mutable. The resource base can change significantly depending on the amount of data available to characterize the resource, the state of knowledge of the geological environment, and how a resource occurs. The estimated world resource base for oil, for example, increased from 600 billion barrels in the early 1940s to over 3900 billion barrels in 2000 (Wood and Long 2000). Similarly, considering the world's proved reserves of oil, the amount changed from ~640 billion barrels in 1980 to more than 1525 in 2012 (US Energy Information Agency 2013). This does not mean that the amount of oil on the planet increased. Rather, the change indicates an improvement in the information available to conduct such an analysis and in the development of improved methods to carry them out. Similarly, the amount of a commodity that can be identified as the resource can change with changes in the economic environment. For example, deposits that could not be economically developed under one set of market conditions may suddenly become economically attractive if the market price increases, as often happens with gold or other metals. Technological improvements that reduce extraction costs or that make it possible to economically develop a deposit or reservoir that was previously only marginally viable are also developments that can change estimates of the resource or the reserve. The development of high-efficiency binary cycle generating systems (discussed in Chapter 10) is an example of such a technological breakthrough that allowed the economic development of moderate temperature geothermal reservoirs, thus changing both the reserve and the resource estimates for geothermal power generation.

In the remainder of this chapter, we will consider primarily the formal resource and how it is estimated. In Chapters 13 and 16, we will discuss potential developments of technologies that might modify the estimates generated using these methods.

DETERMINING THE RESERVOIR VOLUME

Equation 8.1 establishes the importance of accurately knowing the volume of a geothermal reservoir in order to obtain an accurate estimate of the reservoir's heat content. Geothermal reservoirs exist in the subsurface at depths of hundreds to thousands of meters, and it is thus difficult to obtain detailed information about their geometry. In regions where a geothermal reservoir has been identified and developed for use in power generation, there is often sufficient information available from drill holes and production histories to support a relatively precise description of the three-dimensional geometry of the subsurface temperature regime. But even then, substantial uncertain can affect estimates of reservoir volume. As an example, consider the Tiwi geothermal region in The Philippines (Sugiaman et al. 2004). Figure 8.3 is a cross section through that geothermal system, showing the subsurface distribution of the isotherms and the wells upon which they are based. If we consider only the area in the cross section of the reservoir that is at temperatures in excess of 250°C, the area is equivalent to ~42,000 m². If, however, the most eastern well had not been drilled, and the remaining three eastern wells had not exceeded depths of 1400 m, the constrained area of the reservoir could have been estimated to be as much as ~70,000 m². This hypothetical example provides insight into the importance of having data that provide constraints on the extent of the temperature distribution in the subsurface. In the absence of well-defined constraints, it is important to understand what the limitations of a dataset may be and to consider the implications for the accuracy of the assessments. Note, too, that this example represents a two-dimensional section through a three-dimensional volume. To accurately establish the volume of the reservoir, with minimal uncertainty, it is necessary to have an array of wells that delineate the three-dimensional distribution of the relevant isotherms. Because of the cost involved, it is often prohibitive to conduct such an extensive drilling program and estimates of the resource volume will have to rely on best guesses.

In many cases, there may be little or no subsurface data available from a drilling and exploration program, although evidence may exist that a geothermal reservoir is present. The presence of a flowing hot spring is an example of such an occurrence. Strategies for dealing with instances in which subsurface information is inadequate have been developed for some geological settings that allow conservative estimates to be made of reservoir volume. One example of such a strategy comes from the current resource assessment being developed by the US Geological Survey (Williams et al. 2008a). In the Great Basin of the western United States, there occur hot springs that emanate along the so-called range-front faults. Range-front faults occur where a large basin

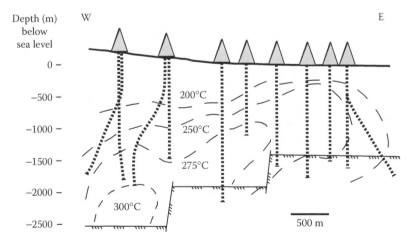

FIGURE 8.3 Cross section through the Tiwi geothermal field, Philippines, showing the distribution of isotherms (dashed lines), well locations (heavy dotted lines), and the crystalline basement rocks (upper limit delineated by hatchured line). Note that several of the wells are deviated. (From Sugiaman, F. et al., *Geothermics*, 22, 57–86, 2004.)

kilometers in width forms by subsidence of a block of continental crust. The block of subsiding crust is bounded by faults that dip at steep angles. The faults are exposed at the surface of the earth along the base of ranges that develop on either side of the subsiding basin. It is because of this topographical arrangement that these are called range-front faults. It is relatively common for springs to form along these faults.

The springs are fed by water that has circulated deep into the earth and emerged along the fault. The faults are high-permeability pathways because repeated movement along the fault crushes the rock into a permeable fault zone. Fluids that circulate deep into the basins usually return upward through these same high-permeability fault zones, emerging at the surface as springs. Often the springs are warm or hot because they have circulated to relatively deep levels where temperatures in excess of 200°C may occur.

Williams et al. (2008a) note that such fault zones have high-permeability regions (damage zones) that may be between 100 and 500 m wide. Using the fluid chemistry, as described in Chapter 6, it is possible to compute a water temperature that should approximate the temperature of the hydrothermal reservoir. Using that temperature and the measured water temperature of the spring, a range for the maximum circulation depth of the spring can be determined, if the local geothermal gradient is approximately known. Because geothermal gradients have been measured in a wide range of settings within the Basin and Range province, usually for academic research efforts, it is possible to compute various possible depths for the reservoir, thus giving bounds on the vertical dimension of the reservoir. The horizontal dimension is less well constrained. In their assessment, Williams et al. (2008a) note that geological evidence and comparison with other geothermal systems that have been developed in this type of setting suggest the horizontal extent, running more or less parallel to the range-front fault, is between 1 and 5 km, with a most likely value of about 2 km. The resulting dimensions allow a range of reservoir volumes to be computed. Statistical analyses or Monte Carlo simulations then allow most likely reservoir volumes to be computed.

Many instances of geothermal springs, however, do not occur in regions where there exists sufficient geological information to allow such an "argument by analogy" approach. For those instances, an assessment can only be based on the knowledge and experience of geologists that has been developed through years of studying such systems. Generally, such ambiguous occurrences are treated as systems deserving of further study and are not directly incorporated into rigorous assessments.

ESTABLISHING THE RESERVOIR HEAT CONTENT

As is evident from Figure 8.3, the heat content and the reservoir volume are intimately related. For example, assuming that Figure 8.3 represented a 1-m-thick slice through the reservoir, the equivalent volume would be approximately 42,000 m^3 within the region existing at temperatures above 250°C. That reservoir volume, however, would more than double if the lower temperature limit for the reservoir were selected to be 150°C.

The decision on what cutoff temperature to use in an assessment will be determined by the technology likely to be available for power generation. As discussed in Chapter 11, power generation at temperatures as low as 150°C or less will use binary power generating systems. However, such systems must also be utilized in settings where there is sufficient cooling capacity to meet the thermodynamic requirement of maximizing the ΔT between production fluid temperature and the cooling capacity of the system. In the assessment performed by Williams et al. (2008a), a temperature of 150°C is the lower temperature bound for defining the reservoir.

Establishing a reservoir temperature structure is rarely possible, except in instances where wells have been drilled and the thermal profile for each well monitored. In the absence of such data, the most useful means for estimating the reservoir temperature comes from geothermometers, such as those described in Chapter 6. Reed and Mariner (2007) and Williams et al. (2008a) conclude that the most reliable geothermometers are the silica, K–Mg, and Na–K–Ca geothermometers. The silica geothermometer provides good results, particularly if the formulation of the geothermometer considers the various effects of different silica polymorphs, as does the method developed by Giggenbach (1992). In the temperature range from 90°C to 130°C, the K–Mg geothermometer (Giggenbach 1988) works well, according to the review of results considered by Williams et al. (2008a). However, it is often difficult to obtain Mg analyses of sufficient quality and consistency, because Mg concentrations can be quite low and are difficult to do at high precision. In the absence of Mg analyses, or in waters that are Cl-rich, the Na–K–Ca geothermometer (see Chapter 6 for a discussion) is utilized.

When using such data, it is important to keep in mind the limitations of computed temperatures. It is generally assumed that the temperature that is obtained from a geothermometer reflects the equilibrium chemical composition of the water at the highest temperature the water experienced and that there was, therefore, very little re-equilibration of the water as it ascended through the rock mass on its way to the surface. The assumptions that equilibrium was achieved and that re-equilibration did not occur are not wholly justified, because reaction rates do not go to 0 once the fluid mass has passed beyond the thermal maximum. Recall Equation 5.8,

$$R = S_A \times k \times T_{fac} \times \alpha \times \phi \times \prod a_i \times \left(\frac{1-Q}{K} \right)^{\omega}$$

where:
   R is the rate (moles/s)
   S_A is the effective surface area exposed to the fluid (cm^2)
   k is the far-from-equilibrium rate constant (moles/cm^2-s)
   T_{fac} is the temperature correction factor for the rate constant k (usually an Arrhenius function)
   α is a power function that accounts for changes in the rate close to equilibrium conditions
   ϕ is a function that modifies the rate for precipitation relative to that for dissolution that is based on experimental data
   a_i is the dependence of the rate on the activities of specific components in solution
   ω is the power dependence based on experimental data that accounts for the particular dissolution or precipitation mechanism

Note that in this formulation, R is temperature dependent and thus will not go to 0 unless equilibrium is achieved at the T of interest. Once equilibrium is achieved, the term k becomes 0 and R goes to 0, but only for that specific T. As the fluid migrates to lower temperatures, k is no longer 0 and increases in magnitude as the ΔT between the equilibrium temperature and the T along the flow path increases. The extent to which re-equilibration will occur will depend upon the residence time of the fluid in the lower temperature regime and is thus dependent on the flow rate—the higher the flow rate, the lower will be the extent to which re-equilibration takes place.

These considerations suggest that the computed temperature from a geothermometer is likely to underestimate the actual reservoir temperature. The extent to which this is the case is not easily determined. For that reason, most geothermometer temperatures used in an assessment should be considered conservative.

SIGNIFICANCE OF HEAT CAPACITY

Once temperature and volume for a reservoir have been established, the only variable left to determine is the heat capacity.

As previously discussed in Chapter 3, the heat capacity is a function of the mineral, or minerals, that make up the rock, the pore volume and whether it is fluid-filled, and the temperature.

The dependence of the heat capacity on temperature for a mineral is usually expressed as a power function of the temperature (Berman and Brown 1985; Berman 1988), such as

$$C_p = k_0 - (k_1 \times T^{-0.5}) + (k_2 \times T^{-2}) + (k_3 \times T^{-3}) \qquad (8.2)$$

where k_i is a fitting coefficient derived from analysis of experimental measurements of heat capacities of individual minerals

Other versions of this relationship have been proposed (e.g., Helgeson et al. 1978; see review by Navarotsky 1995) and used to generate heat capacity values for individual minerals. In most instances, the differences between different fitting methods are on the order of a few percent or less and do not significantly impact a resource assessment. The units of heat capacity are commonly given as kJ/kg-K or kJ/mole-K.

The heat capacity of a rock is determined by the minerals that compose the rock and the pore space or fracture space that may or may not be fluid-filled. Assuming that a rock has minimal fracture or matrix porosity allows calculation of the approximate heat capacity of the rock. The heat capacity of this solid component of the rock can be computed, approximately, as (assuming the units of heat capacity are in kJ/kg-K)

$$C_{p\,rock} = \sum \left(\frac{x_i \times w_i \times C_{pi}}{V_i} \right) \qquad (8.3)$$

where:
  $C_{p\,rock}$ is the constant pressure heat capacity of the rock (J/g-K)
  x_i is the volume fraction of mineral i in the rock
  w_i is the molecular weight (kg/mole) of mineral i
  V_i is the molar volume (m^3 per mole) of mineral i
  C_{pi} is the constant pressure heat capacity of mineral i (J/g-K)

Figure 8.4 shows how the heat content varies for a mixture of minerals, as a function of the proportion of the minerals present at a temperature of 200°C. The hypothetical mineral mixtures, or rocks, are combinations of two minerals—quartz and potassium feldspar, quartz and calcite, and potassium feldspar and calcite. The sensitivity of these calculations to the assumed temperature is not large—if the temperature were increased or decreased by 50°C, the results would shift by less than 5%. These binary combinations are approximations of rock types indicated in the figure. The bulk of the volume of real rocks is generally made up of three to five minerals, but for the types of rocks usually encountered in geothermal systems, the binary combinations provide an indication of the variability to be expected among the main rock types.

There are several key points that are evident from Figure 8.4. First, the difference in heat capacity between the end-member compositions is 10%–15% of the total heat capacity of the rock. This defines the magnitude of uncertainty that can be associated with an assessment of a geothermal reservoir if the mineralogy of the reservoir is poorly known and can only be estimated. Second, most rock types tend to occupy a relatively small range of heat capacity values, reflecting the minerals that dominate the rock volume. Finally, sedimentary rocks composed of complex mixtures of minerals, which are often derived from a variety of different rock types, have a broad range of possible heat capacities, but will tend to overlap the same values as those observed in granitic rocks. This reflects the simple fact that granitic rocks are dominantly composed of various proportions of feldspars and quartz, which are also the erosional remnants of many other rock types. This reflects the resistance to abrasion these particular minerals possess.

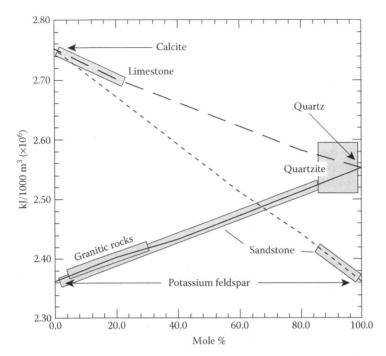

FIGURE 8.4 Heat content (millions of kJ per cubic kilometer) in binary mineral mixes (quartz–potassium feldspar, quartz–calcite, and potassium feldspar–calcite) as a function of the mole proportion of the minerals at 200°C. The rock types indicated by the italicized rock names delineate the regions that approximately correspond to the respective binary mineral mixes.

For highly porous or fractured rocks, the approach outlined above using Equation 8.3 will tend to overestimate the heat capacity of the reservoir. For this reason, it is important that both porosity and permeability be characterized in a reservoir, as well as the mineralogy of the host rock.

EFFICIENCY OF HEAT EXTRACTION

The value of Q_R computed using Equation 8.1 is the total amount of heat contained in the reservoir. This value represents the maximum amount of heat that can be extracted. Realistically, however, only a fraction of this heat is obtainable. There are several factors that contribute to this circumstance. These factors include the ability of wells to access and extract heat and the flow properties of fluids in the reservoir.

As discussed in detail in Chapter 11, the amount of heat that can be used to generate power can be represented as

$$Q_{WH} = m \times (H_{WH} - H_0) \tag{8.4}$$

where:
Q_{WH} is the extractable heat at the wellhead (kJ/s)
m is the mass flow from the wellhead to the turbine (kg/s)
H_{WH} and H_0 are the fluid enthalpy at the wellhead and at a reference end-state temperature, respectively (kJ/kg)

From this relationship, the recovery factor R_R is computed as

$$R_R = \left(\frac{Q_{WH}}{Q_R} \right) \tag{8.5}$$

Values for R_R are important in determining the likely energy that can be obtained from a reservoir. Experience has shown, however, that the recovery factor is difficult to predict. In reservoirs where the permeability is exclusively controlled by fractures, recovery factors are likely to be low, within the range of 0.05–0.2 (Lovekin 2004; Williams 2007; Williams et al. 2008a). In homogeneous porous media, recovery factors can be much higher, potential achieving values as high as 0.5 (Nathenson 1975; Garg and Pritchett 1990; Sanyal and Butler 2005; Williams et al. 2008a). The reason for this wide range in recovery factors reflects the efficiency of fluid extraction from rocks.

For uniformly porous media, a pumped well will reduce the pressure at the bottom of the well. The pressure reduction will be determined by the flow rate at the wellhead. For a liquid reservoir, the mass and volumetric flow into the well at the bottom of the well must equal the mass and volumetric flow out of the top of the well, because the fluid is essentially incompressible. This continuity of mass and volume allows us to use the following equation to describe the difference between the pressure at the well face (or borehole surface) and at a given distance (r) from the well face:

$$P_r = P_R - P_B + \left[\frac{(\mu \times f_v)}{(2 \times \pi \times k \times R_t)} \right] \times \ln\left(\frac{r}{r_w}\right) \tag{8.6}$$

where:
   P_r is the pressure (Pa) at distance r
   P_R is the intrinsic pressure (Pa) of the reservoir
   P_B is the pressure (Pa) at the bottom of the well during pumping
   μ is the viscosity (kg/m-s)
   f_v is the volumetric flow velocity (m^3/s)
   k is the permeability (m^2)
   R_t is the thickness of the reservoir (m)
   r is the distance from the center of the well (m)
   r_w is the distance from the center of the well to the well face (m)

Figure 8.5 shows the variation in pressure as a function of distance from the well for different pumping rates, also known as the draw down. There are several points regarding these results that have significance for geothermal energy extraction.

An important consideration is that the rate of fluid movement through the reservoir is logarithmically related to the distance from the well face. Hence, the rate of heat extraction, in units of J/s, will vary logarithmically with distance from the well. If reinjection is not employed during the time when fluid is extracted for energy production, the pressure in the reservoir will change, resulting in a time-dependent change in the energy flow at the wellhead. If reinjection is employed and is adequate to replenish the volume of fluid withdrawn, this time-dependent effect will not be seen.

A second consideration is that the smoothness of the curves drawn in Figure 8.5 reflects the assumption that the porous medium possesses a uniform permeability. In reality, such a situation is unlikely. Hence, pressure differences will develop throughout the flow field that will affect the flow pattern, diminishing the efficiency with which heat can be extracted uniformly from the reservoir.

Finally, the treatment represented in the figure does not capture the behavior of a fractured medium. If fluid production is primarily through fracture flow, then the main pressure drop in the reservoir will be within the fractures. Perpendicular to the fracture surfaces, there will be an additional pressure differential in the porous medium that is the rock that encloses the fractures,

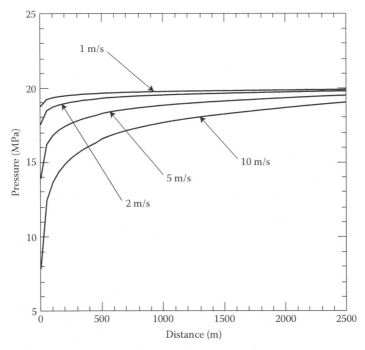

FIGURE 8.5 The pressure "draw down" (in MPa) in a geothermal reservoir as a function of distance (in meters) from a pumped well. The curves are drawn for pumping rates that provide a flow velocity at the wellhead of 1, 2, 5, and 10 m/s. This calculation assumed that the reservoir was 1150 m thick, the permeability was 10 millidarcy, the fluid viscosity was 2×10^{-4} kg/m-s, well diameter was 0.254 m and the reservoir pressure was 20 MPa.

and fluid movement from the rock to the fracture may occur, albeit at a rate significantly lower than in the fracture. This movement of fluid from porous medium to fracture is likely to be small, but will not necessarily be 0. Hence, a certain amount of heat transfer via convection will occur between rock and fracture, in addition to that expected from simple conduction. Nevertheless, the spacing and abundance of fractures will dominate the ability of heat to be extracted from the reservoir. The greater the interface area that is exposed along fracture surfaces in the reservoir, the greater will be the recovery factor. However, the effectiveness of this extraction process generally results in a recovery factor that is modest for fractured reservoirs.

The most successful means for mathematically representing the relationship between fracture flow on the recovery factor was developed by Williams (2007) who used a fractal approach for describing the interaction. Consideration was given to the proportion of flow in various fracture populations in a system. The results of the model were consistent with the observed behavior for well-characterized fracture-dominated systems. Figure 8.6 represents the results using this approach for Dixie Valley and Beowawe, both of which are fracture-dominated geothermal reservoir systems in Nevada. The curves document that a small proportion of the fractures carry a disproportionately high volume of the flow—at Dixie Valley 35% of the flow is accommodated by 10% of the fracture permeability, whereas at Beowawe more than 50% of the flow is accommodated in 10% of the fracture permeability. Because such a small proportion of the total fractures account for a large proportion of the flow, there will be a significant volume of the geothermal reservoir in which interaction with the fracture regime will be minimal. As a result, extraction of heat will not be uniform and the recovery factor will be relatively low.

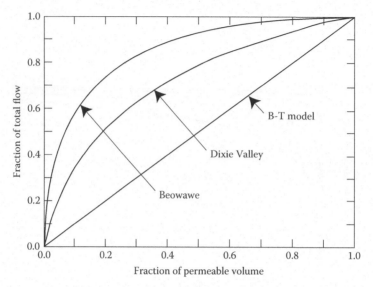

FIGURE 8.6   The curves in this figure compare the proportion of flow for which a given fraction of the permeable volume of a reservoir is responsible. The curves for Beowawe and Dixie Valley are based on the fractal model discussed by Williams (2007) and Williams et al. (2008a). The curve labeled "B-T model" is the relationship based on the Bodvarsson and Tsang (1982) model. (Modified from Williams et al., 2008a.)

SYNOPSIS

Resource assessments attempt to derive statistically meaningful estimates of the amount of energy that could be recovered from geothermal reservoirs. The nomenclature that is used in these efforts identifies the "reserve," which is the energy resource that has already been identified and characterized and is economically viable to develop; the "resource," which considers undiscovered resources or known resources that are currently not economically viable but could be with modest further development; and the resource that is simply not accessible with existing technology or technology that may be developed in the foreseeable future. All of these together constitute the "resource base." For the resource itself, what must be determined in order to develop a rigorous assessment is the volume of each reservoir of interest, its heat content, and the proportion of heat that can be extracted from it (the recovery factor). These parameters are treated statistically, with a distribution that is determined based on experience and history. Once established, estimates are developed using statistical models. The most commonly used method to accomplish a rigorous resource assessment is the Monte Carlo method.

CASE STUDY: ESTABLISHING THE US GEOTHERMAL RESOURCE

Once the heat content and recovery factors have been delineated for specific geothermal reservoirs, as well as the probability distribution associated with them, it is possible to undertake the final part of the resource assessment. This involves calculating the most likely values and their respective ranges for the reservoirs. For undiscovered resources, estimates are developed using an approach in which certain geological settings are given a probability of containing certain types of geothermal resources, based on historical experience and production histories. Once these are determined, the heat contents and recovery factors are similarly estimated.

The resulting database is then used to compute a probability distribution for the geothermal resource. Various statistical approaches can be used to do this. One common method that has been successful in the oil and gas community is the Monte Carlo technique as employed by the US

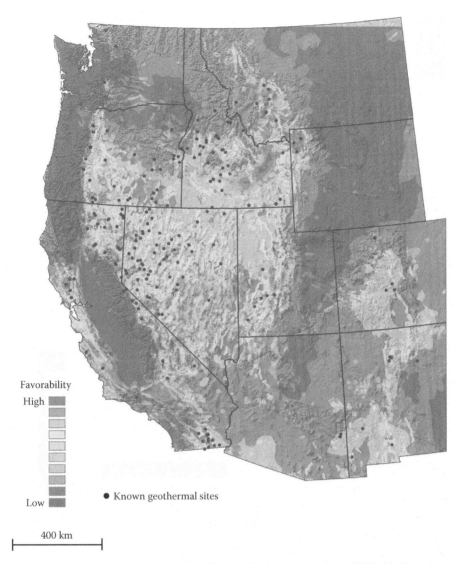

FIGURE 8.7 **(See color insert.)** Western United States showing the locations of individual geothermal sites (black dots). Also shown in color are those regions for which geological evidence suggests the possibility that geothermal reservoirs may be located in the vicinity. The likelihood that a geothermal resource may exist in the region is ranked from low to high "favorability," based on the nature of the geological evidence. (Modified from Williams, C.F. et al., Assessment of moderate- and high-temperature geothermal resources of the United States; US Geological Survey Fact Sheet 2008-3082, 4, 2008b.)

Geological Survey (Charpentier and Klett 2007). The most recent results for geothermal reservoirs that have been derived using this approach are those published by the US Geological Survey in its recent assessment of the geothermal resources of the western United States (Williams et al. 2008b).

The western United States is a complex geological environment within which numerous geothermal sites exist (Figure 8.7). Williams et al. (2008b) defined statistical distributions for the reservoir volumes, heat contents, and recovery factors for the various reservoirs and then used that data to generate estimates of the amount of electrical power that could be generated (Table 8.1). For the reservoir volumes and temperatures (or heat contents), it was assumed that the distributions were

TABLE 8.1

Results of the Geothermal Resource Assessment for the Identified Sites in the Western United States Showing the Potential Electrical Power Production (in MW) at 95% and 5% Confidence Intervals and the Mean, Based on the Monte Carlo Methods Used by Williams et al.

State	Number of Sites	95% CI	Mean	5% CI
Alaska	53	236	677	1,359
Arizona	2	4	26	70
California	45	2,422	5,404	9,282
Colorado	4	8	30	67
Hawaii	1	84	181	320
Idaho	36	81	333	760
Montana	7	15	59	130
Nevada	56	515	1,391	2,551
New Mexico	7	53	170	343
Oregon	29	163	540	1,107
Utah	6	82	184	321
Washington	1	7	23	47
Wyoming	1	5	39	100
Total	248	3,675	9,057	16,457

Source: Williams, C.F. et al., Assessment of moderate- and high-temperature geothermal resources of the United States. US Geological Survey Fact Sheet 2008-3082, 4, 2008b.

triangular, peaking at a most likely value and then decreasing linearly to either side of that value (Figure 8.8) for liquid-dominated geothermal systems. For the recovery factors, it was assumed that the probability of any particular value was the same throughout the possible range of values, which was concluded to lie between 0.08 and 0.2 for systems in which fracture permeability controlled the flow regime and between 0.1 and 0.25 for systems in which porous matrix permeability controlled the flow regime.

Monte Carlo simulations were then carried out using the probability bounds and distributions for the various parameters for each individual site, which are shown in Figure 8.7. The results of the simulations for the known and identified geothermal resources (the "Reserve") are presented in Table 8.1. The results for the undiscovered and enhanced geothermal systems (EGS, which is a geothermal resource discussed in Chapter 13) are presented in Table 8.2. In all cases, the results are presented as the electric power generation potential in megawatts.

The results of the resource assessment provide perspective on the magnitude of the potential contribution geothermal energy that could make to the power needs of the United States. If only the Reserve is considered (Table 8.1 values), nearly 10 GW of power could be produced requiring no new discoveries, assuming the mean value for the estimate. If the mean values for undiscovered and EGS are added (Table 8.2), the total power output could be more than 545 GW. For comparison, total current installed generating capacity in the United States is approximately twice that amount.

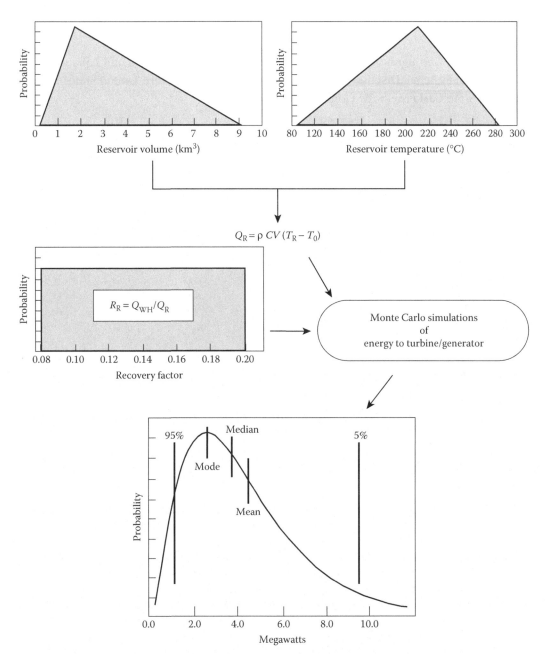

FIGURE 8.8 Representative probability distributions for reservoir volume, temperature, and recovery factor used in the Monte Carlo simulations conducted by Williams et al. (2008b). The probability distributions for the volume and temperature are constrained to be triangular and peak at a specified value that is specific to each reservoir. The probability distribution for the recovery factor is assumed to be uniform. For each site assessed, the combinations of probabilities for these parameters are used in a Monte Carlo simulation to compute a probability distribution for the amount of power that can be generated at that site (lower figure). The "Mode," "Median," and "Mean" are indicated for this particular result, as well as the locations for the amount of power that could be generated with 95% and 5% confidence intervals, respectively.

TABLE 8.2

Results of the Geothermal Resource Assessment for the Undiscovered and Enhanced Geothermal Systems (Discussed in Chapter 13) Sites in the Western United States, Based on the Monte Carlo Methods Used by Williams et al.

	Undiscovered			EGS		
State	95% CI	Mean	5% CI	95% CI	Mean	5% CI
Alaska	537	1,788	4,256	NA	NA	NA
Arizona	238	1,043	2,751	33,000	54,700	82,200
California	3,256	11,340	25,439	32,300	48,100	67,600
Colorado	252	1,105	2,913	34,100	52,600	75,300
Hawaii	822	2,435	5,438	NA	NA	NA
Idaho	427	1,872	4,937	47,500	67,900	92,300
Montana	176	771	2,033	9,000	16,900	27,500
Nevada	996	4,364	11,507	71,800	102,800	139,500
New Mexico	339	1,484	3,913	35,600	55,700	80,100
Oregon	432	1,893	4,991	43,600	62,400	84,500
Utah	334	1,464	3,860	32,600	47,200	64,300
Washington	68	300	790	3,900	6,500	9,800
Wyoming	40	174	458	1,700	3,000	4,800
Total	7,917	30,033	73,286	345,100	517,800	727,900

Source: Williams, C.F. et al., Assessment of moderate- and high-temperature geothermal resources of the United States. US Geological Survey Fact Sheet 2008-3082, 4, 2008b.

Note: Nomenclature is the same as in Table 7.1.

PROBLEMS

8.1. Examine the mean values for known and undiscovered resources in Tables 8.1 and 8.2. Notice that the ratio of known to undiscovered varies considerably between the states. What would be an explanation for this wide variability?

8.2. Answer the same question as above, but for the known and EGS resources.

8.3. Identify three issues that influence whether a geothermal site would be considered a reserve and discuss how those factors can change over time.

8.4. Chemical processes can result in limestone being replaced by calcite. Write the chemical reaction that would represent this process. If 40 volume % of a limestone were replaced by quartz, how would that affect the heat content of the limestone?

8.5. In Figure 8.6, it is demonstrated that fluid flow is often accommodated by only a fraction of the total available permeable volume. Discuss the factors that account for this observation. When exploring for a geothermal resource in fractured rock, what characteristics would you consider favorable for supporting sufficient flow?

8.6. What are the geological processes that have resulted in the concentration of potentially favorable geothermal sites in California, Colorado, New Mexico, Oregon, and Idaho (Figure 8.7)? If you were to undertake a reconnaissance exploration effort, what specific area would you target and why?

8.7. Given the properties of the various geological systems that are implicit in Figure 8.7, what are the classifications that are represented in this geological environment?

8.8. What combination of factors would give the highest probability of being able to develop an 8 MW facility?

REFERENCES

Berman, R.G., 1988. Internally-consistent thermodynamic data for minerals in the system Na_2O-K_2O-CaO-MgO-FeO-Fe_2O_3-Al_2O_3-SiO_2-TiO_2-H_2O-CO_2. *Journal of Petrology*, 29, 445–522.

Berman, R.G. and Brown, T.H., 1985. The heat capacity of minerals in the system K_2O-Na_2O-CaO-MgO-FeO-Fe_2O_3-Al_2O_3-SiO_2-TiO_2-H_2O-CO_2: Representation, estimation and high temperature extrapolation. *Contributions to Mineralogy and Petrology*, 89, 168–183.

Bodvarsson, G.S. and Tsang, C.F., 1982. Injection and thermal breakthrough in fractured geothermal reservoirs. *Journal of Geophysical Research*, 87, 1031–1048.

Charpentier, R.R. and Klett, T.R., 2007. A Monte Carlo simulation method for the assessment of undiscovered, conventional oil and gas. In *Petroleum Systems and Geologic Assessment of Oil and Gas in the San Joaquin Basin Province*, California, ed. A.H. Scheirer. Reston, VA: US Geological Survey, 5 pp.

Garg, S.K. and Pritchett, J.W., 1990. Cold water injection into single- and two-phase geothermal reservoirs. *Water Resources Research*, 26, 331–338.

Gawell, K., 2006. California's geothermal resource base: Its contribution, future potential and a plan for enhancing its ability to meet the states renewable energy and climate goals. California Energy Commission Contractors Report for contract 500-99-13, subcontract C-05-29, California Energy Commission.

Giggenbach, W.F., 1988. Geothermal solute equilibria. Derivation of Na-K-Mg-Ca geoindicators. *Geochimica et Cosmochimica Acta*, 52, 2749–2765.

Giggenbach, W.F., 1992. Chemical techniques in geothermal exploration. In *Application of Geochemistry in Geothermal Reservoir Development*, ed. D'Amore. Rome, Italy: UNITAR/UNDP Centre on Small Energy Resources, pp. 119–144.

Helgeson, H.C., Delany, J.M., Nesbitt, H.W., and Bird, D.K., 1978. Summary and critique of the thermodynamic properties of rock-forming minerals. *American Journal of Science*, 278-A, 229.

Lovekin, J., 2004. Geothermal inventory. *Geothermal Resources Council Transactions*, 33(6), 242–244.

Muffler, L.P.J., 1979. Assessment of geothermal resources of the United States—1978. US Geological Survey Circular 790, p. 163, http://pubs.usgs.gov/circ/1979/0790/report.pdf.

Muffler, L.P.J. and Cataldi, R., 1978. Methods for regional assessment of geothermal resources. *Geothermics*, 7, 53–89.

Nathenson, M., 1975. Physical factors determining the fraction of stored energy recoverable from hydrothermal convection systems and conduction-dominated areas. US Geological Survey, Open-File Report 75-525, p. 50.

Navarotsky, A., 1995. Thermodynamic properties of minerals. In *Mineral Physics and Crystallography*, ed. T.J. Ahrens. Washington, DC: American Geophysical Union, pp. 18–28.

Petty, S., Livesay, B.J., Long, W.P., and Geyer, J., 1992. Supply of geothermal power from hydrothermal sources: A study of the cost of power in 20 and 40 years. Sandia National Laboratory Contract Report SAND92-7302.

Reed, M.J., and Mariner, R.H., 2007. Geothermometer calculations for geothermal assessment. *Geothermal Resources Council Transactions*, 31, 89–92.

Sanyal, S.K. and Butler, S.J., 2005. An analysis of power generation prospects from Enhanced Geothermal Systems. *Geothermal Resources Council Transactions*, 29, 131–137.

Sugiaman, F., Sunio, E., Molling, P., and Stimac, J., 2004. Geochemical response to production of the Tiwi geothermal field, Philippines. *Geothermics*, 22, 57–86.

US Energy Information Agency, 2013. http://www.eia.gov/cfapps/ipdbproject/iedindex3.cfm?tid=5&pid=57&aid=6&cid=regions&syid=1980&eyid=2013&unit=BB.

Western Governor's Association, 2006. Geothermal task force report. Clean and Diversified Energy Initiative Report, p. 66.

White, D.E. and Williams, D.L., 1975. Assessment of geothermal resources of the United States—1975. US Geological Survey Circular 726, p. 155.

Williams, C.F., 2007. Updated methods for estimating recovery factors for geothermal resources. *Proceedings of the 32nd Workshop on Geothermal Reservoir Engineering*, Stanford University, Stanford, CA, p. 6, January 22–24.

Williams, C.F., Reed, M.J., and Mariner, R.H., 2008a. A review of methods by the US Geological Survey in the assessment of identified geothermal resources. US Geological Survey Open File Report 2008-1296, 27 pp.

Williams, C.F., Reed, M.J., Mariner, R.H., DeAngelo, J., and Galanis Jr., S.P., 2008b. Assessment of moderate- and high-temperature geothermal resources of the United States. US Geological Survey Fact Sheet 2008-3082, 4 pp.

Wood, J. and Long, G., 2000. Long term world oil supply (A resource base/production path analysis). Energy Information Administration, US Department of Energy. http://www.eia.doe.gov/pub/oil_gas/petroleum/presentations/2000/long_term_supply/sld001.htm.

FURTHER INFORMATION SOURCES

The Geo-Heat Center at Oregon Institute of Technology (http://geoheat.oit.edu/).
> OIT and the Geo-Heat Center monitor assessments and provide access to reports dealing with the assessments for low- and high-temperature resources.

The Great Basin Center for Geothermal Energy, University of Nevada, Reno, NV (http://www.unr.edu/Geothermal/).
> This research center is an excellent resource for information on geothermal in general and its attributes and availability within the Basin and Range Province in particular. A substantial amount of information is available on the website.

The International Geothermal Association (http://www.geothermal-energy.org/index.php).
> The IGA provides updated summaries of international resource assessments. It is an important resource for accessing global data relevant for geothermal energy use.

The United States Geological Survey (http://www.usgs.gov/).
> The USGS is currently completing the most extensive assessment of geothermal resources in the United States. This website provides access to the results, as they are published, as well as a wealth of information relating to geothermal resources and the geology controlling them.

9 Drilling

Accessing geothermal heat usually requires drilling. Drilling is used to sample the rock and soil in the subsurface in order to determine such things as thermal conductivity, porosity and permeability, type of rock and its physical characteristics, temperature gradients, and other parameters that influence resource evaluation. Drilling is also required to access geothermal fluids and assure a consistent supply of heat at a controlled rate. Finally, drilling is required when reinjection of fluids is necessary. Drilling is usually one of the most expensive single undertakings encountered when developing a geothermal application. For this reason, careful consideration is required when selecting a drill site and the type of drilling technology to be used. The material in this chapter is intended to be an introduction to basic principles that apply when drilling holes for a variety of geothermal applications. The material covered will provide background sufficient to allow further, detailed pursuit of information needed for specific applications. But it is important to understand that drilling remains a skill-intensive enterprise. Although technological advances have been impressive in reducing uncertainty about what is present in the subsurface, detailed knowledge of what will be encountered tens, hundreds, or thousands of meters underground is very rarely available. For this reason, it is important that experienced drilling teams with knowledge of the local and regional drilling lore be used, if at all possible. Such teams should be able to anticipate potential challenges, have access to the materials that would be needed to flexibly respond to them, and act responsibly.

BACKGROUND

A rule of thumb that is often mentioned is that drilling expense accounts for 40%–60% of the costs of a geothermal project. Although the actual costs depend on many variables, there is a general consistency between drilling expenses and depth of the hole drilled. As shown in Figure 9.1, the drilling costs for wells that will be used for power generation run into millions of dollars. For comparison, the costs of completing oil and gas holes by the petroleum industry are also shown. Also shown is the approximate cost curve for completing boreholes for geothermal heat pumps (GHPs), which are discussed in Chapter 11. Although GHP boreholes are generally drilled to depths of only a few hundred feet, an extrapolated curve is shown in the figure in order to facilitate comparison.

The broad range in costs is notable—for any given depth, the drilling costs can vary by up to a factor of 10. Although this pattern holds for both oil wells and geothermal wells, there is a clear tendency for geothermal wells, whether conventional hydrothermal wells or enhanced geothermal systems (EGS) wells, to fall at the more costly part of the overall cost range. These cost patterns reflect the consequences of a variety of technical challenges, the magnitude of which varies from site to site. The remainder of this chapter deals with the various properties that affect these costs. Simpler holes, which are the boreholes for GHPs, are comparable in cost to drilling water wells, currently about $12–$15 per foot. We will begin discussion of the technical challenges faced in drilling boreholes by considering the drilling techniques for the lower-cost GHP projects first.

DRILLING FOR GEOTHERMAL HEAT PUMP AND DIRECT-USE APPLICATIONS

These types of applications require access to low to moderate temperatures. GHP installations need only the stable temperature zone found at depths of a few meters to a few hundred meters. Such systems are best installed where subsurface temperatures fall in the range of about 10°C–25°C (50°F–75°F). Direct-use applications, such as aquaculture, spas, and greenhouses, need access

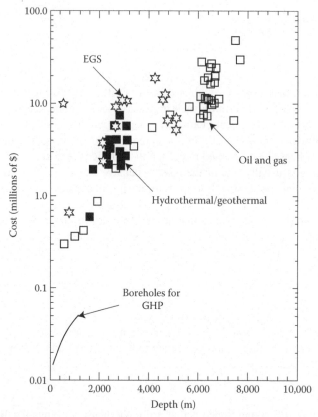

FIGURE 9.1   Completion costs for drilled wells, as a function of the bottom hole depth. (Data from Smith et al. 2000; Bertani 2007; International Geothermal Association 2008; Bolton 2009.)

to similar temperatures. Because these conditions are relatively benign and the drilling depths restricted to relatively shallow levels, the type of drilling equipment that is required is virtually identical to that employed for drilling water wells.

DRILLING EQUIPMENT AND TECHNOLOGY

In regions where the rock is relatively soft, such as in sedimentary basins where the subsurface is composed of sandstone, silt, and other porous, unconsolidated material, drilling is often done with a truck-mounted auger (Figure 9.2). An auger is a helical screw that has a hardened bit for cutting into the soft rock or soil. As the screw rotates, it transports the drill cuttings out of the hole. Such methods usually can reach depths of a few hundred feet in a relatively short time (hours to a few days). Borehole sizes for these applications can be as small as 7 cm (3 inches) to greater than 25 cm (10 inches), depending upon the design of the system and the specific application.

In regions where hard rock will be encountered, it will be necessary to use drilling equipment that can penetrate it. There are several technologies that can be employed, depending upon the size and depth of the hole to be drilled and the budget available for the project. All of the drilling methods excavate the borehole by shattering the rock in the immediate vicinity of the end of the drill string. The specific technology used to shatter the rock, however, varies considerably.

Percussion techniques shatter the rock by repeatedly hammering on it. One method for accomplishing this is to attach the bit and some drill string to a wire line that is repeatedly raised and lowered by the drill rig, allowing the bit to impact onto the rock at the bottom of the hole. Another

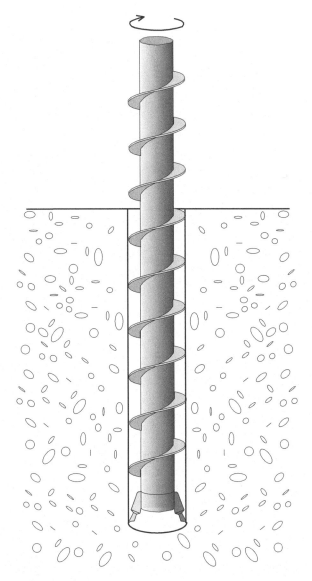

FIGURE 9.2   Auger drilling system.

method uses a mechanical device attached to the drill bit that repeatedly jackhammers the bit into the rock. Such systems are commonly pneumatically driven. High-pressure air is circulated through the hole to remove rock chips and dust. Percussion drilling is useful for relatively small diameter holes of short depth. Rotary methods are generally used for deeper, larger holes.

Rotary drilling techniques shatter the rock by applying pressure on the rock using a rotating bit that is bearing an imposed load. Some bits crush the rock using rolling cutters that have hardened teeth projecting from rollers, whereas other bits grind away at the rock using tungsten carbide or diamond-studded surfaces (Figure 9.3). The roller bits are usually used for soft rocks but they can also be used on harder rocks, if the grain size of the rock and ability of the hardened teeth to shatter the grains is adequate. For hard rock on which the roller bits are inadequate, diamond bits are usually used. These bits come in a variety of sizes and configurations that affect the flow rate of drilling fluids around the bit, the rate of grinding as a function of the applied pressure, and other variables.

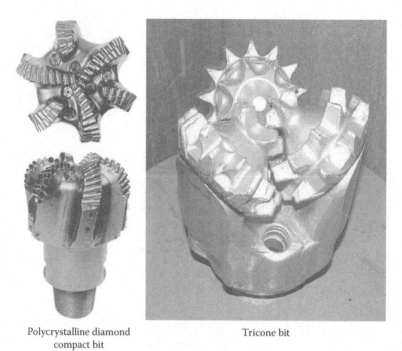

Polycrystalline diamond
compact bit

Tricone bit

FIGURE 9.3   Drill bits for drilling in crystalline rock. The bits on the left have diamond-studded surfaces on the cutting elements. The bit on the right is a tungsten-carbide tricone bit.

Drilling Fluid and Circulation

Once crushed, the rock chips and dust are removed from the hole either by a high-pressure air stream or, more commonly, by circulating a drilling fluid through the hole. The fluid is usually injected down the drill string and returns to the surface via the annulus around the pipe. Reverse circulation, in which fluid moves down the annulus and back up the interior of the drill string, is used in cases where the material being drilled is particularly granular.

Integral to the drilling process is drilling fluid. Although not particularly glamorous, drilling fluid (usually called drilling mud because it is composed of water and clay) performs functions that are necessary to economically obtain a stable, environmentally benign hole. One of those functions is the removal of cuttings from around the bit. If the cuttings from the drilling process are not removed, energy is unnecessarily wasted on grinding the cuttings to a fine powder that will ultimately clog the hole. To remove the cuttings, the drilling mud is pumped down the hole (either inside the drill pipe or outside the drill pipe) at high pressure, and it circulates back to the surface, bringing the cuttings with it. Drilling mud also performs a lubrication function, reducing the friction between the drill string and hole wall as well as lubricating the bit. This reduces wear on the drill pipe and bit while simultaneously reducing the amount of energy needed to drive the rotary motion of the drill string.

Drilling muds also stabilize the hole and act as a barrier to isolate the hole from the environment it penetrates. These functions are particularly important and require careful consideration of the type of mud to use. The importance of this issue can be appreciated if we consider the physical and geological context within which the drilling process occurs.

The rocks that are found at any location on the surface of the earth represent the result of millions of years of geological processes. Rocks of different chemical compositions and physical properties are intimately interlayered with each other. Over time, stresses from a variety of directions may be imposed on that heterogeneous collection of materials, resulting in complex stress fields being stored in the rock, with different orientations of maximum and minimum stresses at different depths. As water infiltrates from the surface, a hydrological regime will evolve that can lead to the formation of one or more aquifers, each with its own flow rates, recharge rates, and direction of flow, reflecting, in part, the sum total of these interacting processes.

Drilling a well into such a geological system perturbs the local balance of forces and processes in complex ways. One possible effect is that poorly consolidated material may collapse into the hole, defeating the drilling effort. Another process that can have the same effect is spallation, in which rock that is under stress may spall off pieces from the borehole wall into the borehole, destabilizing it or jamming the drill string. In instances where multiple aquifers occur, it is likely they will have different water compositions. Drilling a borehole through them provides a potential pathway for cross-communication between the aquifers which, in cases where one aquifer is used as a source for potable water, may result in contamination. Finally, for those instances in which the purpose of the hole is to obtain a controlled volume or mass of geothermal fluid, sealing the hole from the surrounding environment prevents unwanted water from entering the hole or geothermal fluid from escaping from the hole. The use of drilling mud is an important tool for solving all of these problems. How this is accomplished reflects the engineered properties of the mud.

PROPERTIES OF DRILLING FLUIDS

To satisfy the needs described in the preceding section, a drilling fluid must be sufficiently fluid to readily flow through the drill string, around the bit and back to the surface (Figure 9.4) and yet be viscous enough to keep the cuttings from settling back to the bottom of the hole. At the same time, it is important that the mud forms a mechanically intact, gel-like, impermeable layer along the borehole

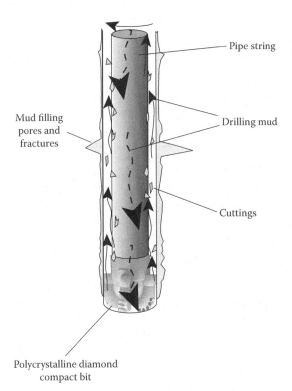

Pipe string

Mud filling pores and fractures

Drilling mud

Cuttings

Polycrystalline diamond compact bit

FIGURE 9.4 Rotary drill string and mud flow. Light gray zone outside the well indicates the extent of penetration of the drilling mud into the surrounding rock. Dashed arrows indicate clean drilling mud flowing to the bit, and solid wiggly arrows indicate mud flowing up the well bore carrying cuttings to the surface.

wall and, to the extent possible, permeate at least a narrow zone of the wall rock. Substances that possess properties that allow flow if mechanically perturbed but remain in a solid or gel-like form if undisturbed are called thixotropic materials. Most drilling muds are designed to be thixotropic.

The most commonly used material for drilling mud is a formulation that consists of a water + clay slurry into which various additives are mixed. Clay mixtures that have a high proportion of montmorillonite are favored. Such mixtures are commonly referred to as *bentonite*. Bentonite is a soft rock that occurs in natural geological deposits in sedimentary environments. Bentonite is a mixture of various clay minerals but its principal characteristic is its high proportion of montmorillonite.

Clay minerals are formed from molecular sheets composed of either silicon atoms linked to four oxygen atoms (the tetrahedral sheets) or aluminum atoms linked to six oxygen atoms (the octahedral sheets). The sheets are bound together by various cations (mainly potassium, sodium, calcium, and magnesium atoms) and are stacked in different ways. Water molecules occupy spaces between the sheets. Different clay minerals have different proportions of the cations, different orders in which the tetrahedral and octahedral sheets are stacked, and different amounts of water in their structure.

The water content of montmorillonite clays has an important impact on the physical characteristic of the mineral. Water is readily added to or removed from the montmorillonite structure via the reaction:

$$(\text{Na}, \text{K}, 0.5 \times \text{Ca})_{0.7}(\text{Al}, \text{Mg}, \text{Fe})_4[(\text{Si}, \text{Al})_8\text{O}_{20}](\text{OH})_4 \cdot n\text{H}_2\text{O} \Leftrightarrow$$

$$(\text{Na}, \text{K}, 0.5 \times \text{Ca})_{0.7}(\text{Al}, \text{Mg}, \text{Fe})_4[(\text{Si}, \text{Al})_8\text{O}_{20}](\text{OH})_4 \cdot (n-1)\text{H}_2\text{O} + \text{H}_2\text{O}$$

(9.1)

Montmorillonite that has the minimum amount of water is so-called dehydrated montmorillonite, whereas hydrated montmorillonite holds a maximum number of water molecules in its structure. The amount of water contained in the clay has a strong effect on the size of the clay molecule—the more hydrated the clay, the larger is its molar volume. Montmorillonite clays are among the clays that can hold the greatest amount of water in their structure in their fully hydrated state. As they absorb water, they visibly swell. Hydrated montmorillonites can expand 10–20 times their dehydrated volume.

This property, as well as their low shear strength, makes them ideal as the primary component in drilling muds—addition of a few percent montmorillonite to water makes a slurry that can be pumped yet still possess adequate viscosity to entrain cuttings while remaining fluid. Such solutions settle into a gel-like state if undisturbed.

Added to this mixture is a material to increase the overall density of the mixture so that it can displace any fluids that may enter the drilled well. Usually barite, a barium sulfate mineral with a high density, is used for this purpose. Organic compounds, many of them unknown because they are part of the proprietary information manufacturers hold regarding their formulations, are added to improve flow, thermal stability, viscosity, or other properties important for the mud.

When this solution enters the pores of the wall rock, it is no longer subject to the perturbations inherent in a pumped fluid. As a result, it becomes gel-like, acting as a fixed mechanical barrier preventing exchange of fluids across the wall of the well and isolating the well from the local environment. The extent to which this slurry penetrates the wall rock will depend on the local permeability and can be highly variable (Figure 9.4). The mechanical strength of the gel is sufficient to hold in place loose material in the wall rock, as well as reduce the incidence of wall collapse.

Occasionally, zones of very fractured rock are encountered. If the extent of the zone is large enough and its permeability high enough, the drilling mud can preferentially flow into it rather than down the hole. To lose circulation of the drilling mud in such a way can be a serious problem. Drilling cannot proceed until such zones are sealed. Sealing is attempted via a variety of means, most commonly by pumping large volumes of mud into the hole or changing the mud formulation. If the lost circulation zone cannot be sealed, there is no choice but to abandon the hole. This is, of course, a costly consequence. As a result, every attempt is made to find the right mud composition for sealing the lost circulation zone.

WELL COMPLETION

Once drilled, the hole must be finished in a way that facilitates its use for the application being pursued. If the hole is to be used for a closed-loop GHP system (see Chapter 11 for a detailed discussion of these systems), the loop or loops of pipe are placed in the hole and the hole is then *grouted*. Grouting is a process in which a permanent material is placed in the hole to keep it sealed and stable. Grouting also performs the function of isolating the borehole from the surrounding environment, thus preventing exchange of chemicals or contaminants between the borehole and any local aquifers, should any of the pipes fail. Another function of grouting is prevention of cross-communication between aquifers—in regions where there exist multiple, isolated aquifers at different depths, it is important that the boreholes do not form a pathway that might allow potable water in one aquifer to be contaminated by water from an aquifer that is made up of non-potable water. Finally, for a GHP application, it is important that the grout have sufficient thermal conductivity in order to facilitate heat transfer between the closed loop pipe and the geological environment and that the grout form a strong bond with the borehole wall in order to prevent water circulation along the wall that would reduce the heat transfer efficiency.

If the purpose of the hole is to obtain geothermal water for use in aquaculture, spas, greenhouses, or other direct-use applications, a liner or *casing* is put in place that will assure the permanence of the hole and protect equipment that may be placed in the hole, such as piping, pumps, and sensors. The casing is grouted into place to assure the same environmental and structural standards are satisfied as in the case of the completely grouted holes for the GHP applications.

Grouting is composed of a mixture of various compounds that can be pumped when initially mixed together but then will harden in place. Various cements, concretes, or special bentonite–polymer mixtures are commonly used. The advantage of these grouts is that they have much lower permeabilities than the surrounding rock, which thus assures that the hole will be physically isolated from the environment. They also have high strength, which helps assure the integrity of the hole.

ENVIRONMENTAL ISSUES

We have discussed the importance of assuring that, as a well is being drilled and once it is completed, it remain isolated from the surrounding environment, particularly if it penetrates an aquifer. Accomplishing this requires careful attention to assuring that the permeability immediately adjacent to a well wall is sealed and that a membrane of grout or cement plus casing, if appropriate, is sound and complete.

Additionally, drilling that requires use of a drilling mud also requires construction of a pit in close proximity to the drilling rig. The pit is used to hold the mixed mud and it is from this pit that the mud is pumped into the well to cool the bit, remove cuttings, and seal the well wall. Mud that returns from circulation carrying the cuttings must also be dumped in a pit. At the completion of drilling, the pits should be cleaned up and the local landscape restored. How this is done and the requirements for meeting local regulations will depend upon the regulatory agency that has jurisdiction over such matters. In some instances, this expense can be significant and must be factored into cost projections for a project.

DRILLING FOR GEOTHERMAL FLUIDS FOR POWER GENERATION

What distinguishes drilling for power generation purposes from that in any other application is the hostility of the environment within which drilling must be done. Although the depth of the wells and the mechanical challenges are similar to those in the oil and gas industry, thus allowing significant technology transfer from that community, accessing and penetrating a geothermal reservoir requires special considerations. The principal challenges that must be addressed are the high pressures and temperatures of the fluid and geological environment and the common presence of fluids that are chemically aggressive. We will consider these challenges by first describing drilling technology in general terms and then describing specific issues that must be addressed in order to complete a geothermal well.

Drilling Rigs

The ability to drill safely and effectively in the type of environment encountered in geothermal systems requires sophisticated, state-of-the-art drilling systems that can handle the unique conditions that will be encountered. Drilling rigs are sized according to the potential load the drill string will impose on them. Hence, they are sized according to the depth to which they will terminate. A sense of the required capabilities can be obtained by considering the requirements for a 6000 m well. Note, however, that most geothermal systems do not currently require drilling to such depths. Should EGS (Chapter 13) be pursued in a consistent fashion, however, such drilling capabilities will be required.

A well extending to 6 km can have a load placed on it that exceeds 350,000 kg. This mass is primarily from the weight of the pipe that the drill rig must control and be capable of retrieving, if necessary. Hence, the structure of the rig must be capable of supporting that load. In addition, the system must be capable of controlling the load on the bit with the appropriate force in order to achieve efficient drilling rates, as described in Section "Confining Pressure and Rock Strength". Finally, since the entire drill string must be rotated in order to drive the bit to excavate the hole, the rig must be capable of applying a torque to the drill string that is on the order of 43,500 newton-meters. The power requirements for such a system will be approximately 4.5 MW.

All drilling rigs have a tower assembly that is used to hoist individual sections of the drilling pipe (usually between 6.1 and 9.1 m long), to hold it in place as new pipe is added to the drill string, and to raise and lower the drill string for various aspects of the drilling process. Drill rigs differ, though, in how they apply torque to the drill string in order to rotate the bit.

Some drilling rigs are designed to apply torque to the drill string using a rotary table. The rotary table is located on the platform or floor of the drill rig where the pipe is lowered into the well. The rotary table is part of the platform that clamps to the pipe and rotates the drill string at a specified rate. Other drill rigs apply torque through what is called a top head drive assembly. The top head drive couples to the end of the drill string and applies torque while being lowered into the well. Both are used in the geothermal industry.

The volume of drilling mud required for such a system must accommodate a minimum of about 135 cubic meters of capacity, in order to adequately lift cuttings from the hole, cool the pipe and bit, and provide lubrication. Obviously, the system must also be capable of pumping that volume of mud in order to maintain flow. For wells that enter geothermal systems, it is common to include in the mud circulation system a mud cooling unit because the return mud can be quite hot as it returns from the bottom of the well. Recirculating hot mud would diminish the cooling capacity of the mud and reduce the lifetime of the bit and the rate of penetration (ROP).

Such systems are generally brought to the drill site in modular assemblies that can be erected and disassembled relatively quickly. These are not truck mounted drilling systems, but rather temporary, complex engineered structures that are assembled on site. The time required to drill such a well is described in the Kakkonda case study see Section "Case Study: Kakkonda, Japan".

Confining Pressure and Rock Strength

One of the important effects that must be considered when drilling holes deeper than a few hundred meters is that the strength of the rock increases as the confining pressure increases. Figure 9.5 summarizes how the fracture strength of a rock increases as the confining pressure increases. Fracture strength is a measure of the stress an intact rock must experience before it fails by fracturing (see Chapter 13 for a detailed discussion of this topic). This parameter provides a means for gauging the

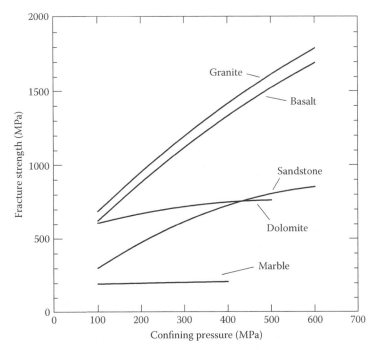

FIGURE 9.5 Rock strength as a function of rock type and confining pressure. (Curves are based on data in Lockner, D.A., Rock failure. In *Rock Physics and Phase Relations: A Handbook of Physical Constants*, vol. 3, ed. T.J. Ahrens, American Geophysical Union, Washington, DC, 127–147, 1995.)

difficulty a drill bit would have grinding its way into the subsurface. Two points are particularly evident in the figure. First, the rock type has a strong influence on the strength of the rock. Even at low confining pressure (<100 MPa, which is equivalent to a depth of a few thousand meters), the fracture strength varies by a factor of three. Second, as the confining pressure increases, the fracture strength of the rock increases, but by different degrees depending upon the rock type. For geological environments where the geology is complex, consisting of different types of rocks that are interlayered, drilling rates will vary and the energy required to drill a deep hole will vary, as the hole is drilled.

The impact of this variation in rock strength is clear when one considers how drilling rate varies with rock type and with increasing confining pressure. Figure 9.6 shows several examples of how drilling rate, or ROP, is affected by rock type and confining pressure. Note that drilling rate can decrease by as much as a factor of 8 over a relatively short depth interval, with most of the change occurring within the first ~15 MPa (~500 m) pressure increase. This effect must be considered when scheduling drilling activities—for relatively deep holes, it is not unreasonable to allocate ~50% of the drilling time to completing the bottom ~10%–~20% of the hole.

A technique employed to accomplish deep drilling is to increase the weight on the bit, which will directly translate into greater force applied to the rock to overcome the rock strength. Figure 9.7 shows how the ROP into the rock varies with the load placed on the bit. Penetration rate can be increased by up to a factor of 2 by approximately tripling the load on the bit. However, Figure 9.7 also emphasizes a crucial aspect that affects penetration rate and that is the nature of the fluid that is used to circulate in the drill hole.

The highest drilling rates obtained in the experimental study shown in Figure 9.7 were obtained using water as the cooling agent. However, water is inadequate for clearing cuttings from the hole—higher viscosity fluids are required to accomplish that. The other fluids used in the study were various mixtures using mineral oil and water, or other compounds. This figure emphasizes the

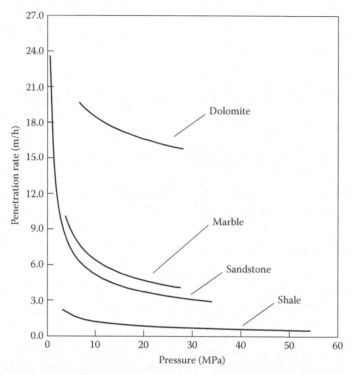

FIGURE 9.6 Penetration rate (ROP, in meters/hour) as a function of the confining pressure. (The curves are based on data and fits from Black, A. and Judzis, A., Topical Report, DE-FC26-02NT41657, http://www.osti.gov/bridge/product.biblio.jsp?query_id = 1&page = 0&osti_id = 895493, 2005; Lyons, K.D. et al., National Energy Technology Laboratory Report NETL/DOE-TR-2007-163, 1–6, 2007.)

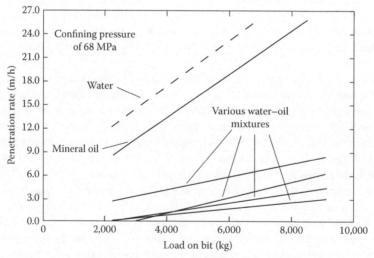

FIGURE 9.7 Penetration rate as a function of the load on the bit and the drilling fluid. Data were generated for a confining pressure of 68 MPa using a laboratory experimental apparatus. The rock being drilled was a sandstone. (The curves are based on data and fits from Black, A. and Judzis, A., Topical Report, DE-FC26-02NT41657, http://www.osti.gov/bridge/product.biblio.jsp?query_id = 1&page = 0&osti_id = 895493, 2005; Lyons, K.D. et al. National Energy Technology Laboratory Report NETL/DOE-TR-2007-163, 1–6, 2007.)

importance of selecting drilling fluids that will minimally diminish penetration rate. Selecting a drilling fluid, however, remains a challenge, as discussed in Section "Temperature and Drilling Fluid Stability".

Rock strength also impacts the ability to maintain the desired orientation of the well. In holes that are intended to be vertical, it is not uncommon for some unintended deviation to occur. This results from the variability in rock strength that can occur on the submeter scale due to the inherent heterogeneous nature of rocks. The impact of this heterogeneity on the bit performance can be appreciated by considering a rock in which there are subvertical zones that are weakened due to alteration or fracturing. When a bit first encounters such a region, the rock strength on one side of the hole being drilled will be less than that of the rest of the hole. If the difference in strength is significant and persists for some distance, the bit will wander into that region that is more readily excavated. The result will be a well that changes alignment with depth. The magnitude of this effect can be quite large, amounting to many meters or tens of meters of deviation from the planned orientation. Changes in drilling rate, drill string abrasion, and other indicators provide the driller with hints as to whether such effects are occurring. In severe cases, downhole measurements are used to determine the actual orientation of the hole with depth. However, these are costly processes because they force the rig to cease the drilling operation for some time period while downhole measurements and assessments are made. It is for this reason, among others, that care is taken when determining how to load the bit, because overweighting it can encourage deviation.

TEMPERATURE AND DRILLING FLUID STABILITY

Although standard drilling muds that are commercially available are adequate for many drilling applications, as temperatures of 200°C are exceeded, the performance of standard drilling muds begins to be compromised. The issues that are encountered include loss of plasticity, flocculation (which affects the viscosity), and shrinkage. The reasons for these effects reflect the thermal and chemical stability of clay minerals.

Consider the reaction for montmorillonite dehydration (refer back to Equation 9.1). It is apparent that this reaction will be very sensitive to temperature, because the right-hand side of the reaction has the high-entropy compound, water, as a product, and high-entropy phases are favored by high temperature. If the reaction occurs as written, with water being removed from the mineral structure, the reaction is a dehydration reaction. Because there are multiple water molecules in the structure of montmorillonite, multiple dehydration steps can affect the montmorillonite phase as temperature increases. Figure 9.8 shows the experimentally determined maximum pressure and temperature conditions for the two main dehydration reactions for montmorillonite. Also shown is the difference between pure Na- and pure K-montmorillonite. The figure indicates that at a given pressure, the temperature at which the dehydration reaction will occur varies by more than 50°C, depending upon the composition.

Empirical evidence from field experience shows that natural montmorillonite, and the natural bentonites within which it occurs, may begin to change properties at temperatures lower than those that have been experimentally determined for the dehydration reactions shown in Figure 9.8. This behavior reflects several effects. Among them is the fact that most montmorillonites are compositionally much more complex than shown in Equation 9.1. The relative proportions of the cations in the structure can vary significantly, reducing the thermodynamic activity of the end-member compositions used in the reaction. The overall result will be a significant decrease in the activity of the product solid mineral, relative to the reactant mineral. This will shift the reaction to lower temperatures than shown in the figure.

In addition, other components in the drilling fluid, whether naturally occurring or synthetic, change their properties with temperature. As a result, the low-temperature behavior of drilling muds that makes them useful in drilling operations are not usually achieved at the elevated temperatures of many geothermal systems.

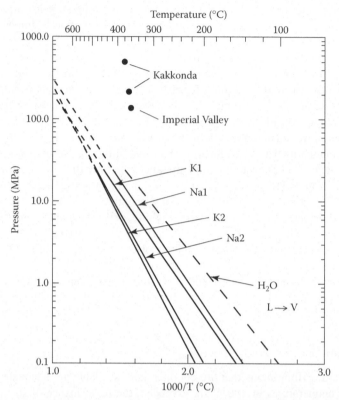

FIGURE 9.8 Experimentally determined dehydration reaction conditions for montmorillonite. The lines labeled K1 and Na1 are the maximum pressure and temperature conditions for the first dehydration reaction for K-rich and Na-rich montmorillonite, respectively. The lines labeled K2 and Na2 are the corresponding conditions for the maximum *P–T* for the second dehydration reactions for the respective compositions. Also shown are the liquid (L) to vapor (V) curve for water and the reported conditions for the geothermal systems in the Imperial Valley, California and the Kakkonda field in Japan. The solid reaction lines are from data in van Groos and Guggenheim. The dashed portions of the lines are linear extrapolations to higher pressure and temperature conditions. (From Zilch, H.E. et al., The evolution of geothermal drilling fluid in the Imperial Valley, *Society of Petroleum Engineers Western Regional Meeting*, Long Beach, CA, March 20–22, 1991; Saito, S. and Sakuma, S., *Journal of the Society of Petroleum Engineers Drilling and Completion*, 15, 152–161, 2000; van Groos, A.F.K. and Guggenheim, S., *Clays and Clay Minerals*, 34, 281–286, 1986.)

To overcome these problems, one approach has been to use a mud cooling system, whereby the returned mud is cooled before being recirculated into the hole. These systems commonly use heat exchangers to remove heat from the returned mud. Such systems, however, add costs to the drilling operation.

Although drilling fluids convey many benefits to the drilling process, they are not without risk. One important factor regarding the use of drilling muds is that they can reduce or eliminate the ability of a rock formation to allow fluid flow. For a geothermal well, this can be a serious problem, because the purpose of the well is to allow hot geothermal fluids to flow into the well for transport to the surface or for reinjecting fluid into a rock formation to recharge the geothermal system. Reduction or loss of flow can happen when a drilling fluid permanently lodges in open spaces, reducing or eliminating permeability. If this *formation damage* occurs, it can jeopardize the success of a geothermal well. Various strategies are employed to flush out material that may reduce permeability, if formation damage occurs. Higher flow rates, changes in fluid composition, flushing, and chemical additives to dissolve the material are options for mitigating formation damage.

These effects have resulted in extensive research into new formulations for drilling muds. Consideration has been given to a variety of polymeric substances and other materials as additives or primary bases for drilling muds to be used at high temperatures. This field remains one of the active research and entrepreneurial activity.

CASING AND GROUTING

Geothermal wells produce hot fluids from deep levels that flow at high rates. The wells often penetrate regions where water is present in aquifers at shallower levels. In order to prevent the inflow of cool waters or leakage of hot fluids out of the well, as well as to maintain the long-term stability of the borehole, the drilled wells are cased with metal casing. In addition, many of these wells are drilled in regions where the pressure of the fluid at depth is higher than the hydrostatic pressure would normally be, resulting in conditions in which hot geothermal fluids can violently escape from the wellhead. To prevent such high-pressure conditions from lifting the pipe out of the ground, and in order to have a sufficiently strong supporting structure to be able to maintain the large mass of the pipe that is held in place in the hole, a series of nested casings of progressively diminishing sizes are grouted or cemented into place as support for the producing well pipe.

Figure 9.9 shows examples of two possible casing methods for a 5000-m well. The drilling process involves drilling a sequence of successively smaller diameter holes. Each hole is drilled with a bit that exceeds the diameter of the casing that will be emplaced in that segment of the hole and the casing is grouted or cemented into place before the next segment of the well is drilled. For example,

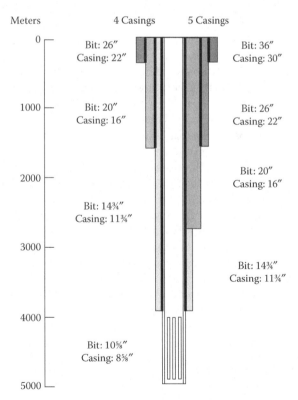

FIGURE 9.9 Examples of two approaches for drilling and casing a 5000-m geothermal well. Casing is shown as the heavy black lines, and cemented or grouted regions are shown by various shades of gray. A 4 casing (left side) technique and a 5 casing (right side) technique are shown. The depths (left side of the figure, in meters) of each section are only illustrative.

in the approach shown for the 4 casing method, the first hole would be drilled using a 26-inch diameter bit to a depth of several hundred meters. Once the selected depth is reached, the drill string would be pulled out and a casing pipe of 22 inches would be lowered into the hole. Grout or cement would then be injected into the 2-inch annulus surrounding the casing and allowed to solidify before the next stage of drilling occurs, which would be done using a 20-inch diameter drill bit. This sequence of staged drilling would continue until the final depth is reached and the producing pipe is emplaced.

It is of the utmost importance that the grouting or cementing process be done with care. Inadequate bonding of the grout or cement to the casing or the enclosing rock can result in fluid flow around the borehole. This situation can destabilize the hole and weaken the casing. In addition, poorly grouted pipe, where voids and inadequate strength result, can ultimately result in casing failure due to corrosion or mechanical failure. Any of these situations are exacerbated if the well is pressurized for testing or stimulation (discussed in Chapter 13), with the possibility that the well will leak or be contaminated by inflowing cold water. Such failures can catastrophically affect the productivity of a geothermal well, ultimately leading to reworking of the well or abandonment.

Whether or not a well is cased all the way to the bottom depends upon the integrity of the rock through which drilling has taken place. Competent rock that has high strength and is not intensely fractured is often drilled and finished as an open hole, without any casing. Such holes are remarkably stable and will maintain their integrity for the lifetime of the power generating facility. If, however, the hole is in highly fractured or otherwise mechanically incompetent rock, various types of liners are used that provide stability so the rock will not collapse into the hole, but which also allow fluid to flow in. Slotted or perforated liners are usually used for this case. An example of a slotted liner is shown in Figure 9.9.

Packers

An important device used to address several issues encountered when drilling is the *packer*. A packer is used to seal off a section of a well for one of several reasons. If it is discovered in the course of testing a well that there are multiple zones with potentially useful or problematic properties, it may be desirable to isolate those zones and conduct tests to evaluate their respective characteristics. Or, if it is desired to seal the porosity in a particular zone of a well, or to increase the permeability of a particular zone of a well, packers can be used to accomplish that.

The required features of a packer are that it can be lowered into the well, hence its diameter must be less than the interior diameter of the pipe or hole; it must be possible to inflate it or in some other way expand its diameter so it can seal the hole at a particular level; and it must be possible to remove it, either by drilling through it or by reducing its diameter once it has been deployed. Several designs have been developed to accomplish these ends.

The most commonly used packer is an inflatable packer that is constructed as a bag from elastomeric materials. This type of device is lowered into the hole and inflated using high-pressure fluid (air or liquid) so that it seals tightly against the pipe or well wall. Once emplaced, it acts as a barrier, preventing communication up and down the well. Emplacement of multiple inflatable packers allows individual regions to be isolated.

Inflatable packers have been used extensively in the oil and gas industry to allow *stimulation* of a particular zone in a well. Stimulation is the process whereby the permeability of a selected region is increased. The most common way through which this is accomplished is by isolating the zone of interest and then injecting into it fluid at very high pressure. If the rock possesses natural fractures that are sealed, these will often slip during the stimulation, a process called *hydro-shearing*, providing increased fracture permeability. In some instances, intact rock will fail and form new fractures if its fracture strength is exceeded by the high-pressure fluid. This process is called *hydro-fracturing*. In either case, this stimulation strategy requires that the injected fluid be contained within the selected region during the stimulation process. Inflatable packers are currently the primary routine means for accomplishing this because they allow containment of the fluid and can then be removed once the process is complete. Both hydro-shearing and hydro-fracturing are discussed in more detail in Chapter 13.

Packers were used successfully in Iceland to stimulate a low-temperature zone that supplied warm (<100°C) water for a geothermal district heating system. In this instance, pressures of about 15 MPa were used, and injection rates of 15–100 kg/s were realized. The result was that fluid flow from the district heating system increased from 300 to 1500 l/s (Axelsson and Thórallsson 2009).

Unfortunately, however, the elastomeric materials that are used in packers generally are unsuitable for applications above ~225°C. As geothermal wells go to deeper levels, extract hotter fluids, and pursue stimulation programs, as is the case for EGS (Chapter 13) new forms of packers will need to be developed. Efforts have been made to employ expandable metal packers and drillable packers, with varying degrees of success (Tester et al. 2006). This remains an important research topic.

LOST CIRCULATION

It has been emphasized that drilling requires constant circulation of drilling mud to cool the bit, lubricate the rotating pipe, and remove the cuttings. For that reason, drillers constantly monitor the rate at which fluid is pumped down the hole and the rate at which it returns. A significant drop in the rate of return of drilling mud signals drilling mud is being diverted into a permeable zone in the wall rock. This can happen when a zone of highly fractured rock is encountered, or if a highly porous region is encountered, both of which are possible at almost any depth and under most drilling conditions. Loss of circulation can be a catastrophic event, in the sense that drilling would have to cease and the hole abandoned if the zone of high permeability cannot be sealed off.

Strategies for sealing such zones vary. In some cases, increasing the rate at which mud is pumped into the hole will ultimately fill the fractures and drilling can continue. In some instances, the fracture network may be so extensive and pervasive that increased pumping will not fill the fractures sufficiently. In that case, other materials may be added to the mud and pumped down the hole. Materials such as cubed alfalfa, granular coal, diesel oil, and sawdust, or any other of a host of materials that have the possibility of clogging the permeability, have been tried. In such cases, there is no guarantee of success, but the investment that has been made in siting and initiating drilling justifies exhaustive efforts to recover circulation before the well is abandoned.

BLOWOUT PREVENTION EQUIPMENT

Fluids in geothermal reservoirs can be very hot and under high pressure. In some cases, it is possible that toxic or combustible gases may be present, as well. In order to prevent the unintentional and rapid release of such fluids, blowout prevention equipment (BOPE) is sometimes required on geothermal wells. The exact conditions that require installation of such equipment vary from place to place, but BOPE installation is likely if a new well is being put in a geothermal field that has not previously been drilled, or if there are known conditions that could lead to hazardous situations if equipment failed. Figure 9.10 shows the main components of a blowout preventer. The rams are the primary device for stopping flow. They consist of steel plates and cutters that can crimp or sever pipe and stop fluid flow at high pressures. BOPE is installed once a well is cemented.

DIRECTIONAL DRILLING

Until the 1970s, nearly all industrial drilling produced wells that were vertical. But the advantages of being able to *deviate* a well at some angle away from the vertical was apparent. One important advantage was that a well that was drilled vertically initially, but which then was deviated toward a particular direction and depth could, if properly guided, greatly increase the length of pipe over which it would be possible to extract a resource. In addition, a single drilling pad could be used to put in numerous wells that were deviated in different directions, thus greatly reducing the acreage needed to access a volume of rock in the subsurface.

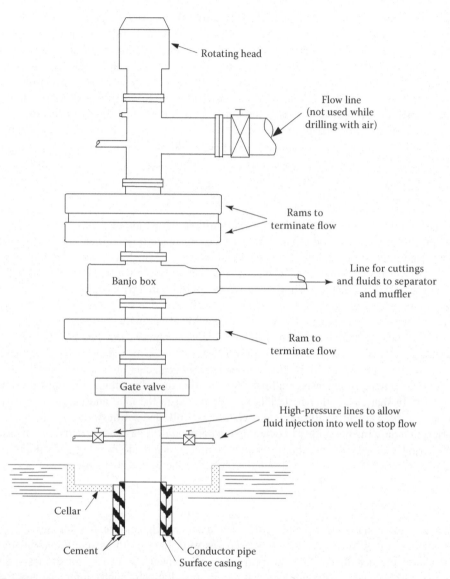

FIGURE 9.10 The components of BOPE. The components of the system, and their common names, are indicated. (Modified from Wygle, P., *Blowout Prevention in California*, 7th edition, California Division of Oil, Gas and Geothermal Resources Publication, Sacramento, CA, 1997; Wygle, P., *Blowout Prevention in California*, 10th edition, California Division of Oil, Gas and Geothermal Resources Publication, Sacramento, CA, 17, 2006.)

In the 1970s, the oil and gas industry began developing tools that would allow deviated wells to be drilled under controlled conditions. This so-called directional drilling has ultimately led to the ability to drill wells horizontally for several thousand meters (Figure 9.11). The ability to do this relies on several technological developments. It also takes advantage of the fact that the drill string, although composed of steel pipe, is remarkably flexible when it extends for hundreds or thousands of meters.

One important development for directional drilling was the invention of the downhole drilling motor. The motor consists of a drill bit in which the cutting rollers or other cutting elements are forced to rotate by the high-pressure flow of drilling mud through a screw-like element inside the pipe at the end of the drill string. This system eliminates the need to rotate the entire drill string at

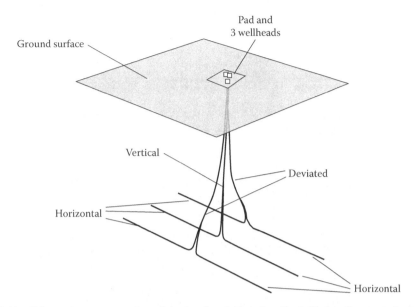

FIGURE 9.11 Schematic representation of deviated and directionally drilled wells, on a single drill pad. In this representation, one vertical well and two deviated wells have been drilled from a single pad. Each of the three wells has been completed with two opposing horizontal runs.

the same rate as the drill bit cutting elements. The cutting assembly of the bit is also inclined about 2° from parallel with the rest of the pipe. This forces the bit to cut preferentially on one side of the hole, thus deviating it. By maintaining control on the orientation of the bit cutting assembly, and the different rate of rotation of the drill string, the well can be made to gradually arc in any desired direction.

More recently, motor assemblies have been developed that allow three-dimensional steering of the rotating bit inside a rotating drill string in real time. These rotary steerable systems provide the ability to more accurately target specific depths and zones. These systems have communication capabilities that allow the location, orientation, depth, and direction to be controlled while drilling is in progress. Currently, the curvature of deviated wells can be as much as 15° per 30 meters. This translates into curvatures of 90° over a distance of less than 200 m.

Use of directional drilling can increase the cost of a well by 20% or more. However, the increased production that can be obtained is usually more than adequate to justify such expenses. This cost–benefit plus the accuracy that can now be achieved while steering the drill bit remotely in real time has resulted in dramatic growth of the directional drilling enterprise since 2010.

Although developed in the oil and gas industry, directional drilling has become standard in the geothermal industry. However, as noted above, the high temperatures of geothermal systems are a challenge for equipment, fluids, and operational tools. The high failure rate of equipment at high temperatures increases the risk of drilling deviated wells, but the benefits still exceed the cost in many cases. Even so, directional drilling and horizontal wells remain a costly challenge, particularly as long as drilling muds, which breakdown at elevated temperatures, are the primary means for driving drilling bits.

Coring

Although costly because of the time it takes, coring during exploration drilling is an important method for acquiring critical information about the rocks that will be encountered when drilling production and injection wells. Coring is usually accomplished by drilling smaller diameter holes

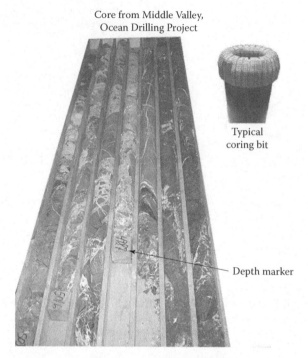

FIGURE 9.12 An example of core (left) obtained from the Ocean Drilling Project laid out in a core box with depth blocks indicating the location of core in the well. On the right is an example of a typical coring bit, with a central hole for the preserved core. (Photograph provided by Robert Zierenberg.)

than normal production holes and by using special drill bits with a central void (Figure 9.12). As the cutting elements of the bit grind away the rock, a central cylinder of rock is kept intact that the bit cuts around. Normally, coring is done in lengths of several meters to tens of meters, depending upon the equipment used. At appropriate intervals for the equipment, drilling is halted and the core is brought to the surface. The system is then redeployed and another section of core is obtained.

SYNOPSIS

Drilling wells for geothermal applications requires strict attention to assuring that environmental protections are in place; especially important is assuring isolation of wells from the environment they penetrate, restoration of drilling sites to an acceptable condition, and preventing releases of unwanted or hazardous compounds. Wells for GHP and direct-use applications are generally shallower than those drilled for power generation purposes. The former are usually drilled with an auger system (if the soil conditions allow) or using rotary drill rigs. Often the same equipment used for water well drilling can be employed for these applications. Wells for power generation are more demanding engineering efforts. Much larger drilling rigs with capabilities to place thousands of meters of casing in hostile conditions are required. Equipment performance and operational issues are challenging because of the extreme temperature and pressure conditions encountered. Nevertheless, technology transfer from the oil and gas industry has allowed the introduction and successful use of such things as directional drilling, enhanced drilling fluids, and trajectory monitoring. These successes need to be expanded upon by the geothermal industry by developing more robust downhole equipment, especially packers and instrumentation, as well as drilling fluids that can sustain high temperatures.

CASE STUDY: KAKKONDA, JAPAN

The Kakkonda geothermal field, located approximately 500 km north of Tokyo, Japan, is one of the world's shallowest, high-temperature geothermal regions. It is located within the active Tohoku volcanic arc, which is part of the subduction zone complex that bounds the Pacific plate in the western Pacific region. This system is of the compressional type and corresponds to the Brophy type B system. In the Kakkonda region, very young granitic rocks, dated at between 110,000 and 240,000 years old (Doi et al. 1995; Matsushima and Okubo 2003), intruded to depths of less than 3 km and supply a prodigious quantity of heat to the area.

More than 70 geothermal wells have been drilled in the area, most reaching depths of between 1000 and 3000 m. The temperature distribution as a function of depth is variable (Figure 9.12). Power has been generated in the area since 1978—currently more than 70 MW power is generated in the region (Matsushima and Okubo 2003).

To better delineate the nature of the geothermal resource, and to evaluate various drilling technologies and strategies, a deep well, designated WD-1 (later WD-1a), was drilled in the region in 1994 and 1995. The target depth was 4000 m. A head drive drilling method was chosen to evaluate, among other things, the ability of such a system to provide adequate cooling during drilling. A mud cooling system was also employed.

The history of events during that drilling program provides insight into the challenges to be expected drilling in such a setting. We will consider these in the following discussion, which is based on events summarized from Uchida et al. (1996) and Saito and Sakuma (2000).

Drilling was initiated on January 5, 1994. The approach was to redrill an earlier hole that had been drilled in 1992 and 1993. The plan was to use a 5 casing design, similar to that depicted in Figure 9.9. The depths, diameters of drill bits, and diameters of the emplaced casing are shown in Figure 9.13.

Previous experience from drilling geothermal wells in the region indicated that the volcanic rock in the region was highly fractured. For this reason, it was anticipated that lost circulation would be an important challenge when drilling the deep hole. Of particular concern was the region where the shallow geothermal reservoir existed. In fact, lost circulation was a major challenge throughout the drilling of the first 2000 m (Figure 9.13). Cement was used successfully to close off many of those zones, but eventually the well trajectory had to be modified at a depth of about 1696 m because of this problem.

In the region where lost circulation was most common due to the intense fracturing of the rock, the average ROP was approximately 6.7 m/day. Beyond that zone, the penetration rate was almost 18 m/day. Although common experience under "normal" conditions suggests that 50% of the cost and time will be committed to completion of the last 10%–20% of the hole, this example demonstrates unequivocally that such a rule of thumb is highly conditional.

The expectation at the time of drilling was that the high-temperature geothermal reservoir would be encountered at about 3000 m depth. Although temperatures in excess of 200°C were encountered initially at that depth, there was no indication of the expected high-temperature reservoir so drilling was continued.

At a depth of 3350 m, elevated CO_2 contents in the drilling mud were detected. Elevated CO_2 can be a health hazard and needs to be controlled. Lime was added to the mud to reduce the CO_2, with the intent of capturing the CO_2 as calcium carbonate. This effort appeared to be successful.

At depths beyond 3451 m, it was discovered that the drilling mud deteriorated because of high temperatures when mud pumping was stopped in order to remove the drill string and change the bit. To overcome this problem, the mud density was lowered to less than 1.1 gm/cm³. It was anticipated that this lower density and higher water content would provide additional time to allow bits to be changed before degradation of the drilling mud could occur. However, beyond a depth of 3642 m, high H_2S contents in the drilling mud were detected. H_2S can be a health hazard and its control is of paramount importance to assure the safety of the drilling crew. It appeared that the H_2S was entering the well from the formation that was being drilled. It was decided to increase the density (and hence the viscosity) of the mud to try to seal off the well from the formation fluids. However,

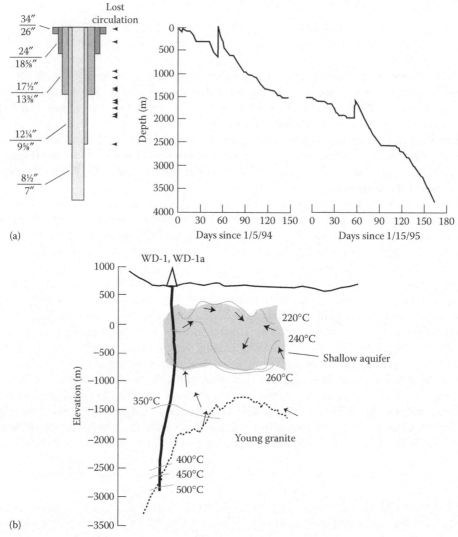

FIGURE 9.13 (a) The drill bit and nested casing specifications for the WD-1 and WD-1a well, as completed. The diameters of the drill bit (numerator in inches) and casing (denominator in inches) are indicated to the left of the figure. The arrows on the right of the well show locations where circulation was lost. To the right is shown the depth drilled as a function of the time because drilling commenced on January 5, 1994. Note that drilling was halted in mid-May, 1994, and recommenced on January 15, 1995. The bottom hole depth of 3729 m was reached on July 6, 1995. (b) A cross section through the region drilled. Shown in the figure is the location of the young, hot granite (upper boundary indicated by the dashed line), generalized flow fields (heavy arrows show inferred flow pattern in the shallow aquifer which has been exploited as a geothermal resource for power generation; lighter arrows show inferred flow path for high-temperature fluids), isotherms (gray lines with temperatures indicated), and the location of well WD-1 and WD-1a. (Data from Uchida, T. et al., *Geothermal Resources Council Transactions*, 20, 543–548, 1996.)

the elevated temperatures at the bottom of the well made it problematic drilling with higher density mud. Drilling was terminated at 3729 m.

Downhole surveys were conducted periodically after drilling. It was found that, after six days, the temperature at the bottom of the hole was 500°C, which was the hottest measured geothermal well at that time.

The experience obtained through the Kakkonda drilling program has provided important lessons for drilling such systems. Overall, the drilling technology, which included monitoring of the hole orientation while drilling, top head drilling, and motor-driven directional controls, can be employed for drilling systems such as this high-temperature field. However, it is also clear from this experience that significant work remains to be done to develop drilling fluids that can function at these elevated temperatures, as well as more robust drilling equipment suitable for prolonged operation in these hostile environments.

PROBLEMS

9.1. What are the purposes of drilling *mud*? What affects the ability of drilling *muds* to perform these functions?

9.2. If you were to hire a driller for installation of a shallow well, what question would you ask to assure you were dealing with a reputable and experienced driller?

9.3. What is the primary component in drilling muds and why is it used?

9.4. What are the environmental issues that need to be addressed when drilling a well for power generation? For non-power applications?

9.5. What is a packer and what is it used for?

9.6. Why is the cost of a deep geothermal well somewhat more expensive than an oil and gas well drilled to the same depth?

9.7. What determines how long it takes to drill a well to any given? Explain why the penetration rate decreases as drilling reaches deeper levels.

9.8. How might directional drilling have affected whether a geothermal site would be considered a reserve or not?

9.9. Using the information in Figure 9.13, make a plot of temperature versus depth. What is the steepest geothermal gradient? The shallowest? At what depth would these two gradients suggest that a temperature of 650°C would be reached? Which of these estimates is reliable and why?

9.10. Why is high temperature a problem for drilling muds?

REFERENCES

Axelsson, G. and Thórallsson, S., 2009. Stimulation of geothermal wells in basaltic rock in Iceland. *IPGT Nesjavellir Workshop*, May 11–12.

Bertani, R., 2007. World Geothermal Power Generation in 2007. *Proceedings of the European Geothermal Congress*, Unterhaching, Germany, May 30–June 1.

Black, A. and Judzis, A., 2005. Optimization of deep drilling performance—Development and benchmark testing of advanced diamond product drill bits & HP/HT fluids to significantly improve rates of penetration. Topical Report, DE-FC26-02NT41657. http://www.osti.gov/scitech/biblio/895493.

Bolton, R.S., 2009. The early history of Wairakei (with brief notes in some unfroeseen outcomes). *Geothermics*, 38, 11–29.

Doi, N., Kato, O., Kanisawa, S., and Ishikawa, K., 1995. Neo-tectonic fracturing after emplacement of Quaternary granitic pluton in the Kakkonda geothermal field. *Japan Geothermal Resources Council Transactions*, 19, 297–303.

International Geothermal Association, 2008. (http://iga.igg.cnr.it/geoworld/geoworld.php?sub=elgen)

Lockner, D.A., 1995. Rock failure. In *Rock Physics and Phase Relations: A Handbook of Physical Constants*, vol. 3, ed. T.J. Ahrens. Washington, DC: American Geophysical Union, pp. 127–147.

Lyons, K.D., Honeygan, S., and Mroz, T. 2007. NETL extreme drilling laboratory studies high pressure high temperature drilling phenomena. National Energy Technology Laboratory Report, NETL/DOE-TR-2007-163, National Energy Technology Laboratory, Morgantown, WV, pp. 1–6.

Matsushima, J. and Okubo, Y., 2003. Rheological implications of the strong seismic reflector in the Kakkonda geothermal field, Japan. *Tectonophysics*, 371, 141–152.

Saito, S. and Sakuma, S., 2000. Frontier geothermal drilling operations succeed at 500°C BHST. *Journal of the Society of Petroleum Engineers Drilling and Completion*, 15, 152–161.

Smith, B., Beall, J., and Stark, M., 2000. Induced seismicity in the SE Geysers Field, California, USA. *Proceedings of the World Geothermal Congress 2000*, Kyushu-Tohoku, Japan, May 28–June 10.

Tester, J.W., Anderson, B.J., Batchelor, A.S., Blackwell, D.D., DiPippio, R., Drake, E.M., Garnish, J. et al., 2006. *The Future of Geothermal Energy*. Cambridge: MIT Press, p. 372.

Uchida, T., Akaku, K., Sasaki, M., Kamenosono, H., Doi, N., and Miyazaki, H., 1996. Recent progress of NEDO's "Deep-Seated Geothermal Resources Survey" project. *Geothermal Resources Council Transactions*, 20, 543–548.

van Groos, A.F.K. and Guggenheim, S., 1986. Dehydration of K-exchanged montmorillonite at elevated temperatures and pressures. *Clays and Clay Minerals*, 34, 281–286.

Wygle, P., 1997. *Blowout Prevention in California*. 7th edition. Sacramento, CA: California Division of Oil, Gas and Geothermal Resources Publication

Wygle, P., 2006. *Blowout Prevention in California*. 10th edition. Sacramento, CA: California Division of Oil, Gas and Geothermal Resources Publication, 17 pp.

Zilch, H.E., Otto, M.J., and Pye, D.S., 1991. The evolution of geothermal drilling fluid in the Imperial Valley. *Society of Petroleum Engineers Western Regional Meeting*, Long Beach, CA, March 20–22.

FURTHER INFORMATION SOURCES

Culver, G., 1998. Drilling and Well Construction. In *Geothermal Direct Use Engineering and Design Guidebook*, eds. P.J. Lienau and B.C. Lunis. Klamath Falls, OR: Geo-Heat Center, Oregon Institute of Technology, pp. 129–164.

Although new developments and advances have been made in the design and use of various components in wells, this remains an excellent overall review of the topics that must be addressed when developing a geothermal well.

Federal Energy Regulatory Commission (http://www.ferc.gov/).

FERC provides important oversight functions for many aspects of the power grid. They maintain an important library of documents, studies, and data useful for understanding power generation and transmission issues.

Geothermal Resources Council (http://www.geothermal.org) and the International Geothermal Association (http://www.geothermal-energy.org/index.php).

The GRC and IGA are organizations that provide information, data, and historical perspective on the geothermal industry. Both hold annual meetings at which recent research results are presented and trade organizations are available for discussion of technology, law, and regulatory issues. The GRC maintains one of the world's premier libraries at its headquarters in Davis, California.

North American Electric Reliability Corporation (http://www.nerc.com/).

NERC develops reliability standards and provides education, training, and certification for the power industry, including courses that deal with power generation and technology. NERC is a self-regulatory organization, subject to oversight by the US Federal Energy Regulatory Commission and governmental authorities in Canada. They also have online courses for all aspects of power generation technology.

10 Generating Power Using Geothermal Resources

The production of electricity using geothermal energy employs technology that is fundamentally indistinguishable from that at most other power generating facilities. Specifically, an electrical generator is powered by a turbine that converts thermal or kinetic energy into electricity. In fossil-fueled power plants, thermal energy drives the turbine, whereas in hydropower plants, the kinetic energy derived from flowing water drives the turbines. However, in two important respects geothermal power production is unique when compared to other power production methods. First, when compared to power generating technologies that supply baseload power, such as fossil-fueled power plants, biomass reactors, or nuclear reactors, there is no fuel cycle required to generate heat because the heat already exists within the earth. Second, when compared to other renewable energy technologies that do not require a fuel cycle for heat generation, such as wind, solar, tidal, or ocean wave technologies, geothermal is not intermittent and provides true baseload capability at a reliability that consistently exceeds 90%. The remainder of this chapter addresses the physics of power generation as it relates to geothermal power production and design issues that are specific to particular types of geothermal resources. For a detailed discussion of geothermal power plant design see the presentation by DiPippo (2008).

HISTORY OF GEOTHERMAL POWER PRODUCTION

The first production of electricity from geothermal steam took place in Larderello, Italy, in 1904. Larderello is in a region of Italy with substantial recent volcanic activity, including explosive steam eruptions that occurred as recently as the late 1200s. With the growth of the Industrial Revolution, interest in exploiting the geothermal resource of the area grew, beginning with the direct use of steam at Larderello to support a local chemical separation industry. However, it was the innovative application of a small steam generator powered by the Larderello geothermal system that gave birth to the geothermal power industry.

The small steam generator that was first used in Larderello lighted four light bulbs in a demonstration project conceived by Prince Piero Ginori Conti. Seven years later, in 1911, industrial-scale production of geothermal power was brought online and used in-house for the commercial chemical applications that were being pursued at the time. In 1916, a 2500 kW generating capacity was brought online to provide electricity commercially to the local communities (Bolton 2009). Larderello remained the world's only geothermal power plant of industrial scale until the 1950s when New Zealand brought online its first geothermal power plant. Between the early 1950s and 1963, New Zealand began generating electricity at Wairakei, with an installed generating capacity of more than 190 MW of electrical power (Bolton 2009). Meanwhile, at The Geysers in California, a utility company, Pacific Gas and Electric, installed a generator with a capacity of approximately 12 MW (Smith et al. 2000). Since that time, the global growth in geothermal power production has been rapid, as documented in Figure 10.1. Table 10.1 lists the global geothermal power generation capability for 1995, 2000, 2007, and 2012. Since 1965, the annual growth rate of geothermal power generation capacity has been approximately 250 MW per year, as indicated by the dashed line in Figure 10.1.

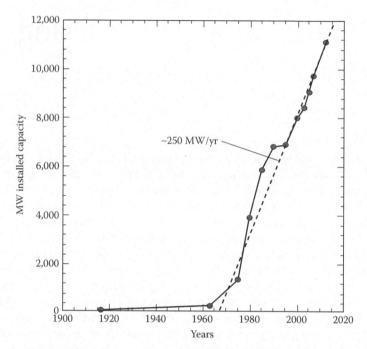

FIGURE 10.1 The installed global geothermal generating capacity, from 1916 through 2012. (Data from Smith, B. et al., Induced seismicity in the SE geysers field, California, USA. *Proceedings of the World Geothermal Congress 2000*, Kyushu–Tohoku, Japan, 2887–2892, 2000; Bertani, R., *Geo-Heat Center Bulletin*, 28, 8–19, 2007; Bolton, R.S., *Geothermics*, 38, 11–29, 2009; International Geothermal Association, 2008. http://www.google.com/url?sa=t&rct=j&q=&esrc=s&source=web&cd=1&ved=0CB8QFjAA&url=http%3A%2F%2Fwww.earth-policy.org%2Fdatacenter%2Fxls%2Fupdate74_2.xls&ei=TF64U9fDGI-hyASh_YDoAw&usg=AFQjCNGjxc8UEE6DK4TZkSvA8sHfXNqPDA&bvm=bv.70138588,d.aWw.)

TABLE 10.1
Geothermal Generation Capacity by Country

Country	1995 (MWe)	2000 (MWe)	2007 (MWe)	2012 (MWe)
Argentina	0.67	0.0	0.0	–
Australia	0.17	0.17	0.2	1.1
Austria	0.0	0.0	1.1	1.4
People's Republic of China	28.78	29.17	27.8	24
Costa Rica	55	142.5	162.5	201
El Salvador	105	161	204.2	204
Ethiopia	0	8.52	7.3	7.3
France	4.2	4.2	14.7	16.2
Germany	0	0	8.4	12.1
Guatemala	33.4	33.4	53	52
Iceland	50	170	421.2	675
Indonesia	309.75	589.5	992	1,333
Italy	631.7	785	810.5	883
Japan	413.7	546.9	535	535
Kenya	45	45	128.8	205
Mexico	753	755	953	983
New Zealand	286	437	471.6	762

(*Continued*)

TABLE 10.1

(Continued) Geothermal Generation Capacity by Country

Country	1995 (MWe)	2000 (MWe)	2007 (MWe)	2012 (MWe)
Nicaragua	70	70	87.4	124
Papua New Guinea	0	0	56	56
The Philippines	1,227	1,909	1,969.7	1,904
Portugal	5	16	23	29
Russia	11	23	79	82
Thailand	0.3	0.3	0.3	0.3
Turkey	20.4	20.4	38	99
The United States	2,816.7	2,228	2,687	3,129
Total	**6,866.77**	**7,974.06**	**9,731.7**	**11,180**

Sources: Smith, B. et al., Induced seismicity in the SE geysers field, California, USA. *Proceedings of the World Geothermal Congress 2000*, Kyushu–Tohoku, Japan, 2887–2892, 2000; Bertani, R., *Geo-Heat Center Bulletin*, 28, 8–19, 2007; Matek, B., 2013 geothermal power: International market overview, Geothermal Energy Association, Washington, DC, 35, 2013; International Geothermal Association, http://iga.igg.cnr.it/geoworld/geoworld. php?sub = elgen, 2008; Bolton, R.S., *Geothermics*, 38, 11–29, 2009; International Geothermal Association, 2008, 2012, IGA Newsletter, v. 89, p.3, IGA Global Geothermal Energy Database.

FLEXIBILITY AND CONSISTENCY

An important attribute of geothermal power generation is that it requires no external fuel infrastructure. Because the heat energy used for power generation is a natural resource in the immediate subsurface vicinity of the power plant, there is no need for transporting fuel or processing fuel in order to extract energy. This property leads to important benefits that make geothermal power generation environmentally and economically attractive.

The absence of a fuel cycle allows power to be generated without refueling. This, and the physical characteristic that heat can be accessed 24 hours a day, every day of the year makes geothermal facilities capable of providing *baseload* power. Baseload power is that which sustains most transmission infrastructure systems and is relied upon to be a consistent source of on-demand electricity. This is in contrast to intermittent power sources, such as wind and solar generation, which are variable due to the fact that wind speeds and solar insolation change with local weather conditions and time of day. Geothermal power generation can thus be an important part of the mix of energy generation technologies employed to sustain a consistent electrical supply.

Recent studies have also shown, however, that geothermal power generation can also be flexible. For a resource to be flexible, a power generator requires that the power output from the generating facility be able to increase or decrease as demand for power changes through the day and as the power input to the grid changes. Changes in power input to the grid occur when generating facilities go offline, such as that which happens when a facility is shut down for maintenance, or due to variable generation as weather or time of day affect the output of wind and solar generation facilities. If designed to provide variable power output, geothermal facilities have the ability to increase or decrease output in response to these changes. Although this flexible capability has been theoretically possible since the inception of geothermal power generation, historically geothermal power production has been exclusively baseload, reflecting the overwhelming need for such a resource in the absence of significant variable generation. However, as wind and solar power have become more widespread, the demand for flexible generation has grown. This suggests that geothermal generating facilities capable of flexible generation are likely to become more common in the future.

The following discussion on geothermal power generation technologies is focused primarily on the designs employed for baseload generation. Flexible generation requires additional design components and operational strategies that are still in the process of being fully implemented. As currently applied, flexible generation is accomplished by varying the amount of hot geothermal fluid that is introduced to the power generating equipment, either through bypassing the generation equipment through the use of secondary piping circuits or by throttling the rate at which geothermal fluid is extracted from the reservoir. The reader is encouraged to follow development of these capabilities, which are posted on the Geothermal Energy Association (GEA) website (http://www.geo-energy.org/).

Additionally, the generation of power from geothermal resources is accomplished at high efficiency and consistency. A typical geothermal plant will operate within 80%–95% of its design capacity over long time periods (weeks to years). This capability provides geothermal power generation with a *capacity factor* of 0.80–0.95. The significance of this high capacity factor is discussed in more detail in Chapter 15.

GENERAL FEATURES OF GEOTHERMAL POWER GENERATION FACILITIES

Producing power from geothermal energy relies on the ability to convert geothermal heat at depth to electricity. To accomplish this requires transferring the heat to a surface power generating facility that can efficiently turn thermal energy into electrical power. The equipment necessary to accomplish this is a piping complex that will bring hot fluids from depth to a turbine facility on the surface where the thermal energy is converted to kinetic energy in the form of a rotating turbine (see Sidebar 10.1). The kinetic energy of the turbine is then converted to electrical energy using an electrical generator. The quantity of electrical energy that can be generated from a geothermal fluid is

$$P_{elec} = \varepsilon_{gen} \times \varepsilon_{turb} \times H_g \tag{10.1}$$

where:
P_{elec} is the number of watts of electrical energy produced, which is equivalent to J/s
ε_{gen} is the efficiency of the electrical generator
ε_{turb} is the efficiency of the turbine
H_g is the rate at which thermal energy is supplied to the turbine

The remainder of this chapter will consider the factors that determine H_g. The factors that influence ε_{gen} and ε_{turb} are the subjects of numerous engineering texts and will not be directly addressed here.

Figure 10.2 is the pressure–enthalpy diagram for water that was previously discussed in Chapter 3. We will use it as a means to examine the conversion of geothermal heat to electrical power.

The energy available for doing work to rotate the turbine shaft that drives the electrical generator is the enthalpy of the fluid. Although the pressure and temperature of the fluid in the geothermal reservoir is the initial energy resource, extraction of fluid from the reservoir and transferring it to the turbine results in some energy loss. The primary sources of energy loss are from conduction of heat from the fluid to the surrounding rock as the fluid flows up the well, frictional losses as the fluid interacts with the well pipe during flow, and the change in gravitational potential energy. All of these losses are relatively small due to the low thermal conductivity of the well pipe and casing, the relatively smooth pipe wall, and the velocity of the rising fluid (on the order of a few meters to tens of meters per second). Collectively, these effects amount to a few percent loss in enthalpy. In the following calculations in which we are considering energy extraction, we will assume these losses to be negligible and that the rising fluid will traverse the length of the well isothermally. Under this assumption, the flow path of the fluid as it ascends to the turbine will follow an isotherm in Figure 10.2.

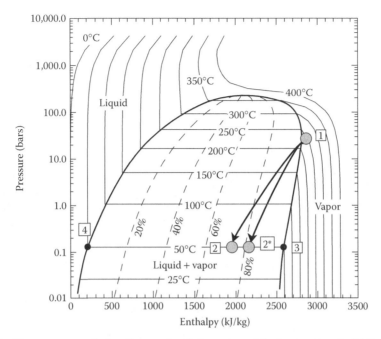

FIGURE 10.2   The pressure–enthalpy diagram for H_2O, contoured for temperature (solid lines). The steam percent in the two-phase liquid + vapor region is indicated by the dashed lines. The critical point temperature is 373.946°C and the critical point pressure is 220.64 bars (from NIST Water Properties). Points 1, 2, 2*, 3, and 4 relate to paths discussed in the conversion of geothermal energy to electricity for a dry steam system.

DRY STEAM RESOURCES

If the reservoir is at a temperature in excess of the critical point (373.946°C) or is at pressure and temperature conditions within the field labeled "Vapor," the fluid will not intersect the two-phase region of "Liquid + Vapor" on its rise to the generating facility. In this case, the high-pressure steam will decompress as it ascends from depth, its enthalpy varying as indicated by the appropriate specific isotherm for that fluid. Such systems are referred to as *supercritical* or *dry steam* systems and are the most sought after.

Why such systems would be attractive becomes obvious when one considers the amount of energy available for conversion to electricity. From a thermodynamic point of view, supercritical and dry steam systems provide the greatest amount of energy per kilogram of fluid extracted. This results from the fact that there is much less separation of liquid from the steam during decompression and the resulting partitioning of the enthalpy between those two phases is minimal. Instead, most of the enthalpy of the fluid remains with the steam as it enters the turbine and becomes available for energy conversion. Such a system requires minimal engineering of the piping between wellhead and the entrance to the turbine. Because there is little liquid to remove from the steam prior to entering the turbine (contrary to most other geothermal systems, as discussed below), design and piping expenses are kept to a minimum. Generally, only a separator that removes particulates from the steam stream is required. Separating particulates that might be present in the steam is important for the prevention of damage to the turbine blades.

The process of steam flow through the turbine can be visualized using Figure 10.3, an entropy–pressure diagram. We will begin at an initial state where the steam falls to the right of the saturation curve, at a temperature of 235°C and a pressure of about 30 bars (the maximum enthalpy point on the saturation curve, Figure 10.2). As the steam enters the turbine, it enters an environment that is less constrained and confined than in the well pipe. As a result, at this lower confining pressure

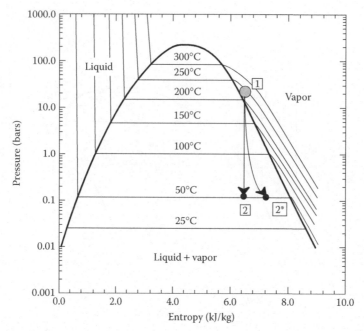

FIGURE 10.3 Pressure–entropy relationships for a dry steam generating facility. Points 1, 2, and 2* correspond to those in Figure 9.2.

condition, it begins to expand. As it goes through its expansion phase, the highest efficiency would be achieved if the expansion were isentropic and reversible (path 1–2 in Figure 10.3). However, as noted previously in Chapter 3, such a process is physically impossible due to the lack of thermodynamic equilibrium, heat losses due to conduction, and frictional effects. The actual pathway followed will be that from 1 to 2*, with a small increase in the entropy of the system and separation of some liquid.

The work that can be produced is the difference in the enthalpy between that of the steam entering the turbine (H_e) and that of the steam exiting the turbine (H_x):

$$w = H_e - H_x \tag{10.2}$$

The enthalpy of the entering steam can be taken as that of saturated steam at 235°C, which is 2804 kJ/kg. For the ideal case in which the work is performed isentropically and reversibly, and with an (assumed) end point of the process at 50°C, the enthalpy of the fluid exiting the turbine would be about 1980 kJ/kg (point 2 in Figure 10.2) and the steam fraction would be about 73%. Total work produced would thus be

$$w = 2804 - 1980 \text{ kJ/kg} = 844 \text{ kJ/kg}$$

However, the actual efficiency of the process must be less than 1.0, resulting in a higher entropy for the overall process. The efficiency of the turbine is the ratio between the realized enthalpy difference achieved during fluid expansion and the ideal, and hence,

$$\text{eff} = \frac{(H_1 - H_{2^*})}{(H_1 - H_2)} \tag{10.3}$$

This efficiency will vary based on how the turbine is designed. Generally, modern steam turbines have efficiencies of 85% or better (see Sidebar 10.1). For such turbines used in geothermal applications, an additional factor affecting performance is the amount of liquid in the steam as it interacts with the turbine. In a classic study, Baumann (1921) demonstrated that the following relationship approximately holds for the performance of steam turbines

$$\text{eff}_w = \text{eff} \times \left[\frac{(1 + x_2)}{2} \right] \tag{10.4}$$

where:
  eff_w is the efficiency of the turbine with wet steam present
  x_2 is the fraction of the steam phase that is pure vapor

This linear relationship shows that for every percent increase in liquid in the steam phase, there will be an approximate half a percent drop in turbine efficiency. Overall performance of a geothermal power plant will thus be sensitive to how condensing liquid is managed before and during passage through the turbine. Modern turbines are engineered with this in mind and invariably have water removal capability integrated into their design at appropriate points along the fluid flow path.

Considering the fluid properties at the end state, along with Equations 10.3 and 10.4, it is possible to show (DiPippo 2008) that the effective enthalpy at 2* can be computed from

$$H_{2*} = \frac{(H_1 - [0.425 \times (H_1 - H_2)] \times \{1 - [H_3 / (H_4 - H_3)]\})}{\{1 + [0.425 \times (H_1 - H_2)] / (H_4 - H_3)\}} \tag{10.5}$$

For the conditions stipulated for the ideal case above, H_{2*} is 2166 kJ/kg, and the work thus performed becomes

$$w = 2804 - 2166 \text{ kJ/kg} = 638 \text{ kJ/kg}$$

The resulting efficiency is then

$$\text{eff} = \frac{(2804 - 2166 \text{ kJ/kg})}{(2804 - 1980 \text{ kJ/kg})} = 0.77$$

For a geothermal site to be useful, it must have the capacity to produce steam at a sufficiently high rate to make generation of electricity economically feasible. Although conditions such as construction costs, exploration investment, access to electrical infrastructure, and other considerations have a large effect on the economic viability of a site (see Chapter 14 for a detailed discussion of the economics of geothermal energy production), it is generally considered a necessity that a site be capable of generating more than a megawatt (MW) of power if it is to be used for power that will be supplied to a transmission infrastructure. If the application will be for local, or *distributed*, generation, such as power for a small community, a few industrial buildings, or a university campus, much smaller units, on the order of a few tens of kilowatts, can be economically employed. For the realistic case considered above in which 638 kJ/kg of fluid were available for work, the flow rate necessary to generate 1 MW would be

$$\frac{1,000,000 \text{ J/s}}{63,800 \text{ J/kg}} = 1.6 \text{ kg/s}$$

Such a flow rate must be sustained for years in order to be viable.

An additional parameter that must be accounted for when generating power from geothermal resources concerns the pressure at the wellhead and its relationship to the flow rate of fluid through the turbine. From Figure 10.2, it is clear that pressure has an effect on performance, because it will establish the conditions for the initial enthalpy of our fluid and the enthalpy difference between initial and final states. The highest possible pressure is that of the reservoir itself. Because most geothermal reservoirs have sufficient pressure such that fluids will exit the well without pumping, the well will pressurize if the wellhead is closed to flow by the wellhead valve. The pressure at the wellhead when the wellhead valve is closed is called the *closed-in pressure* (P_{cl}). The actual closed-in pressure will depend upon the pressure and temperature properties of the reservoir and the depth at the bottom of the well. The minimum pressure possible is that at the cooling tower which cools the condensed fluid at the end of the thermodynamic cycle.

The power output of the turbine (and hence the maximum possible output of the generator) is a product of the mass flow rate times the enthalpy change per unit of mass. The mass flow rate is dependent on the pressure driving the fluid flow. However, the power output will also be a function of the enthalpy drop across the system, which is a function of the pressure drop between the inlet and exit points of the turbine. To understand how these factors interrelate, we will consider an isentropic (i.e., ideal) turbine system. The fluid at the wellhead has an entropy determined by the wellhead closed-in pressure and the temperature. We will assume that the P and T are that of the maximum enthalpy, that is, at 235°C and a pressure of about 30 bars. The enthalpy is 2804 kJ/kg. The mass flow rate can be expressed as a function of the pressure ratio between the closed-in pressure and the effective pressure using (DiPippo 2008)

$$m = m_{max} \times \sqrt{\left[1 - \left(\frac{P}{P_{cl}} \right)^2 \right]} \tag{10.6}$$

The value for m_{max}, the maximum flow rate, can be established by well tests—assume it is 5 kg/s. Table 10.2 shows the relationship between mass flow rate and the ratio of the pressures for the selected set of conditions. Note that the maximum flow rate is achieved with the greatest pressure differential and that it drops to 0 when there is no pressure difference. But maximum flow rate does not correspond to the maximum power generation because the change in enthalpy is 0. Maximum power generation is achieved when there is a balance between flow rate and enthalpy change.

TABLE 10.2

Relationship between Power Generation, Pressure Ratio, and Flow Rate for the Assumed Conditions (P_{cl} = 30 bars; m_{max} = 5.00 kg/s)

Pressure (bars)	Pressure Ratio	Flow Rate (kg/s)	Enthalpy Δ (kJ/kg)	Power (kW)
1.0	0.03	5.00	0	0.00
2.0	0.07	4.99	124.2	619.62
4.0	0.13	4.96	244.2	1210.10
6.0	0.20	4.90	314.2	1539.26
8.0	0.27	4.82	359.2	1730.96
10.0	0.33	4.71	399.2	1881.85
12.0	0.40	4.58	427.2	1957.68
14.0	0.47	4.42	454.2	2008.55
16.0	0.53	4.23	469.2	2005.64
18.0	0.60	4.00	499.2	1996.80
20.0	0.67	3.73	509.2	1897.68
22.0	0.73	3.40	514.2	1764.94
24.0	0.80	3.00	534.2	1602.60
26.0	0.87	2.49	544.2	1357.47
28.0	0.93	1.80	559.2	1000.20
30.0	1.00	0.00	566.2	0.00

Because both the flow rate and the enthalpy change are dependent on the pressure difference, the turbine power output must be computed in a way that takes into account these interdependencies. At each value of P, the enthalpy drop will be

$$\Delta H = H_1 - H^*{}_P$$

where:
H_1 is the wellhead enthalpy (2804 kJ/kg)
$H^*{}_P$ is determined by first noting that the enthalpy of the entering fluid is fixed at 2804 kJ/kg

Because we are considering an isentropic turbine, the entropy at the pressure of interest (P) will be the same for inlet and outlet conditions. Hence, the enthalpy at the condenser pressure will be that at which the entropy is the same as that of the incoming fluid at P. Determining this enthalpy is demonstrated diagrammatically using a simplified enthalpy–entropy diagram for steam (a so-called Mollier diagram) in Figure 10.4. Such a diagram gives the enthalpy and entropy of a fluid as a

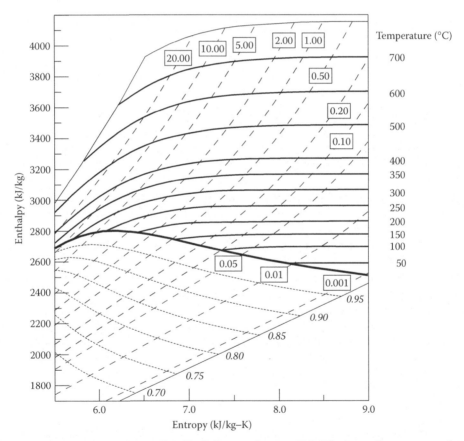

FIGURE 10.4 An enthalpy–entropy "Mollier" diagram for pure H_2O. The heavy line separates the vapor region (above) from the two-phase region (below). Temperature isotherms are shown by the solid lines labeled by temperature on the right; pressure isobars are shown by the long dashed lines with the pressure indicated by the boxed labels (MPa); steam fraction contours are the finely dashed lines in the two-phase region, labeled with italicized stream fraction values. The enthalpy and entropy of the fluid can be found at the intersection of the corresponding pressure and temperature contours for a given set of conditions. In the two-phase region, the percentage of steam can also be determined. (Modified from a diagram developed by Aartun, I., 2001, http://www.nt.ntnu.no/users/haugwarb/Phase_diagrams_and_thermodynamic_tables/PhaseDiagrams/Water.PDF.)

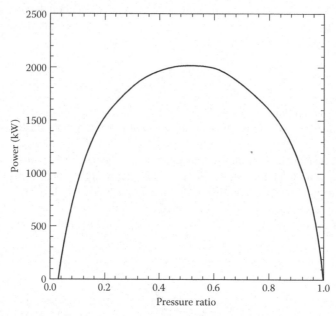

FIGURE 10.5 The relationship between wellhead closed-in pressure and operating pressure and the power output of a generator, as a function of mass flow rate.

function of pressure and temperature and, for the two-phase region, the proportion of vapor to liquid. Such diagrams are calculated using the thermodynamic properties of the fluid being considered and can be either generated using commercially available software or obtained from engineering companies.

The resulting variation in power output, as a function of the pressure ratio between the closed-in pressure and the operating pressure is shown in Figure 10.5. This figure illustrates the importance of assessing the relationship between mass flow rate and the thermodynamic behavior of the entire geothermal resource–generator complex to assure the best performance of a particular facility. As is evident from Figure 10.5, maximum power output is a compromise between maximum fluid flow and the energy available to drive the turbine in a facility and will depend upon the specific conditions at a generating plant.

It is important to remember that the presentation here has relied on the behavior of an idealized turbine that operates isentropically. Engineered systems in the real world will be only approximately represented by this kind of analysis. To accurately evaluate the performance of an engineered facility requires analytical approaches that more realistically represent processes in the actual turbine. An extensive and robust literature is available on this topic; see the references listed at the end of this chapter.

A variety of operational and engineering constraints, beyond the basic thermodynamic limitations of natural systems, must also be evaluated when determining the overall performance of a generating system. Generating facilities invariably have power demands for fluid pumping, monitoring and control, lighting, environmental mitigation, facility power, and the other needs that are parasitic to the generating capacity. These loads will diminish output to the transmission system and must be rigorously catalogued and quantified in order to design and construct efficient facilities that are economically viable.

Although rare, useful dry steam systems are a resource that can provide significant power generation. For example, at The Geysers in California (Figure 10.6), the currently installed capacity is approximately 1400 MW, of which 933 MW of capacity is in operation, making it the world's largest geothermal power generation site, with additional generating capacity currently under development.

FIGURE 10.6 Geothermal dry steam generating facilities at The Geysers, California. The steam in the photograph is from the cooling towers of several generators. The generators are in the buildings adjacent to the cooling towers. (US Geological Survey photograph by Julie Donnelly-Nolan.)

Larderello, in Italy, is the only other operating dry steam facility in the world. An additional dry steam resource has been found in the Republic of Indonesia but has yet to be developed. Initial flow tests of the Indonesian site have produced flow rates of about 3.78 kg/s (Muraoka 2003).

HYDROTHERMAL SYSTEMS

Most geothermal systems currently producing power are wet steam or *hydrothermal* systems. Hydrothermal systems have the common characteristic that their temperature and, hence, enthalpy conditions are on the low enthalpy, liquid-dominated side of the critical point in an enthalpy–pressure diagram (Figure 10.7). As such fluids ascend from depth, they will flash to steam. Whether they flash in the well or on their way to the turbine is a matter of engineering and operational decisions. The pressure and temperature conditions that are most commonly encountered for hydrothermal geothermal resources are enclosed within the shaded region in Figure 10.7.

For illustration purposes, we will consider a geothermal reservoir at 200 bars pressure and 235°C (point 1 in Figure 10.7). The enthalpy at this condition is 1018 kJ/kg (Bowers 1995). This fluid will flash to steam once the pressure has been reduced to the point where the pressure and temperature conditions intersect the two-phase field boundary. In this case where we are considering a temperature of 235°C, the pressure at the two-phase field boundary is 30.6 bars. We will assume, as before, that from this point on the system behaves isenthalpically. In describing the behavior of this fluid as it moves up the well, we will follow the approach described by DiPippo (2008).

The movement of the fluid into and up a well pipe, which is schematically represented in Figure 10.8, is constrained to follow the law of conservation of momentum, which is a fundamental attribute of systems within which hydraulic flow occurs. The momentum equation is

$$m \times a = \sum F_{fl} = -dP - \left(\frac{dF_b}{A}\right) - (\rho \times g \times h) \qquad (10.7)$$

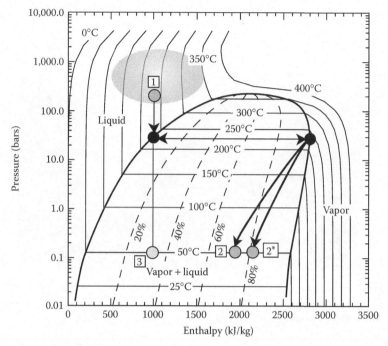

FIGURE 10.7 Pressure–enthalpy diagram for a representative hydrothermal system. Most hydrothermal systems have reservoir conditions enclosed by the shaded field.

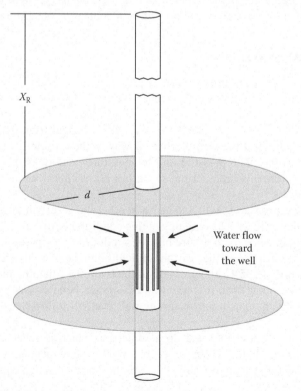

FIGURE 10.8 Schematic representation of the flow field and geometrical relationships for a producing well. The indicated dimensions X_R and d are the depth to the top of the geothermal reservoir and the distance being considered from the well edge, respectively. The shaded regions diagrammatically represent the top and bottom of the reservoir.

where:

m is the mass of the fluid

a is the acceleration as the fluid moves up the well

ΣF_{fl} is the sum of all of the forces acting on the fluid, which resolve into the change in pressure over the depth of the well (*dP*), the frictional forces acting on the fluid in contact with the pipe over the area of flow (*dF_b/A*), and the force due to the overlying column of fluid of density ρ and thickness *h* (*g* is the gravitational constant)

All of these forces are acting in the direction opposite to flow and are therefore given a negative sign.

For liquid only flow, the density is approximately constant. In addition, because water is virtually an incompressible fluid under these conditions, there is no acceleration because flow must conserve both mass and volume. As a result, Equation 10.7 can be rearranged to give the pressure difference between the top and bottom of the well as

$$\Delta P = \left[\frac{(2 \times f \times \rho \times v^2 \times X_R)}{D} \right] + (g \times \rho \times X_R) \quad (10.8)$$

where:

ΔP is the pressure difference between the bottom of the well and the wellhead (MPa)

f is a friction factor

ρ is the fluid density (kg/m³)

v is the velocity of the fluid (m/s)

X_R is the distance from the ground surface to the top of the reservoir (m)

D is the diameter of the well (m)

The friction factor depends upon the roughness of the steel pipe used in the well (Ω), the well diameter, and the Reynolds number (*Re*) which is a dimensionless number that represents the ratio between inertial forces and viscous forces:

$$Re = \frac{(4 \times m_v)}{(\mu \times \pi \times D)} \quad (10.9)$$

where:

m_v is the mass flow rate (kg/s)

μ is the absolute viscosity [kg/(m-s)]

The friction factor is then calculated using the equation

$$f = 0.25 \left[\frac{1}{\left(\log \left\{ [(\Omega/D)/3.7] + \left(5.74/Re^{0.9} \right) \right\} \right)^2} \right] \quad (10.10)$$

Hence, if we assume for purposes of illustration that the pressure is 20 MPa (200 bars) at the top of the reservoir, the fluid density is 837 kg/m³ at 235°C and 20 MPa, the fluid flow rate is 2.0 m/s, the pipe diameter is 0.2 m, the distance to the reservoir top is 2000 m, then the pressure difference from the bottom to the top of the well is

$$\Delta P = 17.5 \text{ MPa}$$

FLASHING

However, assuming that the fluid will remain liquid during ascent of the well is not necessarily a good assumption. As the fluid moves up the well, the physical conditions change sufficiently that, under some circumstances, the fluid may flash while still in the well. The conditions under which that will happen need to be considered in order to understand the processes that must be accounted for when designing and managing a generating facility.

To analyze this situation, a number of characteristics of the system must be established, because they influence how the pressure will change in the well as the fluid moves upward. One of the factors that must be considered is the extent to which pumping of fluid from the well will affect the pressure at the bottom of the well. Fluid movement from the reservoir into the well is constrained by the physical properties of the reservoir, particularly its permeability. As a result, the pressure at the bottom of the well will be less than that in the geothermal reservoir. The magnitude of this effect is expressed by the *draw-down coefficient* (C_D), which must be empirically determined for each well. It is computed from the relationship

$$C_D = \frac{(P_R - P_B)}{m_v} \tag{10.11}$$

where:
   P_R is the reservoir pressure
   P_B is the bottom hole pressure
   m_v is the mass flow rate

P_R, P_B, and m_v are determined by conducting flow tests of the well.

Once the conditions in the reservoir are known, the pressure at which the geothermal fluid will flash to steam can be readily determined from data on the thermodynamic properties of water. For the systems we are considering (water temperature of 235°C), the pressure at which flashing will occur (P_f) is 3.06 MPa (30.6 bars). The distance above the top of the reservoir at which flashing will occur (X_f; also known as the *flash horizon*) can then be computed using

$$X_f = \frac{(P_R - P_f - C_D \times m)}{(\rho \times g) + (\Gamma \times m^2)} \tag{10.12}$$

and Γ is defined as

$$\Gamma \equiv \frac{(32 \times f)}{(\rho \times \pi^2 \times D^5)} \tag{10.13}$$

The depth below the surface is then computed as $X_R - X_f$. Figure 10.9 shows how the flash horizon depth varies as a function of well diameter and mass flow rate.

The relationship between well diameter and the flash horizon is intuitively reasonable. Changes in well diameter affect the factor Γ through an inverse relationship to the fifth power. Hence, increases in well diameter will strongly diminish Γ, resulting in an increase of X_f and a decrease in the depth below the surface. The effects from increasing fluid velocity at a fixed well diameter are more complex, in that fluid velocity is a factor in determining the magnitude of the Reynolds number, draw-down coefficient, and friction factors, all of which figure directly in the flash horizon expression. The net effect is that increasing the fluid velocity increases the mass flux, resulting in an increase in the depth at which flashing will occur.

Whether flashing will occur in the well or between the wellhead and the turbine is determined by the site characteristics and the management strategy for the reservoir. Particularly important in this regard is the rate of pumping and its influence on reservoir quality and lifetime. In either case, however, the quality of steam that is available for useful work strongly influences the overall generating capacity and will be an important consideration when determining how best to manage the well flow and flashing conditions.

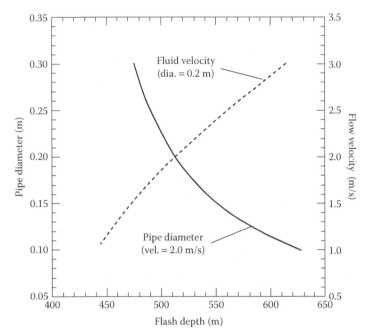

FIGURE 10.9 The dependency of the flash depth on the well diameter and flow rate. The curves are calculated allowing only the indicated variable to evolve.

STEAM QUALITY

One of the important factors influencing the performance of a generating system is the quality of the steam that enters the turbine. Steam quality is defined as the proportion of water vapor to liquid water in the steam phase. In Figure 10.7, the ratio of vapor-to-liquid in the two-phase region is contoured by the dashed lines. As previously noted, Baumann (1921) found that every percent increase in moisture content of the vapor resulted in an approximately half percent drop in efficiency. This raises an important consideration.

Remembering that the process we have been analyzing is an idealized isenthalpic system, it is clear that we will maximize the amount of steam in our system if we reduce the temperature and pressure as much as possible. We will assume that the final state will be 50°C (point 2 in Figure 10.7). Recall that for isentropic conditions the steam enthalpy would be 1980 kJ/kg and the realized enthalpy will be 2166 kJ/kg, as computed from Equation 10.5. From mass balance relationships discussed earlier in Chapter 3, we find that about 33% of the fluid mass is converted to steam (point 3 in Figure 10.7). Hence, for every kilogram of steam that is generated during flashing, we must extract 3 kg of liquid from the reservoir. Because each kilogram of liquid has an enthalpy of 1085 kJ/kg, we must extract 3245 kJ from the reservoir to obtain 638 kJ of work from the available steam. As noted previously, moisture reduces turbine efficiency, and yet in this situation, two-thirds of the mass of the system remains as liquid water. If this were to enter the turbine, the efficiency of our energy extraction process would be diminished by about 30%, which would be an unacceptable loss in generating capacity.

It is crucial to have the ability to separate liquid from vapor in any hydrothermal resource. For this purpose, a cyclone separator (Figure 10.10) is placed between the turbine and the wellhead. The high-pressure mixed vapor–liquid phase that exits the wellhead is piped to the separator where, through a combination of gravity and centrifugal effects, liquid separates from the vapor. The velocity at the inlet pipe is commonly maintained between 25 and 40 m/s (Lazalde-Crabtree 1984), which assures sufficient velocity for efficient separation. The liquid droplets collect on the interior

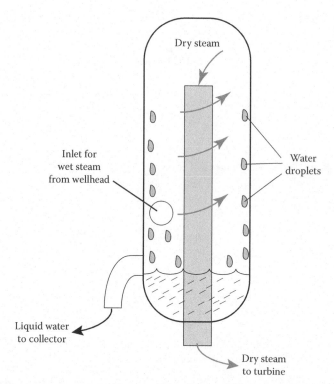

FIGURE 10.10 Schematic representation of a cyclone separator. Mixed vapor and liquid enter at high velocity and flow through the open volume. Centrifugal effects cause the liquid to impact the interior surface and flow down the sides of the separator, collecting and draining at the bottom. The resulting dry steam exits through the high stand pipe and is piped to the turbine.

walls of the separator and flow to a collection basin at the bottom of the collector, from which an outlet pipe allows drainage to a collection facility. Dry vapor exits through a high standing pipe in the interior of the vessel and passes to the steam inlet of the turbine.

From this point on, the performance of the turbine generator system is the same as described for the dry steam resource. The only difference overall remains is the energy available per unit mass of fluid that is extracted from the well, as noted above.

DUAL-FLASH SYSTEMS

An innovation that allowed more complete extraction of energy from geothermal fluids was the development of the dual-flash power plant. Although such plants are more complex, they can be economically justified in many instances because the increased cost of development, operation, and maintenance is more than offset by the income provided by the increased power production.

Dual-flash plants allow 20%–30% more energy extraction than single-flash plants. The thermodynamics of the process are shown in Figure 10.11. The initial step in a dual-flash steam expansion follows the same process as that in a single-flash plant. Beginning at the extraction point from the geothermal reservoir (point 1 in Figure 10.11), the geothermal fluid flashes as it is brought to the wellhead, passes through a cyclone separator, and the separated vapor is introduced to the turbine. The overall path followed is that from point 1 to A and A* (Figure 10.11), which is the initial separation of vapor and liquid. From A*, the vapor expands through the turbine and cools while performing work on the turbine. In a single-flash plant, the turbine stages are designed to extract as much work from the expanding vapor as possible before the vapor exits through the turbine exhaust at

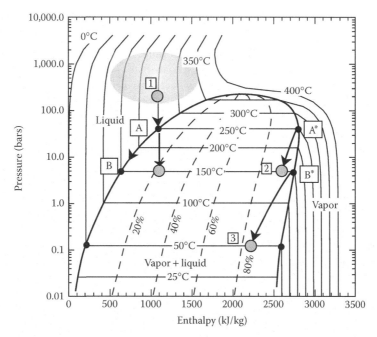

FIGURE 10.11 Pressure–enthalpy relationships for a dual-flash facility. Fluid extracted from the reservoir (point 1, 250°C) is separated into steam that is used for the first-stage flash power generation (points A* to 2). The liquid resulting from that process is separated into a steam and liquid fraction and the resulting steam is used in a second stage of power generation (points B to B* to 3).

a relatively low temperature, somewhere between 50°C and 100°C. Hot water that has condensed from the vapor during the expansion process is removed from the turbine at various points and recycled through either the cooling system or reinjected to the reservoir. Although this process provides substantial energy for power generation, the heat that is in the condensed liquid that separates from the vapor during expansion is lost.

In a dual-flash plant, a high-pressure turbine specifically designed to operate within a relatively narrow pressure and temperature interval is worked upon by the vapor, which is cooled by a relatively modest amount (path A* to 2), perhaps down to 140°C to 160°C. Hot water that has separated from the vapor (A to B) is collected and flashed at lower pressure (point B). This additional vapor is either introduced to a different turbine specifically designed to operate at high efficiency at lower pressure and temperature conditions or introduced to a different part of the turbine that is similarly designed to operate efficiently at lower pressure and temperature. The primary advantage of this dual-flash process is the recovery of a significant fraction of the heat that otherwise would have been lost in the water that was discharged from the turbine complex. The end state for this process would be point 3.

Triple-flash systems have recently been developed and deployed. The turbines in these systems follow the same concept of sequential, incremental extraction of energy as dual-flash systems, but do so using three stages of fluid separation and expansion rather than two.

END STATE: CONDENSERS AND COOLING TOWERS

As noted previously in Chapter 3, the thermodynamic efficiency of a system is ultimately determined by the magnitude of the temperature difference between the initial and final state of the fluid. For the generating systems discussed thus far, an important element for establishing high efficiencies is an effective cooling process that will maximize the temperature drop between the turbine inlet and the exhaust system of the generating plant. To accomplish cooling, the steam that exits the turbine generally flows through a condenser in which cooling water extracts a sufficient amount of heat from the steam to cause the steam to condense to liquid water. This change in temperature also drops the pressure of the system at the exit, which contributes importantly to the overall variation in state variables (pressure and temperature) across the entire turbine flow field. In addition, the pressure in the condenser is maintained mechanically.

Cooling is often accomplished by spraying water into the flowing steam stream from the turbine and collecting the resulting liquid. The mass flow of cooling water needed to sufficiently cool the incoming steam to the point it will condense must be sufficient to decrease the steam enthalpy to that of water on the liquid saturation curve at the target temperature for the system. From our previous discussions, we considered an end-point temperature of 25°C, for which the enthalpy of liquid water is 104.9 kJ/kg. If we assume that the temperature of the steam as it leaves the turbine is 50°C, the enthalpy of the steam when it leaves the turbine is 2592 kJ/kg. This requires that 2487.1 kJ/kg be removed from the steam coming into the condenser to condense all of it to liquid water. The heat capacity of water is approximately 4.2 kJ/kg-K for the conditions we are considering. If we assume the flow rate from the turbine is 2.5 kg/s, the mass flow rate of cooling water can be calculated from the relationship

$$m_{cw} = m_{te} \times \left(\frac{\Delta H}{C_{Pw}} \times \Delta T \right) \tag{10.14}$$

where:

m_{te} is the mass flow rate of vapor from the turbine
ΔH is the required change in enthalpy to achieve the desired end state
C_{Pw} is the heat capacity of water
ΔT is the change in temperature required to reach the end-state condition

For our case, m_{cw} must be

$$m_{cw} = 2.5 \text{ kg/s} \times \left[\frac{2487.1 \text{ kJ/kg}}{(4.2 \text{ kJ/kg} - K \times 25 \text{ K})} \right] = 59.2 \text{ kg/s}$$

This calculated value represents the maximum that would be needed if the cooling cycle finished at the condensing stage. In fact, the condenser is needed simply to convert steam to liquid, in order to drop the pressure across the turbine by reducing the fluid volume from that of steam to that of liquid. The liquid water that usually is condensed in the condenser will therefore not be at the end-state temperature of 25°C, but at some other higher design temperature. In addition, although converting 100% of the steam to liquid would provide the greatest efficiency, conversions of 80%–95% would still provide a sufficiently large pressure change. Realistically, under these operating conditions, m_{cw} will be a small fraction of the 59.2 kg/s. The actual value will depend on the operating conditions of the condenser, such as its pressure and the design temperature for the condensed water.

Following the condenser phase, the warm water from the condenser is cooled to the end-state condition by passage through a cooling tower. In the cooling tower, the liquid water is sprayed into a volume of moving air, allowing for evaporative cooling. In this process, approximately half to three-quarters of the mass of fluid supplied to the turbine is transferred to the atmosphere.

Regardless of the actual amount of water used in the condensing phase and during evaporative cooling, this result is significant because it emphasizes the importance of treating water as an additional resource requiring careful management. This result also emphasizes that, in order to achieve high efficiency in the generation process, a significant amount of cooling water is needed. We will return to this point when we discuss the environmental issues associated with the utilization of geothermal energy in Chapter 14.

BINARY GENERATION FACILITIES: ORGANIC RANKINE CYCLE

Up to this point in this chapter, we have discussed power generation strategies that are currently employed to produce power directly from geothermal fluids powering steam turbines. However, at temperatures below about 150°C–180°C, the power that can be extracted from such systems is limited by the efficiencies that can be achieved (remembering, again the importance of the temperature difference between the resource and the end state) and the energy density of the fluid, that is, the number of joules per kilogram of fluid that can be extracted. Although there are numerous sites around the globe that have hot water that is readily available in the temperature range of 90°C–180°C, converting that thermal energy to electricity was not possible until the 1960s.

In 1961, Harry Zvi Tabor and Lucien Bronicki developed a method for utilizing a low-boiling temperature organic fluid as the working medium to power turbines for electrical generation. Shown in Figure 10.12 is a schematic of a basic binary system. Using a high-efficiency heat exchanger, the heat from a geothermal fluid is transferred to a fluid with a low boiling point. This heating process vaporizes the fluid and the resulting vapor is used to power a turbine using the thermodynamics of an expansion cycle. Thus, the basic concepts that apply in a binary electrical generating facility for powering the turbine are very similar to that in a steam turbine.

The fluid that is used to power the turbine must be one that not only has a boiling point significantly below that of water, but also is relatively nonreactive with those materials used in the piping and turbine system of a generating plant. One commonly used liquid is isopentane (C_5H_{12}) that has a boiling point of 28°C, a heat capacity of about 2.3 kJ/kg-K (Guthrie and Huffman 1943) and a heat of vaporization of about 344 kJ/kg (Schuler et al. 2001). Geothermal fluids with moderate temperatures are thus capable of boiling relatively large masses of isopentane.

In Figure 10.12, the values that are shown for the temperatures along the flow path are more or less typical of binary plants that use isopentane. These values are useful for evaluating the energetics of such systems. For the flow rate and temperatures shown, and using the constant pressure heat capacity for isopentane, we can calculate approximately the power transferred to the turbine. The enthalpy drop per unit time across the turbine during the expansion is given by

$$\Delta H = m \times C_p \times \Delta T \tag{10.15}$$

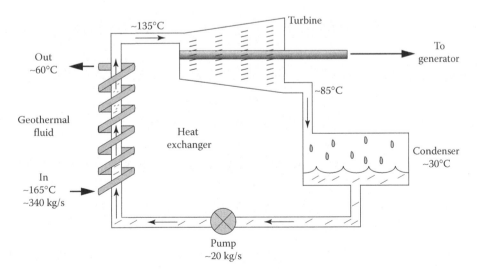

FIGURE 10.12 Schematic of a basic binary geothermal power plant. The values shown for each part of the system are generalized. (Data from Kanoglu, M., *Geothermics*, 31, 709–724, 2002.)

For the conditions in the figure, this becomes

$$\Delta H = 20 \text{ kg/s} \times 2.29 \text{ kJ/kg-}K \times 50 \text{ K} = 2288.4 \text{ kJ/s} = \sim 2.3 \text{ MW}$$

For the efficiencies of turbines used previously (85%), the maximum possible power output for this system would be about 1.95 MW. This example provides an indication of the size of generating facility that can be built for systems with these characteristics. By combining a variety of engineering enhancements and coupling together several units with generating capacities similar to that calculated here, facilities that produce 10–20 times that amount of power can be developed. Indeed, binary plants with power outputs from 2 MW to more than 50 MW are common. Larger generating facilities are also in place.

Binary generating facilities that are intended to provide power to local industries, facilities, and communities are also available. A pioneering application of small binary power plants has successfully been developed at Chena Hot Springs Resort in Alaska (Erkan et al. 2007). Two 210-kW generators have been installed using the local hot spring resource, providing low cost power to a relatively remote setting. At Oregon Institute of Technology in Klamath Falls, Oregon, a single unit has been installed that provides 280 kW (gross power output) from a single binary generator. Figure 10.13 shows the arrangement of this particular generator and its dimensions.

An important environmental aspect of binary plants is their absence of emissions. The flow path of the fluid powering the turbine forms a closed loop (Figure 10.12). In other words, the isopentane or other fluid continuously cycles from condenser to heat exchanger to turbine and back to the condenser. In addition, the fluid from the geothermal reservoir is pumped through the heat exchanger and back into the reservoir, without any venting. Unlike flash plants that inevitably exhaust some proportion of the geothermal steam to the atmosphere during the cooling and condensation process, the closed loop design of a binary plant results in no atmospheric discharge during the equivalent part of the cycle. In addition, because the boiling point of the fluid in the closed loop is so low, air

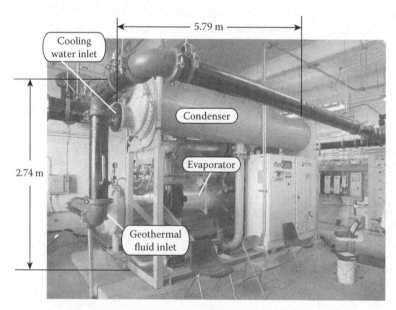

FIGURE 10.13 A 280-kW binary power system as installed at the Oregon Institute of Technology. The length dimension is from geothermal fluid inlet to geothermal fluid outlet (not shown). The height dimension is from the base of the unit to the center point for the inlet for the cooling water. The maximum width of the unit is 2.29 m. Total unit weight is 12,519 kg. The power specification is for the gross power output. The net output is between 225 and 260 kW and depends on the ΔT between the incoming geothermal fluid and the cooling water. This particular unit is produced by Pratt & Whitney. (Photograph by the author.)

cooling often is sufficient to accomplish the necessary cooling and condensation. Hence, binary plants produce no atmospheric discharge.

The absence of water cooling, however, makes binary plants susceptible to changes in ΔT due to the diurnal and seasonal fluctuation of the air temperature. In locations where the outside air temperature remains relatively low, such as during the winter months in most parts of the United States, northern Europe, Canada, and other moderate- to high-latitude settings, binary plants can generate power close to their rated outputs. During the summer months, however, high midday temperatures reduce the efficiency of the plant by reducing the ΔT. The result is a reduction in generating capacity by as much as 20%. Research efforts are focusing on development of cooling systems that can overcome such effects by allowing for a scaled cooling capacity that fluctuates with the ambient air temperature and humidity, thus allowing maintenance of a near-constant ΔT.

SYNOPSIS

Generating electrical power from geothermal resources requires no fuel while providing true base-load energy at a capacity factor that is consistently above 0.9. Recent engineering and operational strategy developments also allow geothermal power to be generated flexibly. Extracting geothermal energy requires the ability to efficiently transfer heat from a geothermal reservoir to a power plant where the heat is converted to electrical energy. This process is accomplished using production wells that penetrate hundreds to thousands of meters into the subsurface, providing a flow path for the reservoir fluid to ascend from depth. Injection wells are used to replenish the reservoir by recycling condensed water, supplementing it (if needed) with other water sources. Reservoirs are of several types. Dry steam reservoirs have sufficient enthalpy to vaporize all available water. Such systems are the simplest to engineer and have the highest energy availability of all geothermal resources but they are geologically uncommon. Hydrothermal systems are more common. Such systems usually possess sufficient heat (temperatures in the range of about 160°C–250°C) at elevated pressures to allow water to flash to steam as it ascends the wellbore and approaches the turbine. Power is generated in dry steam and hydrothermal power generating plants by the expansion of the steam as it expands and cools as it traverses through a turbine facility designed to extract as much energy from the fluid as efficiently as possible. Several designs have been developed to take advantage of the steam enthalpy, including single-flash and dual-flash plants and multistage turbines. Lower temperature geothermal resources can be used to generate power using binary plants that employ a working fluid (usually an organic compound such as isopentane or propane, or an ammonia–water solution) that has a boiling temperature significantly below that of water. In a binary plant, the geothermal water flows through a heat exchanger, transferring its heat to the working fluid, and is then reinjected into the reservoir. Binary plants are becoming the fastest growing part of the geothermal energy market. They have the advantage of requiring lower temperature resources, they emit no gases to the atmosphere, and they can be built in modular form.

CASE STUDY: THE GEYSERS

As a case study of geothermal power production, we will examine The Geysers in northwestern California. Although The Geysers is one of the rare dry steam fields and cannot be considered a typical geothermal resource for that reason, its development history highlights important issues that affect many geothermal power operations, whether the resource is a dry steam system or a hydrothermal system. It is from that perspective that we will use this system for study.

Geology

The Geysers is located in a complex geological environment that remains the topic of considerable research. The Geysers region lies just south of the landfall of the Mendocino transform fault, which is part of a triple junction plate boundary involving the Cascadia subduction zone off the coast of Washington and Oregon, and the San Andreas Fault (Figure 10.14). It has been argued that this triple junction has migrated northward from Baja California to its present position over the last ~25 million years. As it migrated, it allowed the mantle to come in contact with the overlying continent without an intervening slab of subducting oceanic crust (Furlong et al. 1989) at the edge of this

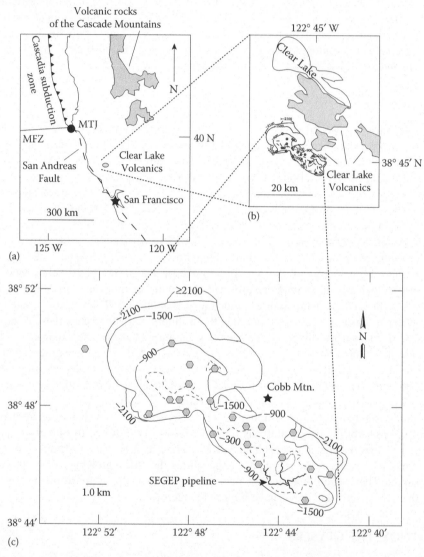

FIGURE 10.14 Geologic setting and depth to the top of the steam horizon at The Geysers, California. (a) Regional tectonic framework showing the location of The Geysers relative to the Cascade Mountains, the San Andreas fault, the Mendocino fracture zone (MFZ), the Cascadia subduction zone, and the Mendocino triple junction (MTJ). (b) The location of the steam field relative to the surface location of the Clear Lake Volcanics, which are volcanic rocks erupted within the last 700,000 years. (c) Contour map showing the depth to the top of the steam reservoir (dashed contour is the −300 m depth). Elevations are in meters relative to sea level. Also shown are the locations of the power plants (shaded hexagons) and the SEGEP pipeline. (Data sources from Geothermal Resources Council. *Monograph on The Geysers geothermal field*, Special Report No. 17, 1992; Lawrence Berkeley National Laboratory, Calpine, Inc., and Northern California Power Agency, Integrated high resolution microearthquake analysis and monitoring for optimizing steam production at The Geysers geothermal field, California, California Energy Commission, Geothermal Resources Development Account Final Report for Grant Agreement GEO-00-003, 41 pp., 2004.)

migrating system. The window allowed hot mantle to interact with crustal rocks at a relatively shallow level, resulting in high heat flow and the formation of a string of volcanic centers that are progressively younger as one proceeds from south to north in California.

This tectonic framework instigated the development of the thermal anomaly, of which The Geysers is a manifestation. Today, the region is the site of very recent volcanic activity that formed eruption centers as recently as 10,000 years ago (Donnelly-Nolan et al. 1981; Hearn et al. 1995). Heat flow in the area is about 500 mW/m^2 (Walters and Coombs 1989), which is about 60 times the global average. Steaming ground and hot springs were present in the area when it was first considered for development in the early 1900s. These surface manifestations require the presence of magma bodies at a relatively shallow depth below the surface. Modeling of the thermal structure and evolution of the area, along with analysis of fluid inclusions in rocks from the region, suggest that magma bodies that have solidified are the primary heat source, existing at depths as shallow as 3 km (Dalrymple et al. 1999).

The primary geothermal resource in the region is a subsurface zone of dry steam that is trapped below a very low permeability rock mass. The process that formed this system is not well understood, but the basic elements of the system can be described. Shallow-level heat sources commonly cause the development of vigorously convecting groundwater environments. The combination of hot groundwater and freshly erupted volcanic rocks can readily result in extensive recrystallization and alteration of the original rock. Such processes commonly lead to sealed fractures and clogged pore spaces in the rock, with the ultimate consequence that rocks that were once permeable become much less permeable, reducing or eliminating fluid flow pathways. Such a process likely occurred relatively quickly in the subsurface above the primary heat sources in rocks that probably had low intrinsic permeability when they were first emplaced. Once the fluid flow pathways were reduced or eliminated, convection of the hot fluid upward would be reduced. Because convection is an effective heat transfer process, sealing of the fluid flow pathways would effectively trap heat above the heat source. With sufficient energy input, the groundwater would boil and vaporize in the subsurface. The impermeable barrier of rock that formed during water–rock interaction would then act as a cap, trapping a zone of superheated or *dry* steam in the area that is now being utilized for geothermal power production.

The volume of this system is large. Drilling has extended to well over 3000 m and no liquid zone has yet been found. This implies the total steam reservoir must be over a thousand meters thick in places. The dry steam temperature is between 225°C and 300°C, depending upon location within the field (Dalrymple et al. 1999; Moore et al. 2001; Dobson et al. 2006).

POWER GENERATION HISTORY

The first attempt to generate power using geothermal steam at The Geysers was in 1921 when a 35-kW generator was installed to supply power to a local hotel. Although power was successfully generated, the steam energy conversion system used for power production was damaged by particulates in the fluid and the chemically aggressive nature of the fluid and was abandoned after a relatively short time.

In 1961 power production was again pursued, this time with more robust technology and a better understanding of geothermal power production principles. The first plant was an 11-MW plant that was built near the site of the original 35-kW power plant. Today, there are a total of 21 power plants with an installed generating capacity of about 1400 MW. The locations of the power plants are shown in Figure 10.14c.

Drilling in the region can be a complex undertaking. On the positive side, the hard, impermeable cap rock that sealed in the resource allowed the wells to be completed without casing. These open holes reduce costs because liners or other metal pipe need not be emplaced in the hole to keep it stable. However, the terrain is rugged and the ground unstable, partly due to the geology and partly due to the presence of aggressive gases and fluids that have reacted with the rock, reducing the cohesion of the soils

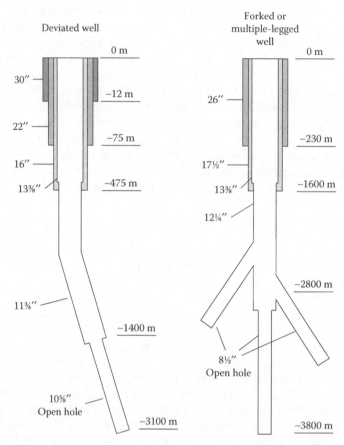

FIGURE 10.15 Deviated and branched wells. The diameters and respective depths are approximate and are generalized. (From information presented by Pye, D.S. and Hamblin, G.M., *Monograph on The Geysers Geothermal Field.* Special Report No. 17, Geothermal Resources Council, Davis, CA, 229–235, 1992 and Henneberger, R.C. et al., Advances in multiple-legged well completion methodology at The Geysers geothermal field, California, *Proceedings of the World Geothermal Congress*, vol. 2, Florence, Italy, 1403–1408, 1995.)

and shallow bedrock. These conditions have combined to make it economically expedient to utilize a drill pad located in a stable location for wells that access resources some distance to the side of the drill site. One drilling method that has been employed to tap the reservoir is *deviated* wells, which are wells that angle off in a predetermined direction – (Figure 10.15). Recently, such methods have become more common in the geothermal industry, building on a long history of such drilling methods in the oil and gas industry (see discussion in Chapter 9). Another drilling innovation that has been utilized is the branched or forked or multiple-legged well (Figure 10.15), which will have several wells deviate off of one vertical well. Such methods gather the steam from several places in the reservoir and pipe it through a single wellhead, increasing the per well production and simplifying the above-ground piping.

EMISSIONS

In the course of exploration, it was discovered that the composition of the steam system varied from place to place. Common *noncondensable* gases include CO_2, H_2S, and HCl. In the northern part of the field, especially, it was found that the HCl concentration is particularly high, causing the steam to be extremely aggressive. This acidic, low pH resource is currently not economical to develop because exotic, costly materials would be required for the piping and turbine components. As a result, the northern part of the resource has yet to be developed.

The noncondensable gases can be problematic. Emission standards for H_2S vary from region to region, depending upon local restrictions established by appropriate agencies. The Occupational Health and Safety Administration recommends that H_2S in the workplace not exceed 20 parts per million, with a maximum of 50 parts per million for 10 minutes. Typical H_2S emissions from geo-thermal power plants are less than 1 part per billion (Energy Efficiency and Renewable Energy Office of the Department of Energy, http://www1.eere.energy.gov/geothermal/geopower_cleanair.html). However, that is not the case at The Geysers, where the steam can contain up to 0.15% H_2S by weight. As a result, an abatement program was developed to remove H_2S from the emission stream. That pro-cess oxidizes H_2S to elemental sulfur with an efficiency in excess of 99.9%. The sulfur by-product that is produced in this process has become a useful commodity provided to the local agricultural industry.

Emissions of CO_2 are also important to consider, because they have an impact on greenhouse gas releases to the environment. These are discussed in detail in Chapter 15 and will not be further addressed here.

SUSTAINABILITY AND REINJECTION

The sustainability of geothermal reservoirs for power production is an issue that deserves careful con-sideration. The Geysers provide a particularly interesting example of this issue. The most basic means for establishing sustainability of a resource is to consider the historical record of energy production per well or to use some proxy for that metric. With that data, one can model the long-term behavior of the resource and establish whether it can be used in a sustainable manner. For the context within which we are considering this question, we will define sustainability as the ability to maintain power generation at some specified level for at least 20 years, which is a reasonable design lifetime for a power plant.

The history of power production varies from plant to plant, and at The Geysers, over the nearly 50-year history of operations, some plants have been dismantled or replaced. If we consider a subset of the plants, however, a clear picture emerges. Figure 10.16 shows the power production history from the entire geothermal complex and the per well energy production (Figure 10.16a). Also shown (Figure 10.16b) is the power production for five specific power plants (Calpine Power Plant Units 13, 16, 18 and NCPA Plants 1 and 2) over the period 1996 through 2007 (Khan 2007).

The rate of steam production throughout The Geysers grew from 1970 to 1983, at which point overall steam production began a long decline until 1995, when steam production stabilized. When considered on a per well basis, steam production dropped between 1970 and 1978 at which point it stabilized for a few years, then declined again between 1984 and 1995. Since 1995, production has been stable. If we assume that the initial steam properties were those of point 1 in Figure 10.3 and provided enthalpy down to a temperature of a little over 50°C (typical for turbines utilized in systems like this), the enthalpy contribution is approximately 300 kJ/kg of steam. From the peak of steam production in 1987 of about 110 billion kg of steam to 1995 when production was about 60 billion kg of steam, the average annual decline in available enthalpy would be approximately

$$\frac{(1.1 \times 10^{11}\,\text{kg} - 6.0 \times 10^{10}\,\text{kg}) \times 300\,\text{kJ/kg}}{8\,\text{years}} = 1.875 \times 10^{12}\,\text{kJ/year}$$

The loss of this energy is due to the hydrologic framework of the system. In essence, natural liquid groundwater cannot flow into the reservoir and be vaporized to steam at the same rate at which steam is being extracted. Dry steam systems have low rates of natural recharge because liquid water must flow around the impermeable zone that caps the system and then migrate into the reservoir region. In order to maintain steam production at a constant level, the steam extraction rate must be managed so that it occurs at the same rate as the natural system is recharged, or liquid water must be injected into the reservoir at a sufficient rate to produce extractable steam to balance that which is removed.

Predicting the rate at which steam can be removed during the early stages of development of a dry steam field must be based on experience rather than theoretical models, at least for the current

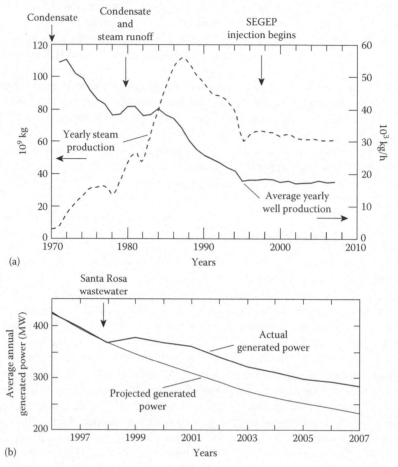

FIGURE 10.16 The effects of extraction and reinjection. (a) Total steam production, by year, for the entire field (dashed line) and the yearly per well average of steam production (solid line). The approximate times when various reinjection protocols were started are shown along the top of the figure. (b) The response of a subset of generating units in the southeast part of the field to the SEGEP injection program. (Modified from Khan, A., The Geysers geothermal field, an injection success story, Ground Water Protection Council, Annual Forum, 2007.)

state-of-the-art. This reflects the fact that the volume of the reservoir is usually unknown, the permeability of the rock is usually insufficiently characterized to portray realistic steam flow paths in three dimensions, and the actual natural recharge rate of the system is unknown. The historical record for The Geysers documents that steam was extracted from the system faster than it was being replenished during its early development, as indicated by the prolonged drop in per well steam production, even though the overall steam production was increasing. These facts made it apparent that injection of water would be a necessity if power production were to be maintained at The Geysers.

Beginning in 1969, injection of condensate from the generating complex was initiated. Wells that were marginally productive were reconfigured as injection wells and the condensate was injected through them into the reservoir. On a mass base, the injected fluid was equivalent to 25%–30% of the extracted steam (Brauner and Carlson 2002). Additional volumes of water were obtained from local streams to supplement the condensate, increasing by a few percent the total liquid mass injected into the reservoir. However, these fluid volumes were insufficient to prevent further decline in steam production.

In 1995, a project, called the South East Geysers Effluent Pipeline (SEGEP) Project, was initiated to inject treated wastewater from the city of Santa Rosa, 26 miles to the east, into the steam field. A pipeline (see Figure 10.14 for location of the pipeline) was constructed that was capable of providing 29.5 million liters (7.8 million gallons) a day to the southern part of the field for injection into a complex of 7–10 wells (Brauner and Carlson 2002). As is clear from Figure 10.16, wastewater injection stabilized steam production. Additional wastewater injection capability, amounting to an additional 41.6 million liters (11 million gallons) per day, is currently under development.

Water injection also had the benefit of reducing H_2S in the fluid. Prior to the injection program in 1995, the concentration of H_2S in the steam was increasing at various wells in the area. This reflected the impact of reduced dilution by steam in the reservoir. However, within a few months of initiating the injection of wastewater, the H_2S concentration dropped by nearly 20% due to the dilution of the noncondensable gas component in the reservoir.

It is important to note, however, that there is not a simple relationship between the mass of water injected and the mass of steam recovered. The total rate of water injection into The Geysers in 1999 was approximately 3.83 million kg/h, most of which was distributed among a few wells in the southeastern area. At the same time, approximately 6.6 million kg/h of steam was produced from the entire field. In other words, a mass equivalent to 56% of the total mass of steam produced in the entire field was injected into an area accounting for less than 20% of the geothermal resource area. Such a situation suggests that the subsurface flow paths of water and steam must be complex.

Figure 10.17 provides a graphic demonstration of the subsurface behavior of fluid injected into the system. The figure shows the distribution of *microseismic* events around one of the injection wells. Microseismic events are short releases of elastic energy by the rock, due to a number of possible mechanisms. Cold water injected into hot rock can cause rapid cracking due to contraction,

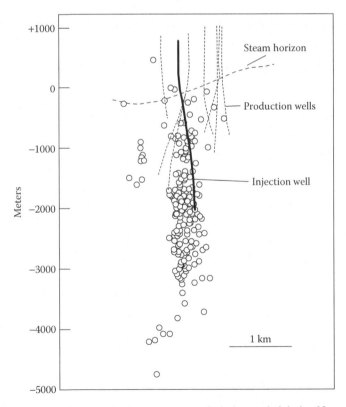

FIGURE 10.17 Distribution of microseismic events (open circles) recorded during November 2007, related to injection at well 42B-33 (solid line labeled "Injection well"). The approximate location of the steam horizon is indicated and the production wells are shown as the dashed lines. (Data are from Smith, B. et al., Induced seismicity in the SE geysers field, California, USA. *Proceedings of the World Geothermal Congress 2000*, Kyushu–Tohoku, Japan, 2887–2892, 2000.)

which can be detected by sensitive seismometers. Other detectable events that can occur in these settings are the forced opening of old fractures in the rock, formation of new fractures, and the release of preexisting stresses in the rock that develop in response to tectonic forces. Whatever the specific mechanism, the location of the microseismic events provides a picture of the possible migration path of the injected water. As is clear in the figure, much of the microseismic activity occurs well below and away from the production wells. This implies that at least some portion of the injected fluid escapes from the immediate vicinity of the volume of rock from which the production wells are accessing steam. However, the fact that steam production stabilized over the long term suggests that repressurization of the reservoir was successful. The mass balance between injected liquid and steam formation that contributed to recharging the reservoir cannot be rigorously determined, however. In addition, although production has been stabilized, the overall reservoir performance remains below that achieved during the late 1960s and early 1970s. It remains to be determined the extent to which the new injection program will bring production back to those earlier levels.

PROBLEMS

10.1. What is a dry steam resource and why is it favored over other types of geothermal resources for power generation?

10.2. Assume that steam entering a turbine is at 235°C, has a flow rate of 5 kg/s, leaves the turbine at 35°C, and has 84% efficiency. What is the power output of the turbine, assuming an ideal process?

10.3. At what pressure will the fluid in problem 10.2 flash to steam?

10.4. Assume that the pressure at the top of a reservoir 3000 m underground is 35 MPa, and it contains a fluid at 235°C. Assume that the fluid is flowing in the well at 3 m/s and that the fluid density is 850 kg/m³. At what depth in the well will the fluid flash?

10.5. How many kg/s of cooling water is necessary to cool 60°C water that is exiting a turbine to 25°C if the flow rate of the hot water is 5 kg/s?

10.6. What factors determine the sustainability of a geothermal resource? What operational strategies can be employed to enhance the sustainability?

10.7. What is a Rankine cycle? Why is it important for the use of geothermal resources for power generation?

10.8. What is the approximate minimum temperature for a useful geothermal power generation system that employs binary technology? What determines this temperature?

10.9. If the deposits on turbine blades shown in Figure 10S.3 were calcium carbonate, what would be a chemical method that could be used to prevent them from forming?

REFERENCES

Baumann, K., 1921. Some recent developments in large steam turbine practice. *Journal of the Institution of Electrical Engineers*, 59, 565–663.

Bertani, R., 2007. World geothermal energy production, 2007. *Geo-Heat Center Bulletin*, 28, 8–19.

Bolton, R.S., 2009. The early history of Wairakei (with brief notes on some unforeseen outcomes). *Geothermics*, 38, 11–29.

Bowers, T.S., 1995. Pressure-volume-temperature properties of H_2O-CO_2 fluids. In *Rock Physics and Phase Relations*, ed. T.J. Ahrens. Washington, DC: American Geophysical Union, pp. 45–72.

Brauner, E. Jr. and Carlson, D.C., 2002. Santa Rosa Geysers recharge project: GEO-98-001. California Energy Commission Report 500-02-078V1. California Energy Commission, Sacramento, CA.

Dalrymple, G.B., Grove, M., Lovera, O.M., Harrison, T.M., Hulen, J.B., and Lanphyre, M.A., 1999. Age and thermal history of The Geysers plutonic complex (felsite unit), Geysers geothermal field, California: A $^{40}Ar/^{39}Ar$ and U-Pb study. *Earth and Planetary Science Letters*, 173, 285–298.

DiPippo, R., 2008. *Geothermal Power Plants*. 2nd edition. Amsterdam, The Netherlands: Elsevier. 493 pp.

Dobson, P., Sonnenthal, E., Kennedy, M., van Soest, T., and Lewicki, J., 2006. Temporal changes in noble gas compositions within the Aidlin sector of The Geysers geothermal system. Lawrence Berkeley National Laboratory Paper LBNL-60159, p. 12.

Donnelly-Nolan, J.M., Hearn, B.C. Jr., Curtis, G.H., and Drake, R.E., 1981. Geochronolgy and evolution of the Clear Lake Volcanics. In *Research in The Geysers-Clear Lake Geothermal Area, Northern California*, eds. R.J. McLaughlin and J.M. Donnelly-Nolan. Washington, DC: US Geological Survey, pp. 47–60.

Erkan, K., Holdman, G., Blackwell, D., and Benoit, W., 2007. Thermal characteristics of the Chena Hot Springs, Alaska, geothermal system. *Proceedings of the 32nd Workshop on Geothermal Reservoir Engineering*, Stanford University, Stanford, CA, January 22–24.

Furlong, K.P., Hugo, W.D., and Zandt, G., 1989. Geometry and evolution of the San Andreas Fault zone in Northern California. *Journal of Geophysical Research*, 94(B3), 3100–3110.

Guthrie, G.B. Jr. and Huffman, H.M., 1943. Thermal data XVI. The heat capacity and entropy of isopentane. The absence of a reported thermal anomaly. *Journal of the American Chemical Society*, 65, 1139–1143.

Hearn, B.C., Donnelly-Nolan, J.M., and Goff, F.E., 1995. Geologic map and structure sections of the Clear Lake volcanics, northern California. US Geological Survey Miscellaneous Investigations Map I-2362, http://pubs.usgs.gov/imap/2362/.

Henneberger, R.C., Gardner, M.C., and Chase, D., 1995. Advances in multiple-legged well completion methodology at The Geysers geothermal field, California. *Proceedings of the World Geothermal Congress*, vol. 2, Florence, Italy, pp. 1403–1408.

International Geothermal Association, 2008. http://www.google.com/url?sa=t&rct=j&q=&esrc=s&source=web&cd=1&ved=0CB8QFjAA&url=http%3A%2F%2Fwww.earth-policy.org%2Fdatacenter%2Fxls%2Fupdate74_2.xls&ei=TF64U9fDGI-hyASh_YDoAw&usg=AFQjCNGjxc8UEE6DK4TZkSvA8sHfXNqPDA&bvm=bv.70138588,d.aWw.

International Geothermal Association, 2008, 2012. *IGA Newsletter*, v. 89, p.3, IGA Global Geothermal Energy Database.

Kanoglu, M., 2002. Exergy analysis of a dual-level binary geothermal power plant. *Geothermics*, 31, 709–724.

Khan, A., 2007. The Geysers geothermal field, an injection success story. Ground Water Protection Council, Annual Forum, http://pakistanli.com/papers/FtF.htm

Lawrence Berkeley National Laboratory, Calpine, Inc., and Northern California Power Agency, 2004. Integrated high resolution microearthquake analysis and monitoring for optimizing steam production at The Geysers geothermal field, California. California Energy Commission, Geothermal Resources Development Account Final Report for Grant Agreement GEO-00-003, 41 pp.

Lazalde-Crabtree, H., 1984. Design approach of steam-water separators and steam dryers for geothermal applications. *Geothermal Resources Council Bulletin*, 13(8), 11–20.

Matek, B., 2013. 2013 geothermal power: International market overview. Geothermal Energy Association, Washington, DC, 35 pp.

Moore, J.N., Norman, D.I., and Kennedy, B.M., 2001. Fluid inclusion gas compositions from an active magmatic–hydrothermal system: A case study of The Geysers geothermal field, USA. *Chemical Geology*, 173, 3–30.

Muraoka, H., 2003. Exploration of geothermal resources for remote islands of Indonesia. *AIST Today*, 7, 13–16.

Pye, D.S. and Hamblin, G.M., 1992. Drilling geothermal wells at The geysers field. *Monograph on The Geysers Geothermal Field*. Special Report No. 17, Geothermal Resources Council, Davis, CA, pp. 229–235.

Schuler, L.D., Daura, X., and van Gunsteren, W.F., 2001. An improved GROMOS96 force field for aliphatic hydrocarbons in the condensed phase. *Journal of Computational Chemistry*, 22, 1205–1218.

Smith, B., Beall, J., and Stark, M., 2000. Induced seismicity in the SE geysers field, California, USA. *Proceedings of the World Geothermal Congress 2000*, Kyushu–Tohoku, Japan, pp. 2887–2892.

Walters, M. and Coombs, J., 1989. Heat flow regime in The Geysers-Clear Lake area of northern California. *Geothermal Resources Council Transactions*, 13, 491–502.

FURTHER INFORMATION

Federal Energy Regulatory Commission (http://www.ferc.gov/).

    FERC provides important oversight functions for many aspects of the power grid. They maintain an important library of documents, studies, and data useful for understanding power generation and transmission issues.

Geothermal Resources Council (http://www.geothermal.org) and the International Geothermal Association (http://www.geothermal-energy.org/index.php).

The GRC and IGA are an organizations that provide information, data, and historical perspective on the geothermal industry. Both hold annual meetings at which recent research results are presented and trade organizations are available for discussion of technology, law, and regulatory issues. The GRC maintains one of the world's premier libraries at its headquarters in Davis, California.

North American Electric Reliability Corporation (http://www.nerc.com/).

NERC develops reliability standards and provides education, training, and certification for the power industry, including courses that deal with power generation and technology. NERC is a self-regulatory organization, subject to oversight by the US Federal Energy Regulatory Commission and governmental authorities in Canada. They also have online courses for all aspects of power generation technology.

SIDEBAR 10.1   Turbines

The most common means for producing electrical energy is through the use of a turbine to power an electrical generator. A turbine is a mechanical device that converts the energy of a fluid to mechanical rotational energy.

In the case of geothermal power production, the working fluid is steam. The turbine is used to convert the enthalpy of the steam to useful power. The arrangements of the components of a basic steam turbine are shown in Figures 10S.1 and 10S.2.

Energy conversion is accomplished by allowing high temperature and pressure geothermal steam to expand into a region of lower temperature and pressure that is produced using a condenser and cooling system that cools the steam. As the high P–T steam expands, it is directed to flow through a set of stationary blades (the stator) that focus the fluid flow toward a set of blades mounted on a rotating shaft (the rotator). The energy of the fluid is partially converted to rotational energy as the fluid moves over the rotating blades of the rotator. During this expansion process, the fluid transfers its thermal enthalpy to kinetic energy of the rotating turbine shaft and the pressure and temperature of the steam drop.

The efficiency of this process depends critically on the shape and dimensions of the blades. Because the enthalpy content of the fluid is constantly dropping as it moves through the turbine, blade form and size must be different along the length of the flow path in order to maximize energy transfer from the fluid to the shaft. Modern steam turbines consist of a series of stator–rotator pairs arranged in stages that are designed specifically to extract, as efficiently as possible, energy at different P–T conditions. The high-pressure stage usually has smaller diameter blade wheels than the lower pressure stages, reflecting the decreasing energy density of the fluid at lower pressures—at lower energy densities (and pressures) a greater blade area is required for efficient energy extraction. Figure 10S.1 shows the configuration for a single stage turbine while Figure 10S.2 shows the schematic relationships and flow path for a single stator–rotator pair in a turbine.

Modern steam turbine design has benefited from decades of engineering refinement and scientific analysis. The result is that today's multistage steam turbines are capable of operating at high efficiencies, commonly in excess of 85% of the theoretical maximum for extracting energy from the available steam resource.

Any modification of the flow path of the expanding steam or the blade geometry, or change in the steam properties of the incoming steam will diminish the efficiency of the turbine, because its design is optimized for specific conditions. Of particular concern in geothermal power production is the common occurrence of components other than H_2O in the steam supply. Although achieving the highest steam quality from the geothermal fluid is assiduously pursued, thermodynamics prevent the formation of pure steam from a solution that contains solutes. Regardless of how small the partitioning may be for a component between steam and liquid (see Chapter 5 for a discussion of vapor–fluid partitioning) some amount of all species in a solution will exist in the fluid phase. If the concentrations in the fluid are significant, as for example the Salton Sea brine in Table 5.1, deposition of some solid will inevitably form on the turbine blades. This is apparent if we consider the solubility of silica (Figure 5.3), which is one of the most common minerals to be deposited on turbine blades in geothermal power systems. A 5°C drop in temperature results in a decrease in the solubility concentration by a factor of approximately 0.1–5.0, depending upon the temperature interval considered. Calcium carbonate is another solute component that commonly impacts turbine efficiency via the same mechanism. For systems with typical flow rates through the turbines (1.0–5.0 kg/s), it becomes only a matter of time before turbine blade form is compromised sufficiently by deposits of minerals to reduce the efficiency of turbine performance to unacceptable levels. When that happens costly turbine overhaul is required. It is for this reason that it is important to remove

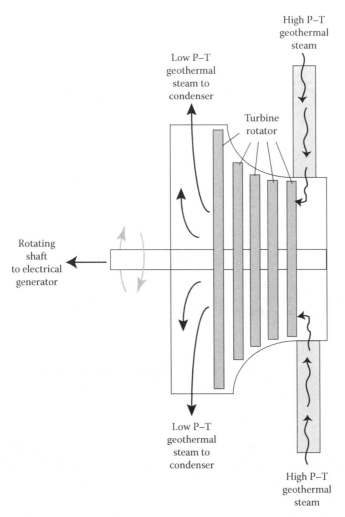

FIGURE 10S.1 Schematic of a single-stage turbine. High-pressure steam that has been piped from a cyclone separator and particulate trap enters the turbine on the right and flows through a series of rotating turbine blades. Each set of blades forms a wheel that is of increasing diameter along the flow path. The diameter of the blade wheel is a design parameter based on the expected pressure and temperature of the steam at that location along the flow path. The steam is exhausted from the turbine to a lower temperature and pressure condenser and cooling system. For multistage turbines, the steam is dehumidified and reinjected into another set of turbine blades after it is exhausted from the preceding stage before it finally enters the condenser-cooling system. The turbine blades are attached to a shaft that is directly connected to the electrical generator.

from the fluids a sufficient amount of the dissolved loads to allow the longest possible time between turbine overhauls (Figure 10S.3).

Also affecting efficiency is the presence of gas phases that do not condense. These noncondensable gases (NCG) diminish the energy that can be transferred to the turbine, which is specifically defined for the thermodynamic properties of pure steam. The principle NCGs in geothermal steam are CO_2, H_2S, and NH_4. Depending upon the chemistry of the fluid, removal of NCGs may be accomplished before the steam enters the turbine.

Steam turbines used in geothermal power plants differ from those in conventional power plants powered by fossil fuels or nuclear power because the steam properties are different. Geothermal systems provide steam at temperatures of 200°C–350°C, in contrast to the >1500°C steam in other power plants. As a result, the turbine design for the relatively low-pressure geothermal systems is optimized for these less severe conditions.

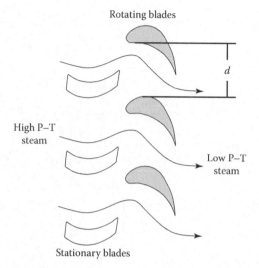

FIGURE 10S.2 Schematic geometry of the stator–rotator couple along the steam flow path within the turbine. The geometry of each set of blades is designed for specific temperature and pressure conditions as well as flow rates, to maximize extraction of energy from the expanding steam. The distance d is one of the critical parameters that influences turbine efficiency.

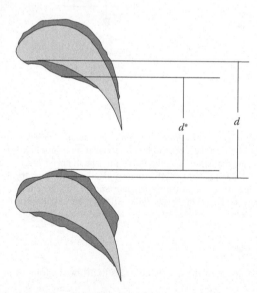

FIGURE 10S.3 Deposits (dark, irregular material coating the light gray turbine blades) on the turbine blades modify the design distance, d. The presence of the deposits adds roughness to the blades, which diminishes the smooth flow of the fluid over the shaped blade and reduces the energy input to the blade. In addition, the deposits diminish the distance d to d^*, which reduces the flow volume and hence the total rate of energy transfer.

11 Low-Temperature Geothermal Resources

Geothermal Heat Pumps

Heat pump technology is one of the most sophisticated engineering accomplishments to come out of the twentieth century. Heat pumps are simple devices that operate at the highest efficiency levels accomplished by heat-transporting systems. They are, in essence, close analogs to actual Carnot cycles, operating on precisely the same principles as that which Carnot imagined in his thought experiment. They transfer heat in ways that can accomplish both heating and cooling, while consuming a small fraction of the amount of energy they move. Their great advantage comes from the fact that they move heat that already exists, using basic thermodynamic principles and thus do not require that heat be generated. Because generating heat (through combustion, friction, or some other means) is usually inherently less efficient than moving preexisting heat, a heat-transfer engine provides efficiencies that are difficult to match through other means. In this sense, they provide the ideal method to satisfy energy demands for heating and cooling buildings and spaces. This chapter considers the principles that need to be addressed when coupling heat pumps to the earth's near-surface heat reservoir, design concepts for such systems, and issues that need to be considered to successfully implement such an application.

BASIC HEAT PUMP PRINCIPLES

Heat pumps that utilize the earth's heat energy are fluid-mediated mechanical devices that transfer heat from one location to another, relying on the thermodynamics of fluid systems. The first known use of a heat pump was a device invented by Peter Ritter von Rittenger in the mid-1850s in Austria that employed exhaust steam for heat in salt mines. But it was not until the early 1900s that heat pumps, mainly as the heart of household refrigerators, became commonplace.

In a refrigerator, a heat pump is used to remove heat from the interior air of the refrigerator (heat source) into the air of the room in which the refrigerator is located (the heat sink). This process is exactly analogous to a typical Carnot cycle, with the air of the refrigerator interior being the heat source used in the initial isothermal expansion phase of the cycle, and the room air being the heat sink into which heat is expelled from the cycle during the isothermal compression phase, as depicted in Figure 11.1. The fluid through which heat transfer occurs is the refrigerant, which has long been an organic compound with a very low boiling temperature. For building heating, ventilation, and air conditioning (HVAC) purposes geothermal energy is utilized, where the heat transfer principles and processes are the same, and the only difference being the heat source and sink.

Geothermal heat pumps (also called ground-coupled or ground-source heat pumps [GSHPs]; sometimes abbreviated as GHPs) take advantage of the immense thermal mass of the earth and its moderate, constant near-surface temperature. In regions where summer cooling is required, exterior daytime temperatures generally are above 26°C (80°F). Where winter heating is required, exterior temperatures generally are below 18°C (65°F). At most locations, the temperature in the subsurface at depths of a few tens of meters is a consistent 4°C–13°C (40°F–55°F). This consistent temperature means that the subsurface is a heat reservoir from which heat can be extracted for heating or deposited for cooling, provided efficient heat transfer between the earth and a building can be accomplished.

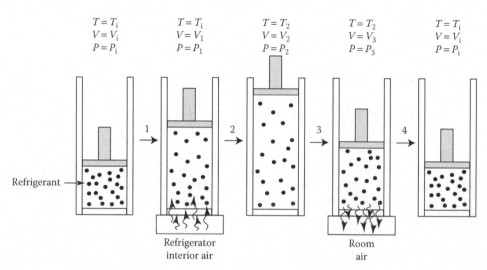

FIGURE 11.1 Schematic representation of the thermodynamic cycles used to remove air from the interior of a refrigerator and transfer it to the room within which the refrigerator is housed. This process is indistinguishable from that characteristic of a Carnot cycle, as presented in Chapter 3.

The basic configuration of an earth–building couple is depicted in Figure 11.2. The design shown is that of a vertical loop of pipe that continuously circulates a working fluid between a heat pump inside the building and the earth. Such a system is called a *closed loop* because the working fluid circulates within the loop of pipe and does not exit it. Numerous other designs have also been employed for closed-loop systems, including single and multiple horizontal closed loops that are emplaced in trenches or other excavated settings (Ochsner 2008). Open-loop designs also exist that use groundwater that is pumped through a heat pump from a water well and then reinjected back into the groundwater system. The basic principles of heat exchange of an open-loop system are the same as those of a closed-loop system. However, management of the resource is different and will be treated separately in a later section.

Figure 11.3 is a schematic representation of a heat pump coupled to the earth. The principle elements of the heat pump are a coil containing a refrigerant fluid that has a boiling point that is lower than that of the local subsurface, a compressor, a pressure reduction valve, and a capability to exchange heat with a room (left of figure) and with the earth (right of figure). The refrigerant that is used varies by manufacturer (see Table 11.1 for characteristics of some commonly used refrigerants). These are now stipulated to be nonozone-depleting compounds. An additional component required for the successful operation of this system is a pump that will circulate the water between the heat pump and the earth-coupled loop.

A complete *heating* cycle of the pump involves the path from A through D in Figure 11.3 (a *cooling* cycle could be accomplished by allowing heat deposition from the room side into the closed-loop fluid). At A, the cool liquid refrigerant passes into a heat exchanger where it acquires heat from the working fluid that has circulated through the earth's thermal reservoir. If the thermal exchange efficiency between the earth and the circulating fluid is sufficiently high, the working fluid will have a temperature close to that of the earth at the depth of the pipe in the outside borehole. Because the refrigerant has a boiling temperature substantially below that of the local subsurface, the refrigerant boils when it flows through the heat exchange coil that links it to the external fluid. As this heating of the refrigerant occurs, it goes through a phase transition from liquid to gas via boiling. At B, the gas pressure is increased by a compression pump, resulting in an increase in the gas temperature, reflecting the fact that work has been done by the compressor on the gas. The hot gas then passes through another heat exchanger in the building, where its temperature drops as it exchanges heat with the room (point C), heating the room air. The warm

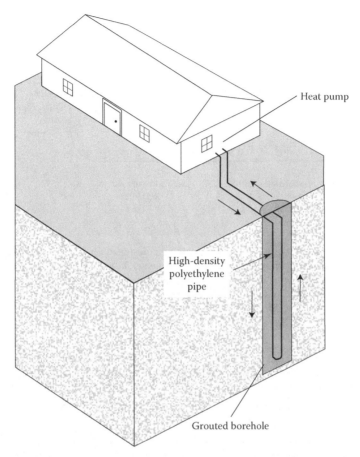

FIGURE 11.2 Schematic of a GHP plumbing system utilizing a single borehole. Although single borehole installations are common, multiborehole systems are usually required for most installations. The borehole depth is typically about 90 m, with the piping grouted in place using a cement–clay mixture that has high thermal conductivity. The heat pump and HVAC unit are typically located in a service closet in the house.

gas then passes through a pressure reduction valve (D), which results in a pressure drop and the gas condensing back to a liquid. At this point, the cycle has been completed and the working fluid is again available for heating.

THERMODYNAMICS OF HEAT PUMPS

Heat pump technology employed in ground source systems is based on the same thermodynamic principles as that employed by binary geothermal power generation systems using an organic Rankine cycle. The obvious differences between the systems are the conditions and energy densities that are utilized and the fluid used in the heat transfer process.

The nonaqueous fluids used in heat pumps have boiling points well below that of cool ground waters. In Figure 11.4, the pressure–enthalpy diagram for propane is shown, contoured for temperature. Also shown is the boundary between the liquid and liquid + vapor field for water under these pressure–enthalpy conditions, for comparison. Generally, the working fluid heated by interaction with the ground will be between 5°C and 15°C (the light gray field in Figure 11.4). The working fluid will transfer its heat to the refrigerant fluid in the heat pump at the heat exchanger on the right side of Figure 11.3. The fluid in the heat pump at that point will be a liquid at

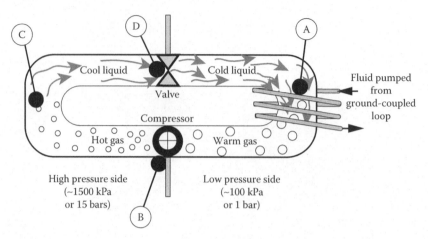

FIGURE 11.3 Schematic of a ground-coupled heat pump designed for heating. In heating mode, fluid that has circulated through the subsurface enters a heat exchange coil on the right of the figure and heats the circulating refrigerant that is pumped through the heat pump. Heat pumps are also available that can accomplish heating and cooling, in which case the circulation scheme for the refrigerant and the heat exchange configuration allow the refrigerant to switch between modes that either accept heat or deposit heat in the circulating fluid in the closed loop.

TABLE 11.1

Thermodynamic Properties of Some Compounds Potentially Useful as Refrigerants

Name	Formula	Molecular Weight (g/mol)	Density (kg/m³)	Melting Temperature (°C)	Boiling Temperature (°C)	Heat of Vaporization (kJ/kg)	Constant Pressure Heat Capacity (kJ/kg-K)
R134a	H_2FC-CF_3	102.03	1206	−101	−26.6	215.9	0.853
Propane	C_3H_8	44.096	582	−187.7	−42.1	425.31	1.701
Isopentane	C_5H_{12}	72.15	626	−160	28	344.4	2.288

approximately 0°C and about 100–400 kPa (the actual temperature and pressure will depend on the heat pump design and the fluid used in the heat pump). Hence, the fluid in the heat pump will increase in temperature at constant pressure, following the arrow pointing to the right in the dark gray field in Figure 11.4. When the liquid in the heat pump reaches the temperature of the liquid–vapor two-phase region, the liquid will absorb heat at constant pressure and temperature while the liquid vaporizes and will continue until vaporization is complete. The amount of energy absorbed will depend upon the heat of vaporization of the fluid. At that point, if the vapor remains in contact with the heat exchanger, it will become *superheated*, meaning it will acquire additional heat beyond the heat of vaporization.

Once the vapor has formed, it is compressed to a pressure about 10 times that at which it is vaporized (arrow in the gas field in Figure 11.4). The vapor then moves through a heat exchanger in the room or building, losing heat at constant pressure, crossing back through the two-phase liquid + gas field. As this happens, the fluid gives up its heat of vaporization and the vapor condenses back to a liquid (long arrow in Figure 11.4 pointing to the left).

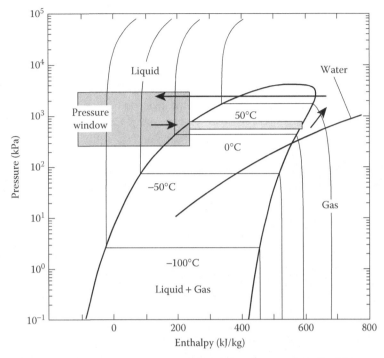

FIGURE 11.4 Enthalpy–pressure diagram for propane showing the two-phase liquid–gas region. For reference, the low-temperature limb of the liquid boundary for the water system is also shown (see Figure 3.8 for details of the water system).

COEFFICIENT OF PERFORMANCE AND ENERGY EFFICIENCY RATIO

The efficiency of the heat pump process is measured by comparing the energy required to drive the system to the amount of heat transferred. Consider the thermodynamic properties of the system in Figure 11.4, again. To vaporize the propane from a liquid to a gaseous state takes approximately 425 kJ/kg of propane. The heat capacity of water is about 4180 kJ/kg-K. Normally, a temperature drop of about 10°C can be expected between the inflow and outflow of the heat exchanger for ground loop on the right-hand side of the heat pump in Figure 11.3. Hence, assuming the fluid flowing in the ground loop is simply water, 1 kg of geothermally heated water in the closed loop feeding the heat pump can vaporize approximately 10 kg of propane by adding just over 4 kJ of energy to the propane. The compressor in a heat pump is driven by a small electric motor, with power ratings on the order of 1.5 kW. The power consumed by the motor does work on the fluid. Assuming the motor has an efficiency of 0.8, and the flow rate for the geothermal fluid is 1 kg/s, the rate of input of energy to the working fluid in the heat pump is

$$E_{\text{Tot}} = 4180\,\text{J/s} + 0.8 \times 1500\,\text{J/s} = 5380\,\text{J/s}$$

A measure of the efficiency of this system can be obtained by comparing the total heat input, that is, 5380 J/s, to the amount of energy consumed, that is, 1.5 kW to run the compressor. The ratio of these values is the efficiency,

$$\frac{E_{\text{Tot}}}{E_{\text{consumed}}} = \frac{(4180\,\text{J/s} + 0.8 \times 1500\,\text{J/s})}{1500\,\text{J/s}} = 3.59$$

For heating cycles, this measure is called the coefficient of performance (COP) and is defined as

$$COP = \frac{Delivered\ heat\ energy}{Compressor\ electrical\ demand}$$

Common values for the COP for GHPs are between 3.0 and 5.0, meaning that 300%–500% of the energy used to run the heat pump is delivered to the space to be heated. For comparison, the most efficient gas-fueled furnaces convert 90%–95% of the energy that is potentially available in the gas to useable heat for heating and have a COP of about 0.9.

The thermodynamic significance of the COP can be understood by considering the path described in Figure 11.4. The compression cycle increases the enthalpy of the gas from about 600 kJ/kg to about 680 kJ/kg. This is the work done by the compressor on the gas phase and is the electricity demand of the system. The heat delivered to the building is the heat of vaporization that is released when the fluid condenses, which is where the long, left-pointing arrow crosses the two-phase boundary at about 390 kJ/kg in Figure 11.4. Hence, the heat delivered to the room, compared to the energy consumed in the compression, is

$$COP = \frac{(680 - 390\,kJ/kg)}{(680 - 600\,kJ/kg)} = 3.6$$

which is essentially the same as that computed above using a slightly different method. Actual COP values depend upon the design of the heat pump components and its operating parameters as well as the temperatures of the end points of the cycle.

The cooling efficiency is measured in units of energy efficiency ratio (EER) and is the cooling capacity (in Btu/hour) of the unit divided by its electrical input (in watts) at standard peak rating conditions. EER values for GHPs are generally in the range of 15–25.

NEAR-SURFACE THERMAL RESERVOIR

The soil and rock that make-up the top few hundred feet of the earth act as a heat reservoir that evolves in response to two heat sources. As previously noted in Chapter 2, heat flow from the earth's interior averages 87 mW/m². The source of this heat is the slow cooling of the earth's interior and radioactive decay in the crust. In deep mines around the world, temperatures at depths of 1–2 km can be as high as 60°C (140°F), with local thermal gradients as high as 0.5°C–1.5°C per 100 feet. Such gradients would result in a surface temperature of between 5°C–42°C (40°F and 107°F), in the absence of an atmosphere or sun. Other regions have lower heat flows and would be cooler. Hence, solely on the basis of heat flow, there would be a wide range of temperatures in the shallow subsurface, with local hot spots and cold spots.

SOLAR INSOLATION

The atmosphere and solar insolation mitigate that extreme temperature variability at the surface that would result solely from heat flow. The average daily influx of solar energy that reaches the surface is about 200 W/m² for the earth as a whole (Wolfson 2008). This energy input, along with that which interacts with the atmosphere to generate weather patterns, moderates the variation in surface temperature throughout the world. This effect heats the top tens of centimeters of the land surface during the day and allows it to slowly cool at night, all within a restricted temperature range that is strongly dependent upon latitude. This diurnal effect slowly propagates into the subsurface.

Added to this is the effect of precipitation which, when coupled with the diurnal effect of solar insolation, significantly adds to the thermal energy content of the subsurface. As rainfall occurs

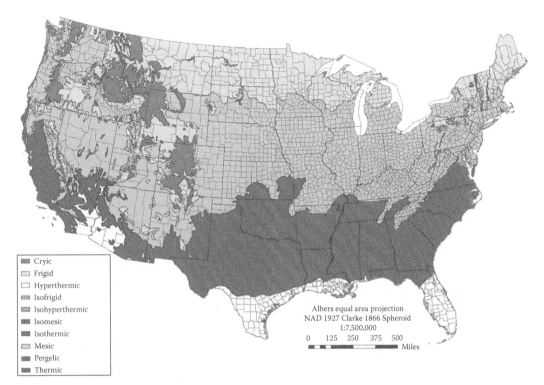

Cryic
Frigid
Hyperthermic
Isofrigid
Isohyperthermic
Isomesic
Isothermic
Mesic
Pergelic
Thermic

Albers equal area projection
NAD 1927 Clarke 1866 Spheroid
1:7,500,000
0   125   250   375   500
                              Miles

FIGURE 11.5 Near-surface soil temperatures throughout the United States. (Natural Resources Conservation Service, http://soils.usda.goc.)

(or snow melts), the fate of the resulting meteoric water is complex. Some of the water immediately returns to the atmosphere via evaporation, some is taken up by plants and either incorporated into the plant material or respired back into the atmosphere, some runs over the surface to local drainage systems, and some percolates into the ground where it slowly flows in the subsurface to the local water table and enters the aquifer system. The percolating water most significantly impacts the subsurface thermal reservoir by absorbing heat from the surface solar energy and transporting it to deeper levels in the soil and rock. The high heat capacity of water assures that solar energy is relatively efficiently transported to the subsurface. The result is that the energy content of the subsurface reflects the sum of both solar and geothermal energy inputs, with the former dramatically moderating the geothermal variability. Figure 11.5 shows the mean annual soil temperature around the United States, obtained from the Natural Resources Conservation Service (http://soils. usda.goc). Note how the temperatures strongly correlate with latitude as well as regional climate patterns.

It is important to note, however, that the soil temperature map shown in Figure 11.5 reflects conditions at shallow (<5 m) depth. The variability of temperatures in the subsurface as a function of depth from region to region has an important influence on geothermal heat pump performance, a topic we will discuss later in this chapter when we consider system designs.

SOIL CHARACTERISTICS

Figure 11.6 shows idealized and observed temperature variability as a function of moisture content, depth, and season for a study site in Finland. There are two idealized curves (thinnest lines) for each season shown. The outer curves represent temperature variation with depth for wet soils, whereas the inner curves are for dry soils. The difference between these curves is due to the fact that, for a

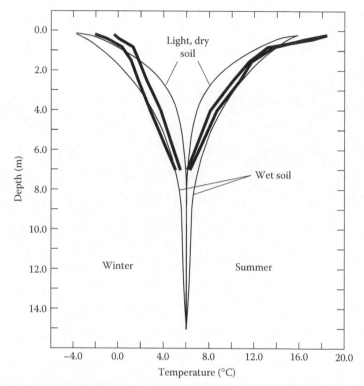

FIGURE 11.6 Temperature profile annual extremes in soil, to a depth of 15 m. The thin lines are idealized temperature profiles for summer maximum and winter minimum temperatures. There are two curves for each seasonal extreme, the more extreme curve representing the idealized variation for wet soil and the less extreme curve showing the variation of dry soil. The idealized curves assume that the variation is symmetrical about a mean temperature. The heavy solid curves are actual measurements from a site in south Finland, representing two different years. (From Lemmelä, R. et al., *Geophysica*, 17, 143–154, 1981.)

given heat flow over a specified distance, the temperature gradient will be a direct inverse function of the thermal conductivity. This is apparent by rearranging Equation 2.3:

$$\frac{(\nabla x \times q_{th})}{\nabla T} = k_{th} \tag{11.1}$$

As the thermal conductivity decreases, as it must if the saturation in a rock decreases, the thermal gradient must increase.

The idealized curves in Figure 11.6 are drawn for the case where the annual surface temperature swing is 10°C around the local mean temperature. However, the behavior of real systems is more complex. Note that the summer swing from the mean in the Finland observations shown in the figure is about 14°C, but the winter swing is about half that amount, reflecting the effects of local weather patterns and latitude. At lower latitudes, the swing is likely to be more symmetrically distributed about the mean annual soil temperature. Adding to this complexity is the fact that temperature swings can range from as little as 5°C to values exceeding 20°C, depending upon local weather and seasonal variability.

Regardless of the actual conditions at a site, two important conclusions can be drawn from the information in Figure 11.6. First, shallow closed-loop systems that are installed in trenches must be designed with careful attention paid to the local soil properties and conditions. Use of standard curves of thermal properties for generalized soil types is sufficient as a guide to estimate design parameters, but optimal performance will only be achieved by conducting a careful and thorough survey of soil thermal conductivity and variability (see Case Study, below). When conducting such measurements, attention must also be paid to the local history of weather patterns, because regional precipitation patterns can strongly influence how thermal conductivity may change monthly. Local agricultural research stations and soil conservation service offices can be a good resource for obtaining maps and data on local conditions. But once excavation or drilling is undertaken, soil samples must be taken and thermal properties measured, if optimal designs are to be obtained. Second, seasonal variability is much less of a design factor if a borehole system is installed. Once depths of about 10 m are exceeded, the effects of fluctuating solar insolation and weather are damped out, resulting in a virtually constant thermal state.

THERMAL CONDUCTIVITY AND HEAT CAPACITY OF SOILS

The success of GHP systems results from the high heat capacity and high thermal conductivity of soil and rock. Table 11.2 provides a summary of the thermal conductivities and constant pressure heat capacities of some common geological materials. The importance of these properties can be appreciated by considering the amount of heat that must be transferred between buildings and the earth when utilizing GHPs for HVAC purposes.

Consider a case in which the rate of heat lost from a building in winter was 7 kW. If the building is to maintain a constant temperature for a workday that goes between 7:00 AM and 5:00 PM, the total heat loss over that period is 252,000 kJ. From Table 11.2, the amount of heat available from the subsurface, per cubic meter of material, can be calculated from

$$Q = C_p \cdot \Delta T \cdot V \tag{11.2}$$

where:

Q is the heat that can be obtained
C_p is the constant pressure heat capacity
ΔT is the temperature change for the material of interest
V is the number of moles per cubic meter of the material of interest

---

TABLE 11.2
Thermal Conductivity (W/m-K) and the Constant Pressure Heat Capacity (C_p [J/mole-K]) of Some Common Materials at 25°C

Material	k_{th}	C_p	Q	m³
Quartz[a]	6.5	44.5	1960	128.5
Alkali feldspar[a]	2.34	203	2000	130
Calcite[a]	2.99	82	2103	120
Kaolinite[a]	0.2	240	2408	105
Water	0.61	75.3	4181	60

Source: [a] C_p computed from Helgeson, H.C. et al., *American Journal of Science*, 278-A, 229, 1978.

Note: The amount of heat, Q (kJ/m³-K), that must be supplied or removed, per cubic meter of material, for 1°C of temperature increase or decrease at about 25°C is shown. The number of cubic meters of each material needed to supply 7 kW of heat is shown in the column m³ (see text for details).

Shown in Table 11.2 is the amount of heat per cubic meter that can be obtained from each material, for a 1°C temperature change at near-surface conditions. Also shown in the table is the total number of cubic meters of material that would be needed to supply the heat required to account for the heat loss. As discussed below, boreholes in many GHP applications are drilled to depths of about 250–295 m. Thus, a single borehole could easily supply the needed heat, in principle.

However, whether or not the heat can be supplied at the needed rate depends on the thermal conductivity of the material in the vicinity of the closed loop. To establish the rate at which this will occur requires solving the following equation and comparing the results to the demand:

$$\frac{\Delta Q}{\Delta t} = C_v (T_f - T_i)/(t_f - t_i) \times V \tag{11.3}$$

where:

ΔQ is the amount of heat that is required to be added or removed from the system (J)
Δt is the time over which the heat is added or removed (s)
C_v is the constant volume heat capacity (J/m^3-K)
$(T_f - T_i)$ is the temperature change between the initial (i) and final (f) states
$(t_f - t_i)$ is the time duration (s)
V is the volume of interest (m^3)

For a borehole system operating under the conditions described above, and using a value of 2225.2 kJ/m^3-K for the C_v for kaolinite, and assuming a volume of 1 m:

$$\frac{\Delta Q}{\Delta t} = 61.81 \text{ J/s}$$

In Table 11.2, it is evident that kaolinite can only provide about 0.2 J/s per meter length of borehole under optimal conditions. As a result, a borefield in such material would require substantially more boreholes to provide the rate of heat transfer needed, even though the available heat in one borehole is adequate. Comparing the thermal conductivity of kaolinite to that of other materials in Table 11.2, it is evident that most other geological materials are much better suited to support a GHP system than is a site composed solely of dry kaolinite. Indeed, for most materials, the thermal conductivity is sufficiently high such that the closed-loop systems usually have the capacity to accommodate loads at the rate of approximately 3520 watts per 180 meters, which is approximately equivalent to 1 ton of cooling capacity from a 295 foot borehole.

DESIGN CONSIDERATIONS FOR CLOSED-LOOP SYSTEMS

HEATING AND COOLING LOADS

Despite these myriad considerations, the principles involved in designing a ground loop are relatively straightforward. The principle challenge is establishing the length of underground piping needed to satisfy a cooling and/or heating load for a given building. The important parameters that must be quantitatively taken into account are the heating and cooling load of the building or space to be controlled, the available heating and cooling capacity of the local thermal reservoir, and the rate and efficiency with which heat must be transferred from or to the reservoir.

CALCULATING LOOP LENGTH

Calculating building energy demands (heating and cooling loads) will not be treated in this book. Excellent software packages are available to complete such calculations using state-of-the-art data from a variety of sources, including those listed at the end of this chapter in "Further Information

Sources." In the following discussion, it will be assumed that the required heating (C_H) and cooling (C_C) capacities needed for the building loads are about 13,400 and 11,750 W, respectively.

There are several approaches used for calculating the length of piping needed for a given system. The primary variables such calculations must consider are the load, the efficiency of the heat pump, the rate at which heat transfer can occur, the available heat in the thermal resource, and the amount of time the load demand will be imposed on the system. Because sizing requires consideration of the seasonal variability at the site being considered, most of these parameters are considered on an annual basis. The approach presented here is that described by the International Ground Source Heat Pump Association (Oklahoma State University 1988). To determine the length of underground piping needed for a heating loop, the equation is

$$L_H(\text{m}) = \frac{\{(C_H) \times [(\text{COP}-1)/\text{COP}] \times [R_P + (R_S \times F_H)]\}}{(T_L - T_{\min})} \tag{11.4}$$

For a cooling loop, the corresponding equation is

$$L_C(\text{m}) = \frac{\{(C_C) \times [(\text{EER}+3.412)/\text{EER}] \times [R_P + (R_S \times F_C)]\}}{(T_{\max} - T_H)} \tag{11.5}$$

where:
  R_P is the resistance to heat flow of the pipe (which is equivalent to 1/thermal conductivity of the pipe)
  R_S is the resistance to heat flow of the soil (which is equivalent to 1/thermal conductivity of the soil)
  F_H (F_C) is the fraction of time the heating (cooling) system will be operating
  T_L (T_H) is the minimum (maximum) soil temperature at the depth of installation
  $T_{\min}$ ($T_{\max}$) is the minimum (maximum) fluid temperature for the selected heat pump

In Figure 11.7, the horizontal lengths of pipe required for heating and cooling purposes for a pair of soil temperatures are plotted as a function of the soil thermal conductivity. It is obvious that there is a strong dependence on soil thermal conductivity and temperature conditions. This dependence is greatest under dry conditions, becoming less pronounced, but nevertheless very significant, at higher saturations and thermal conductivities. For illustrative purposes, consider the effect on loop length for the variously saturated quartz sands depicted in Figure 2.4. The difference in pipe length that would be required for the least and most saturated sands in that figure is nearly a factor of 4. Clearly, it would be easy to undersize or oversize a ground loop system if care is not taken to adequately characterize the thermal properties of the soils at a site.

Establishing the design length importantly constrains the design implementation. For a vertical system that will utilize boreholes, less land is required but because drilling costs can be significant, the initial investment expenses can be high. For a horizontal loop system, the amount of land required for excavation and loop emplacement can be significant, but the installation costs are less than for vertical boreholes. If there is sufficient land available for a horizontal loop system, Equations 11.4 and 11.5 can be used to size the loop length and assist in determining what loop configurations would be appropriate.

Open-loop systems differ from closed-loop designs in that the working fluid is directly extracted from the groundwater system, passed through the heat pump heat exchanger, and either reinjected into the subsurface or released into the surface environment. Such systems rely only on the fluid temperature of the groundwater that is extracted and thus do not depend on knowledge of the thermal properties of the soil or bedrock at the site. Instead, it is necessary to obtain reliable data on the temperature variation that the groundwater system historically has exhibited. In general, such temperature swings will be minor, compared to that seen in the near-surface soils, usually ranging

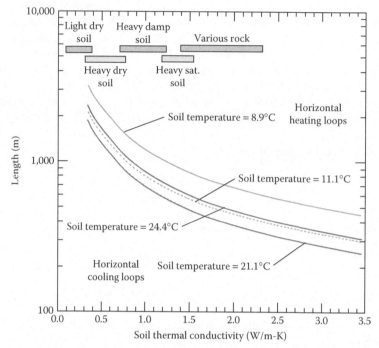

FIGURE 11.7   Computed loop length for heating and cooling, closed-loop GHP systems. In these calculations, it was assumed that the COP of the heat pump was 3.24 and the EER was 7.8. Pipe thermal conductivity was assumed to be 14.8 W/m-K, the heating and cooling run time fractions were 0.5 and 0.6, respectively, and the heat pump fluid T_{max} and T_{min} were 37.8°C and 4.4°C, respectively. For reference, the range of thermal conductivities for light, dry soil (Light dry soil); heavy, dry soil (Heavy dry soil); heavy, damp soil (Heavy damp soil); heavy, saturated soil (Heavy sat. soil); and crystalline rocks (Various rock) are also shown.

within 5°C of some median value. These systems do not require a specified loop length, thus making sizing of the system easier. However, disposal of the groundwater, whether by reinjection or by release into a surface catchment system, requires special consideration and care in order to avoid potential water contamination. Because of that concern, special permits are often required to install such systems.

LOCAL VARIABILITY: WHY MEASUREMENTS MATTER

We have discussed the importance of having accurate subsurface information in order to design GHP systems that function as planned. Although some geothermal heat pump systems have been designed using rule-of-thumb approaches for subsurface properties, the natural variability of geological systems makes such approaches highly risky. Consider, for example, the subsurface temperature.

Figure 11.8 shows the temperature versus depth variability for several metropolitan areas in the United States. For Boston and Dallas, the temperature in the subsurface closely follows a linear gradient that allows for simple relationships to be considered. For the Los Angeles Basin region, however, a wide range of subsurface temperatures are observed, spanning a temperature range of as much as 30°C (~40°F) at depths of only a few hundred feet. As shown in Figure 11.9, the temperature of the water from a geothermal loop that enters the heat pump significantly affects the efficiency (COP) of the system (Davis 2013). As a result, systems that do not have the heat pump and the loop length carefully designed to match the subsurface temperature variability can be over- or underdesigned for the local conditions.

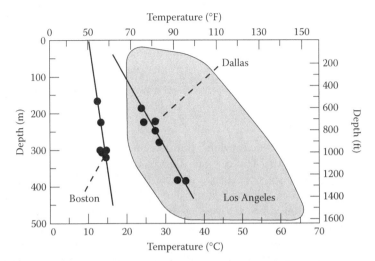

FIGURE 11.8 Variation in temperature versus depth for three metropolitan areas in the United States. (Data obtained from Battocletti, E.C. and Glassley, W.E., Measuring the costs and benefits of nationwide geothermal heat pump deployment, US Department of Energy, Final Report for Award DE-EE0002741, 319, 2013.)

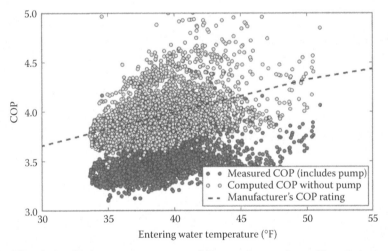

FIGURE 11.9 The relationship between temperature of the water in a geothermal loop that enters a heat pump and the resulting efficiency (COP) of the system. The light gray dots are the computed COP, not accounting for the pump energy use, and the dark gray dots include the impact on efficiency of including energy consumption of the pump. (From Davis, M., Ground source heat pump system performance: Measuring the COP. Ground Energy Support, 12, http://www.groundenergy.com, 2013.)

Figure 11.9 also emphasizes additional points. As noted when discussing subsurface temperature conditions (Figure 11.6), seasonal weather and climate patterns can cause temperature fluctuations of 20°C or more. The manufacturer's reported COP of a heat pump is usually referenced to a specified temperature. Adequately sizing a heat pump must consider what the seasonal variability will be in order to install the most efficient system.

Finally, Figure 11.9 also makes clear that the COP, although useful for comparison and reference purposes, must be treated with intelligence. Because loop size is a design feature that is determined by the engineer responsible for the system, the COP reported by a manufacturer will not consider the energy used to circulate water through the geothermal loop, because the loop size is unknown to the manufacturer. Actual observed performance, that is, the COP, of an installed system should approach that of the ideal design, but will not match it. Properly designed and installed geothermal heat pump systems will invariably outperform any other equipment used for HVAC purposes, but the realized energy savings will not exactly match that predicted solely on the basis of a "name-plate" COP.

SYNOPSIS

GHP systems take advantage of the low-temperature geothermal resource that exists in the shallow subsurface (< 300 m depth). The constant flux of heat from the interior of the earth plus the influx of solar energy to soils provides a reliable thermal reservoir from which heat can be extracted. High-efficiency heat pumps that are coupled to this reservoir can move the thermal energy from the earth to buildings by use of simple heat pumps. The heat pump technology currently available allows COP values in the range of 3–5, which makes them the most energy efficient means for HVAC purposes. Sizing such systems is a straightforward and well-developed practice, but relies on the availability of high-quality information regarding subsurface properties and the properties of the building to assure that a functioning system reliably meets the demand placed on it.

CASE STUDY: WEAVERVILLE AND A US COST–BENEFIT ANALYSIS OF GHP INSTALLATION

Weaverville Elementary School is a 50-year-old public school located in northwestern California. In July and August, the hottest months in Weaverville, the average daytime high temperature is about 34°C (93°F). In winter, the coldest months are December and January, when the daytime high temperature averages about 8°C (46°F). Heating and cooling are thus required for both time periods. For reasons of economy and building maintenance efficiency, it was decided to retrofit the building complex with geothermal heat pump heating and cooling capability.

Prior to installation of the geothermal heat pump system, the building complex utilized 20 air source heat pumps, two diesel boilers, one propane boiler, two propane heaters, 11 evaporative coolers, and four dual-fuel units.

For purposes of sizing the GHP system for the building complex, it was calculated that the necessary heating capacity needed for the site was about 47 kW. The calculated cooling capacity that was needed was estimated to be 35 kW. The amount of time required for HVAC was calculated to be 700 hours/year.

The computed borehole length for the stipulated geothermal heat pumps and ground loop piping was 1005 m (3320 ft). In this calculation, it was assumed that the subsurface temperature would be about 11.5°C (53°F), the COP of the heat pumps would be about 4.8, and the EER would be 23.5. In the final configuration, 11 separate borefields were put in place, with between 8 and 22 individual boreholes per field, each borehole being 90 m (295 ft) deep. A total of 38 GHPs were installed in classrooms and office spaces. The system was brought online in September 2003. At the time the system was completed, an additional 371 sq. m (4000 sq. ft) of classroom space was added to the school, and all of the classrooms were provided with year-round HVAC, which was an upgrade for 14 classrooms.

In Figure 11.10, the amount of oil and propane consumed before and after installation of the GHP system is shown. Figure 11.11 shows the change in electricity consumption over the same time period, and the total dollar cost for energy production is shown in Figure 11.12. These changes in energy source amount to a reduction in the emissions of greenhouse gases as shown in Figure 11.13. By eliminating oil consumption and reducing propane and electricity consumption, CO_2 emissions were reduced by almost 50% and SO_2 emissions were nearly eliminated. From both an economic and environmental perspective, the investment in the system was beneficial.

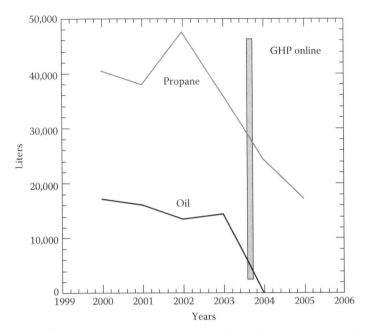

FIGURE 11.10 Oil and propane use (in liters per year) at Weaverville Elementary School, from 1999 to 2005. Only the total annual consumption is shown. Therefore, the drawn lines only connect year-end values and are not meant to represent monthly values. The September time period when the GHP system was energized is shown for reference.

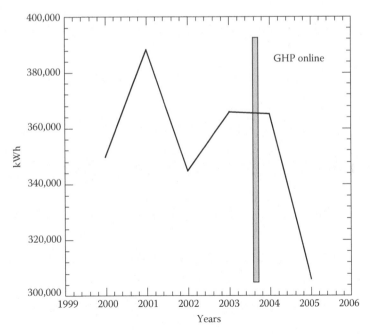

FIGURE 11.11 Electricity consumption (in kWh) at Weaverville Elementary School, from 1999 to 2005. Only the total annual consumption is shown. Therefore, the drawn lines only connect year-end values and are not meant to represent monthly values. The September time period when the GHP system was energized is shown for reference.

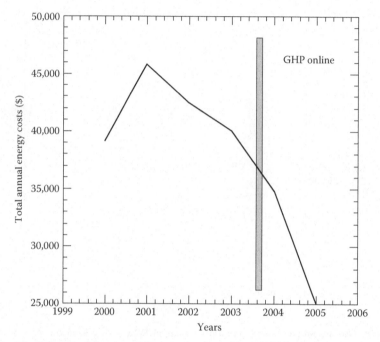

FIGURE 11.12 Total annual energy costs ($) at Weaverville Elementary School, from 1999 to 2005. Only the total annual consumption is shown. Therefore, the drawn lines only connect year-end values and are not meant to represent monthly values. The September time period when the GHP system was energized is shown for reference.

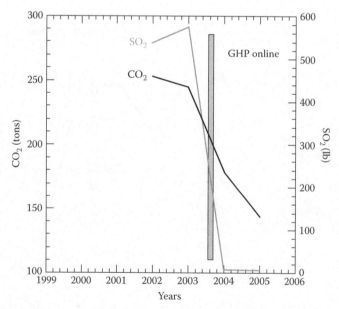

FIGURE 11.13 CO_2 (tons) and SO_2 (lb) annual emission at Weaverville Elementary School, from 2002 to 2005. Only the total annual consumption is shown. Therefore, the drawn lines only connect year-end values and are not meant to represent monthly values. The September time period when the GHP system was energized is shown for reference.

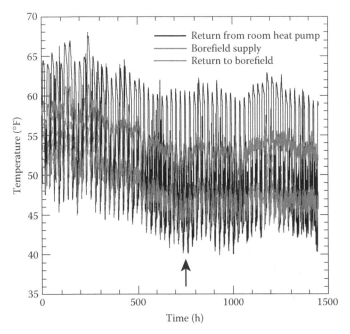

FIGURE 11.14 **(See color insert.)** Temperature variation of the working fluid coming from the borefield (red line labeled "Borefield supply"), returning to the borefield (blue line labeled "Return to borefield"), and exiting the heat pump from a specific room (black line labeled "Return from room heat pump"). The time period covered is from November 4, 2005, to January 4, 2006. The arrow points to the period represented in Figure 11.15.

Temperatures of the borefield supply and return water were monitored for several months, as was the outlet temperature of water from several of the rooms in the school. In Figure 11.14, the recorded temperatures are shown for a two-month period from early November through early January, which was a period of high heating demand. The spiky nature of each curve reflects the diurnal changes due to the timed thermostat used to control the system, as discussed in the reminder of this section, and the cycling of the heat pump system.

The supply temperature for the borefield can be assumed to be within a few degrees of the average ground temperature for the borefield taken as a whole. At the beginning of the recorded period, the supply temperature was approximately 14°C (57°F), but it dropped to about 11°C (52°F) by the beginning of January. Such seasonal fluctuation is relatively common for regions with a strong seasonal variation in ground temperature. This reflects the fact that some infiltration of cold surface water and cold air transfers to the deeper regions, resulting in a seasonal cooling pattern. Despite this variability, note that the design temperature of 11.5°C (53°F) is very close to the low end of the range, allowing for adequate heating capacity of the system.

Close examination of the diurnal change reveals some important characteristics of systems using geothermal heat pump technology. Figure 11.15 shows the temperature variation over one daily cycle in the middle of winter. The temperature of the fluid circulating through the borefield and heat pump was recorded every few minutes at numerous locations. Shown in the figure is the temperature at the outlet of the heat pump that supplied heat to one of the classrooms ("Return from room heat pump"), the temperature in the main supply line from the borefield that connected to the room ("Borefield supply"), and the temperature in the main return line to the borefield ("Return to borefield"). During the period shown, the thermostat that controlled the heat pump cycling was programmed to set to 65°F at 17:00, when the classrooms would be unoccupied, and reset to 72°F at 06:30 before the classrooms would be occupied.

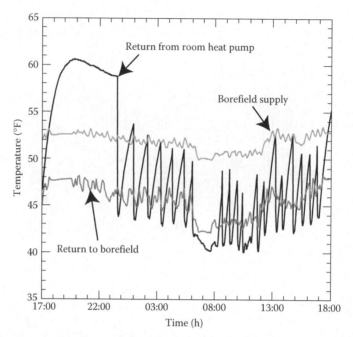

FIGURE 11.15   **(See color insert.)** Detailed daily behavior of the temperature of the fluids. The time period covered in the figure is from Monday, December 5, 2005, 17:00, to Tuesday, December 6, 2005, 18:00. The curves are colored as in Figure 11.14.

The temperature recorded for the heat pump return shows that the fluid progressively increased in temperature from 17:00 until approximately 19:30, when the temperature began a slow decline. This "hump" in the temperature curve reflects the fact that there was no fluid circulating in the system because the thermostat was set back to 65°F at 17:00, and the room temperature at that point was at least 72°F. Between 17:00 and approximately 19:30, the motionless fluid in the heat pump heated as it approached thermodynamic equilibrium with the room in which the heat pump was located. As the room began to cool in the evening, the temperature in the pump then also began to cool. At around 23:00, the room temperature dropped below 65°F, causing the heat pump compressor to activate and the fluid to circulate in the system. This activation of the system caused the fluid temperature in the heat pump return to drop to values close to that of the main borefield return temperature. Because several separate rooms and their respective heat pumps fed into the borefield return line, the borefield return line represents the average of those return flows from the individual heat pumps. When the pump cycled off when the room reached the thermostatically controlled 65°F, the heat pump circulation ceased and the temperature climbed. Fortuitously, the room temperature dropped below 65°F and cycled the heat pump back on just about when the fluid in the heat pump had approached the borefield supply temperature. It is this cycling that causes the spiky oscillation of the heat pump return temperature. At 06:30, the thermostat reset to 72°F and the heat pump remained on for about 2 h while the room temperature increased. From 08:30 until 17:00, the heat pump cycled on and off, and the return temperature stayed close to the borefield return temperature. At 17:00, the cycle repeated.

Note that the borefield supply temperature increases slightly, from 52°F to 53°F, during the hours after 17:00 when the system was shut off. This represents the recovery process of the natural system to the condition in the subsurface when there is no heat removed from it. When the heat pump cycles on, fluid circulation results in a drop in the supply temperature by about a degree or two, which is especially noticeable during the time when the heat pump was in continuous operation between 06:30 and 08:30. From 08:30 onward, the system slowly recovers because the heat pump was cycling on for only brief periods. These variations in the supply temperature occurred in response to the impact the return flow to the borefield had on the overall heat balance in the system. The relative stability of the supply from the borefield is noteworthy, providing evidence that the geothermal resource is stable.

The spiky nature of the curve showing the "Return from room heat pump" reflects the on–off cycles of the heat pump. At the time this system was installed, heat pumps were generally *single-stage*, meaning that energy input to the room during a heating cycle was at a single, constant rate. Since that time, *two-stage* and *multistage* systems have been developed that reduce even further the energy consumption of these heat pump systems, thus smoothing out the spiky nature of the temperature curves and reducing impact on the subsurface thermal reservoir.

In Figure 11.16, the long-term change in the ratio of the supply temperature to the return temperature is plotted. Also shown is a linear least squares fit to the data. The slight positive slope to this trend indicates that there was some residual heat being added to the subsurface, thus increasing the overall thermal budget available for use. This was happening because warm fluid was circulated into the subsurface each time the heat pump turned on. The net result is that the heat transferred between the subsurface and the buildings actually cycled back and forth. That behavior is typical of most GHP installations. If the installed system is used for heating and cooling, the energy budget is affected by this heat effect and can affect performance, although usually the effect is minimal. In systems where there is relatively rapid movement of groundwater within the depth interval penetrated by the borehole, such an effect will be minimized or eliminated.

The implications of installing such geothermal heat pump systems on energy consumption and greenhouse gas emissions at a national scale were analyzed by Battocletti and Glassley (2013) for the United States. In that study, the 30 largest metropolitan areas were considered and installation of geothermal heat pumps in residences and commercial buildings were modeled. The results (Figure 11.17) show that the energy savings would vary from one climate zone to another—the coldest regions would see the greatest savings in energy consumption—up to nearly 75%. However, due

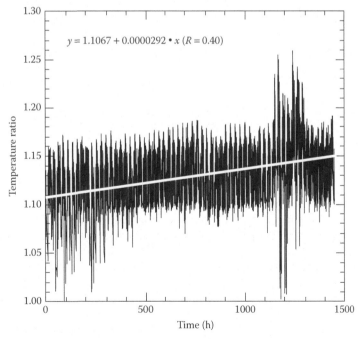

FIGURE 11.16 Variation in the ratio of borefield supply temperature to return temperature to the borefield for the time period from November 4, 2005, to January 4, 2006. The white line is a linear least squares fit to the data. The equation for the line is shown in the figure, as is the goodness of fit.

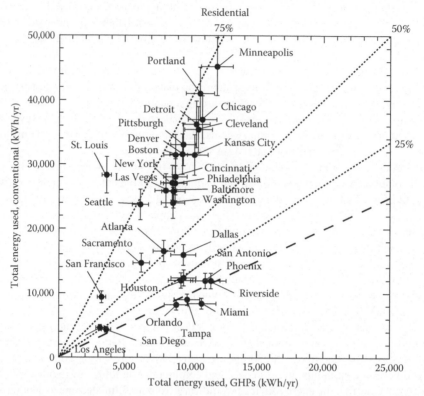

FIGURE 11.17 Comparison of the energy consumed to heat a standard home using a GHP system (horizontal axis, in kWh/yr) in 30 metropolitan regions in the United States, compared to the energy used for a conventional system. The heavy dashed line indicates those values for which the energy use is identical. The labeled lightly dashed lines indicate the percent of energy savings for a GHP system relative to a conventional system. The amount of energy for the conventional systems is based on the energy technology mix used to generate power in the specific metropolitan area (i.e., oil, coal, gas, and nuclear). (From Battocletti, E.C. and Glassley, W.E., Measuring the costs and benefits of nationwide geothermal heat pump deployment, US Department of Energy, Final Report for Award DE-EE0002741, 319, 2013.)

to issues associated with heat transfer efficiencies, the most temperate regions would see minimal impact on energy use.

PROBLEMS

11.1. Describe how a geothermal heat pump works. What determines its efficiency?

11.2. What does COP measure? What variables go into computing COP?

11.3. How might you change the performance of a heat pump so that its COP went from 3.5 to 5.0?

11.4. At 25°C, how much heat is available, per degree temperature drop, from a soil composed of equal parts quartz, alkali feldspar, and calcite, with a water-filled porosity of 15%?

11.5. For a heating system employing a heat pump with a COP of 3.8 for a building heating load of 15 kW, calculate the length of the heating loop needed. Assume that the soil properties are those in problem 11.4, the thermal conductivity of the pipe is the same as the soil, the fractional heating time is 0.5, the minimum soil temperature is 10°C, and the heat pump fluid minimum temperature is 3°C.

11.6. From the Weaverville case study, explain, hour by hour from Figure 11.14, the factors that are controlling the temperature for the return from the room heat pump.

11.7. What would be the advantages and disadvantages of running the heat pump 24 h a day, if the pattern in Figure 11.14 is being considered?

11.8. In Figure 11.17, the three cities for which energy savings are minimal or negative are from Florida. What factors minimize the advantage of geothermal heat pump systems in this setting?

REFERENCES

Battocletti, E.C. and Glassley, W.E., 2013. Measuring the costs and benefits of nationwide geothermal heat pump deployment. US Department of Energy, Final Report for Award DE-EE0002741, 319 pp.

Davis, M., 2013. Ground source heat pump system performance: Measuring the COP. Ground Energy Support, 12 pp. http://www.groundenergy.com.

Helgeson, H.C., Delany, J.M., Nesbitt, H.W., and Bird, D.K., 1978. Summary and critique of the thermodynamic properties of rock-forming minerals. *American Journal of Science*, 278-A, 229.

Lemmelä, R., Sucksdorff, Y., and Gilman, K., 1981. Annual variation of soil temperature at depths 20 to 700 cm in an experimental field in Hyrylä, south-Finland during 1969 to 1973. *Geophysica*, 17, 143–154.

Ochsner, K., 2008. Geothermal Heat Pumps. London: Earthscan, p. 146.

Oklahoma State University, 1988. Closed-loop/ground-source heat pump systems: Installation guide. National Rural Electric Cooperative Association Research Project 86-1. International Ground Source Heat Pump Association, Stillwater, OK, p. 236.

Wolfson, R., 2008. Energy, Environment, and Climate. New York: W.W. Norton & Company, p. 532.

FURTHER INFORMATION

American Society of Heating, Refrigeration, and Air-Conditioning Engineers (ASHRAE) (http://www. ashrae.org/).

ASHRAE provides data, information, training, and resources for all aspects of building HVAC interests. They have established standards useful for building design, sustainability, and use of renewable energy resources, including GHPs. They provide standard methods for, among other things, computing building energy demand and heating and cooling loads

Ground Energy Support LLC (http://groundenergysupport.com/wp/).

Ground Energy Support LLC is a company that provides monitoring equipment for geothermal heat pump systems. Research they have conducted on GHP performance provides useful objective insight into the actual performance of installed GHP systems. Reports on their research can be obtained at their website.

International Ground Source Heat Pump Association (IGSPHA). Stillwater, OK (http://www.igshpa.okstate.edu/).

IGSPHA is an important resource for materials related to GHP systems. The training programs and materials are excellent. The software packages they provide for design and load calculations (e.g., LDCAL5) are recommended.

US Department of Agriculture, Natural Resources Conservation Service (http://soils.usda.goc).

The NRCS maintains a website containing numerous databases and links for information regarding soil properties and conditions. It is a useful resource for initially determining soil conditions that can influence the design of GHP systems.

12 Direct Use of Geothermal Resources

Unlike geothermal power generation, in which heat energy is converted to electricity, direct-use applications use heat energy directly to accomplish a broad range of purposes. The temperature range of these applications is between ~10°C and ~150°C. Given the ubiquity of this temperature range in the shallow subsurface, these types of applications of geothermal energy have the potential to be installed almost anywhere that has sufficient fluid available. In this chapter, the basic principles that underlay these applications will be detailed and examples of the applications will be discussed. We will begin the discussion by examining the estimates of the magnitude of the resource.

ASSESSING THE MAGNITUDE OF THE DIRECT-USE RESERVOIR

Approximately 5.4×10^{27} J (Dickson and Fanelli 2004) of thermal energy is available in the continents of the earth, of which nearly a quarter is available at depths shallower than 10 km (Lund 2007). To be useful directly, the heat must be significantly above ambient surface temperatures and capable of being transferred efficiently. Such conditions have traditionally been satisfied in places where warm or hot springs emerge at the surface or in locations where high thermal gradients allow shallow drilling to access heated waters. Such sites are relatively restricted in their distribution, being concentrated in regions where volcanic activity has occurred recently or where rifting of continents has occurred (Chapter 2). For these reasons, a relatively small fraction of the large amount of heat contained within the continents can be economically employed for geothermal direct-use applications.

The proportion of the heat that is readily available is not well known because thorough assessment efforts to quantitatively map the distribution of such resources have thus far been limited. If it is assumed that active plate margins and collision zones are the most likely sites where elevated, near-surface temperatures are sufficient to drive the circulation of warm water, approximately 1%–10% of the area of continents has the potential to support direct-use applications using available technology for accessing subsurface fluids. As drilling technology improves and fluid circulation to support heat harvesting at depth improves, the proportion of the continental thermal resource that can be accessed will significantly expand.

As of 2010, approximately 122 TWh/yr of thermal energy was used for direct-use purposes, worldwide, which was derived from an installed capacity of 50,583 MW (Lund et al. 2011). For comparison, global consumption of electricity in 2006 was 16,378 TWh/yr (Energy Information Administration 2009). The installed capacity of direct-use applications nearly doubled between 2000 and 2005, growing at a rate of about 13.3% per year (Lund et al. 2005). Between 2005 and 2010, the change was almost 79%, with an annual growth rate of about 12.3% per year (Lund et al. 2011).

The growth in installed capacity of direct-use applications reflects a rapid growth in international development of this type of system. In 1985, 11 countries reported using more than 100 MW of direct-use geothermal energy. In 2010, that number had increased to 78. Shown in Table 12.1 is a summary of installed capacity of direct-use geothermal energy systems. The global distribution of these systems reflects the diversity of applications for which they have been engineered.

TABLE 12.1
Capacity of Installed Direct-Use Applications in 2005 and 2010

Country	Capacity (MWt) 2005	Capacity (MWt) 2010	Change in Installed Capacity	Change (%)
Albania	9.6	11.5	1.9	19.6
Algeria	152.3	55.6	−96.7	−63.5
Argentina	149.9	307.5	157.6	105.1
Armenia	1.0	1.0	0.0	0.0
Australia	109.5	33.3	−76.2	−69.6
Austria	352.0	662.9	310.9	88.3
Belarus	2.0	3.4	1.4	71.1
Belgium	63.9	117.9	54.0	84.5
Bosnia and Herzegovina	−	21.7	21.7	−
Brazil	360.1	360.1	0.0	0.0
Bulgaria	109.6	98.3	−11.3	−10.3
Canada	461.0	1,126.0	665.0	144.3
Caribbean Islands	0.1	0.1	0.0	3.0
Chile	8.7	9.1	0.4	4.7
China	3,687.0	8,898.0	5211.0	141.3
Colombia	14.4	14.4	0.0	0.0
Costa Rica	1.0	1.0	0.0	0.0
Croatia	114.0	67.5	−46.5	−40.8
Czech Republic	204.5	151.5	−53.0	−25.9
Denmark	330.0	200.0	−130.0	−39.4
Ecuador	5.2	5.2	0.0	−0.8
Egypt	1.0	1.0	0.0	0.0
El Salvador	−	2.0	2.0	−
Estonia	−	63.0	63.0	−
Ethiopia	1.0	2.2	1.2	120.0
Finland	260.0	857.9	597.9	230.0
France	308.0	1,345.0	1,037.0	336.7
Georgia	250.0	24.5	−225.5	−90.2
Germany	504.6	2,485.4	1,980.8	392.5
Greece	74.8	134.6	59.8	79.9
Guatemala	2.1	2.3	0.2	10.0
Honduras	0.7	1.9	1.2	176.1
Hungary	694.2	654.6	−39.6	−5.7
Iceland	1,844.0	1,826.0	−18.0	−1.0
India	203.0	265.0	62.0	30.5
Indonesia	2.3	2.3	0.0	0.0
Iran	30.1	41.6	11.5	38.2
Ireland	20.0	152.9	132.9	664.4
Israel	82.4	82.4	0.0	0.0
Italy	606.6	867.0	260.4	42.9
Japan	822.4	2,099.5	1,277.1	155.3
Jordan	153.3	153.3	0.0	0.0
Kenya	10.0	16.0	6.0	60.0
South Korea	16.9	229.3	212.4	1,256.8

(Continued)

TABLE 12.1

(Continued)   Capacity of Installed Direct-Use Applications in 2005 and 2010

Country	Capacity (MWt) 2005	Capacity (MWt) 2010	Change in Installed Capacity	Change (%)
Latvia	1.6	1.6	0.0	1.9
Lithuania	21.3	48.1	26.8	125.8
Macedonia	62.3	47.2	−15.1	−24.3
Mexico	164.7	155.8	−8.9	−5.4
Mongolia	6.8	6.8	0.0	0.0
Morocco	−	5.0	5.0	−
Nepal	2.1	2.7	0.6	29.4
The Netherlands	253.5	1,410.3	1,156.8	456.3
New Zealand	308.1	393.2	85.1	27.6
Norway	600.0	3,300.0	2,700.0	450.0
Papua New Guinea	0.1	0.1	0.0	0.0
Peru	2.4	2.4	0.0	0.0
Philippines	3.3	3.3	0.0	0.0
Poland	170.9	281.1	110.2	64.5
Portugal	30.6	28.1	−2.5	−8.2
Romania	145.1	153.2	8.1	5.6
Russia	308.2	308.2	0.0	0.0
Serbia	88.8	100.8	12.0	13.5
Slovak Republic	187.7	132.2	−55.5	−29.6
Slovenia	49.6	104.2	54.6	110.0
South Africa	−	6.0	6.0	−
Spain	22.3	141.0	118.7	532.5
Sweden	3,840.0	4,460.0	620.0	16.1
Switzerland	581.6	1,060.9	479.3	82.4
Tajikistan	−	2.9	2.9	−
Thailand	2.5	2.5	0.0	1.6
Tunisia	25.4	43.8	18.4	72.4
Turkey	1,495.0	2,084.0	589.0	39.4
Ukraine	10.9	10.9	0.0	0.0
United Kingdom	10.2	186.6	176.4	1,729.6
United States	7,817.4	12,611.5	4,794.1	61.3
Venezuela	0.7	0.7	0.0	0.0
Vietnam	30.7	31.2	0.5	1.6
Yemen	1.0	1.0	0.0	0.0
Total	**28,268.0**	**50,583.1**	**22,315.1**	**78.9**

Sources: Lund, J.W. et al., *Geothermics*, 34, 691–727, 2005; Lund, J.W. et al., Direct utilization of geothermal energy: 2010 Worldwide review. *Geothermics*, 40, 159–180, 2011.

Figure 12.1 summarizes some of the types of direct-use applications that have been installed that utilize warm geothermal fluids. The applications have been grouped according to broad categories that share certain engineering approaches. Underlying all of these are the basic principles of heat transfer. The following Section "The Nature of Thermal Energy Transfer" describes these principles and how they are employed to address specific design needs.

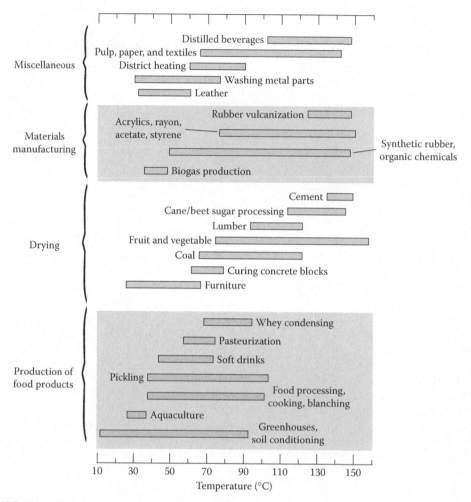

FIGURE 12.1 The temperature range for various direct-use applications. Specific processes are grouped by the way in which heat is utilized or by the type of industry. Food production that utilizes heat is grouped as one category of direct-use application, although heat in this industry is utilized in a variety of ways (space, water, and soil heating to support agriculture and aquaculture; cooking, blanching, and sterilizing food; and processing food products). Drying primarily reflects the use of heat to remove moisture from materials. Materials manufacturing utilizes heat to drive chemical reactions. Miscellaneous represents specific types of applications unique to particular industries.

NATURE OF THERMAL ENERGY TRANSFER

Thermal equilibrium is achieved when coexisting systems or components of a system reach the same temperature, as discussed in Chapter 3. If they are not at the same temperature, and in the absence of imposed thermal barriers or perturbations, heat will spontaneously transfer from the warmer body or bodies to the colder body or bodies until all of the bodies in the system are at the same temperature. It is this fundamental principle upon which all direct-use applications rely. It is, however, the driving force that also results in unwanted heat losses as heat is moved from one place to another. The ability to manage heat movement by minimizing unwanted losses and maximizing useful heat transfer is required to construct and operate an efficient direct-use application. The physical mechanisms that accomplish heat transfer are conduction, convection, radiation, and evaporation. Each of the applications shown in Figure 12.1 is affected by one or more of these heat transfer mechanism. For this reason the transfer mechanisms are discussed in the following subsections. Although these were also covered in Chapter 2, this discussion emphasizes materials and processes that will be encountered in the development of direct-use applications, rather than the bulk earth processes.

HEAT TRANSFER BY CONDUCTION

Heat transfer by conduction occurs when atoms and molecules exchange vibrational energy. At the macroscopic level, this process is manifested as changes in temperature when two bodies, each at a different temperature, are placed in contact with each other. This process is shown schematically in Figure 12.2 in which the temperature trajectory over time is depicted for two bodies brought into contact at time t_1. T_1 and T_2 represent the initial temperatures of the bodies 1 and 2, respectively, and T_3 is the equilibrium temperature they eventually achieve at time t_2. Note that T_3 is not half way between T_1 and T_2, reflecting the effect of heat capacity. In this example, the heat capacity of body 1 must be higher than that of body 2.

Conductive heat transfer is described by the following relationship:

$$Q_{cd} = k \times A \times \frac{dT}{dx} \qquad (12.1)$$

where:

Q_{cd} is the rate at which heat transfer occurs by conduction over the area A
k is the thermal conductivity (W/m-K)
dT/dx is the temperature gradient over the distance x (m)

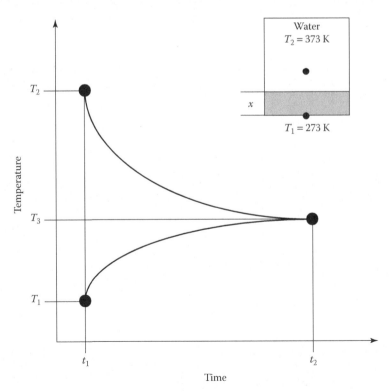

FIGURE 12.2 Schematic representation of the change in temperature over time when two bodies that are at different temperatures are placed in contact. Heat transfer is assumed to take place solely by conduction. The box in the upper right schematically represents the conditions for which the calculations were conducted that are shown in Figure 12.3.

Equation 12.1 is sometimes called Fourier's law of heat conduction. As Equation 12.1 indicates, the rate of heat transfer can be increased by increasing the area over which heat transfer will occur or decreasing the distance.

As an example, consider a container of water at 373 K that is placed on a copper plate. It is assumed that heat is only lost from the base of the container and that the thermal conductivities for the materials are constant. Plotted in Figure 12.3 is the rate at which heat will be lost from the container (J/s) for different thicknesses of the copper plate. It is this rate of heat loss that determines what the thermal gradient will be. The thinner the plate, the greater is the temperature gradient across the plate, and hence, the greater the heat loss rate.

For comparison, the heat loss curves for chrome–nickel steel and sandstone are also shown. Table 12.2 lists the thermal conductivities of a range of materials that may be used in direct-use applications. Note that, for the materials considered in the figure, there is a difference of more than 2 orders of magnitude in the rate of heat loss, but that, for many common materials, the difference can be more than 4 orders of magnitude. Note, too, the sensitivity of the heat loss rate to the thickness of the material being considered.

The example results shown in Figure 12.3 are simplified versions of the behavior of real materials. Although large temperature changes are not associated with most direct-use applications, it must be appreciated that thermal conductivity is a temperature-dependent characteristic of a substance or material. Thus, accurate computation of heat transfer rates must take into account the temperature dependence of k, which was not the case for the calculation carried in the construction of Figure 12.3. The results emphasize the importance of knowing the thermal conductivities and spatial geometry of the materials that may be used for engineering a direct-use application, or which may be encountered when constructing a facility. Incorrect data on these parameters can result in seriously undersizing thermal insulation, inadequately sized piping, and underestimating the rate

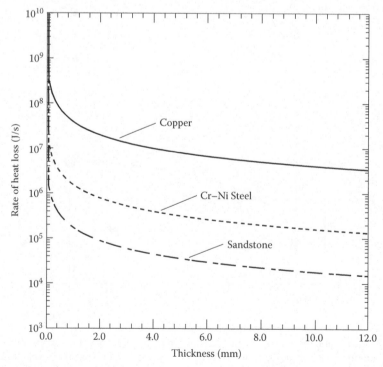

FIGURE 12.3 Rate at which heat is lost (J/s) from a container of water as a function of the thickness of the material (mm) across which the heat is transferred.

TABLE 12.2
Thermal Conductivities (k) of Selected Materials

Material	k_{th}(W/m-K)
Silver (pure)	410
Copper (pure)	385
Aluminum (pure)	202
Nickel (pure)	93
Iron (pure)	73
Carbon steel, 1% C	43
Chrome–nickel steel (Cr—18%, Ni—8%)	16.3
Quartz	6.5
Magnesite	4.15
Marble	2.9
Sandstone	1.83
Glass, window	0.78
Glass wool	0.038
Concrete	1.40
Mercury (liquid)	8.21
Water	0.556
Ammonia	0.054
Hydrogen	0.175
Helium	0.141
Air	0.024
Water vapor (saturated)	0.0206
Carbon dioxide	0.0146

Source: Holman, J.P., *Heat Transfer*, 7th Edition, McGraw-Hill, New York, 1990.

of heat loss to the environment, all of which can seriously compromise the efficient operation of a direct-use system.

HEAT TRANSFER BY CONVECTION

Heat transfer by convection is a complex process that involves the movement of mass that contains heat. Previously, this topic was discussed in Chapter 2, in terms of convective flow of hot rocks deep in the mantle, where viscous forces strongly influence the behavior of materials. Convective heat transfer also occurs at interfaces between materials, as when air is in contact with a warm pool of water or is forced by a fan to flow at high velocity through a heat exchange unit. In such cases, bouncy effects, the character (e.g., turbulent versus laminar) of the flow, the development of boundary layers, the effects of momentum and viscosity, and the effects of the surface properties and shape of the geometry of the flow pathway influence heat transfer.

Figure 12.4 schematically shows some of the influences on convective heat transfer for the simplest possible case, namely, fluid flowing over a flat surface. This geometry is grossly similar to that which is typical of many situations in direct-use applications. In this instance, airflow is depicted by the arrows as laminar (not turbulent) and the surface at the interface between the flowing air and the body of warm water is assumed to be perfectly smooth. Cool air at temperature T_2 moves over a body of warm water that is at temperature T_1. Viscous and frictional forces act to slow the movement of air near the surface of the body of water, forming a *boundary layer*, which is a region where a velocity gradient develops between the interface and the main air mass that has velocity v.

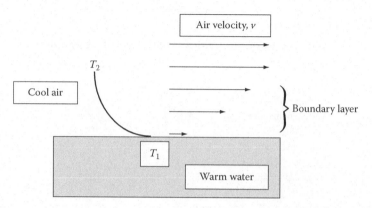

FIGURE 12.4 Schematic representation of convective heat transfer at an air–water interface. The length of the arrows is proportional to the air velocity at the distance from the interface. The boundary layer in the air adjacent to the interface is that region for which the velocity is affected by the presence of the interface. The temperature gradient in the air is indicated by the solid line labeled T_2. T_1 is the temperature of the water.

At the interface, the velocity approaches 0. The characteristics of the boundary layer are dependent upon the fluid properties, velocity, temperature, and pressure. Heat is transferred by diffusive processes from the water surface to the fluid at the near-zero velocity boundary layer base, causing its temperature to approach that of the water, T_1. As a result, a temperature gradient forms, in addition to the velocity gradient, between the main mass of moving air and the water–air interface. This thermal gradient becomes the driving force behind thermal diffusion that contributes to heat transfer through this boundary layer. Advective transport of heated molecules provides an additional mechanism for heat to move through the boundary layer and into the main air mass, resulting in an increase in the temperature of the air.

The rate at which convective heat transfer occurs follows Newton's "law of cooling," which is expressed as

$$Q_{cv} = h \times A \times dT \tag{12.2a}$$

where:

Q_{cv} is the rate at which heat transfer occurs by convection
h is the convection *heat transfer coefficient* (J/s-m²-K)
A is the exposed surface area (m²)
dT is the temperature difference between the warm boundary and the overlying cooler air mass, $(T_1 - T_2)$

Values for h are strongly dependent on the properties of the materials involved, the pressure and temperature conditions, the flow velocity and whether the flow is laminar or turbulent, surface properties of the interface, the geometry of the flow path, and the orientation of the surface with respect to the gravitational field. As a result, h is highly variable and specific to a given situation. Table 12.3 lists values of h for some geometries and conditions.

Determining values for h requires geometry-specific experiments or access to functional relationships that have been developed for situations that are closely analogous to those of a given application. For example, convective heat loss from a small pond over which air is flowing at a low velocity can be reasonably well represented by (Wolf 1983; Rafferty 2006)

$$Q_{cv} = (9.045 \cdot v) \times A \times dT \tag{12.2b}$$

where:

v is the velocity of the air (m/s)
The effective unit of the factor 9.045 is kJ-s/m³-h-°C

TABLE 12.3

Selected Convection Heat Transfer Coefficients (h)

Condition	h(J/m²-s-K)
Static environment with T1–T2 = 30 K	
Vertical plate in air (0.3 m high)	4.5
Horizontal cylinder in air (5 cm diameter)	6.5
Horizontal cylinder in water (2 cm diameter)	890
Flowing air and water	
2 m/s airflow over 0.2 m square plate	12
35 m/s airflow over 0.75 m square plate	75
10 m/s airflow in 2.5 cm tube at 0.2 MPa	65
0.5 kg/s water flow in 2.5 cm diameter pipe	3500
50 m/s airflow across 5 cm diameter pipe	180

Source: Holman, J.P., *Heat Transfer*, 7th Edition, McGraw-Hill, New York, 1990.

As an example of how this relationship can be applied to a real-world situation, consider how quickly a pool of water might lose energy if a light breeze were flowing over it. For a 5 m × 5 m external pool of water that is at 30°C (303 K), with a 3 m/s wind and an external temperature of the air of 0°C (273 K), the heat loss would amount to

$$Q_{cv} = (9.045 \times 3.0 \text{ m/s}) \times 25 \text{ m}^2 \times 30°C = 20{,}351.25 \text{ kJ/h} = 5{,}653.1 \text{ W} \qquad (12.2c)$$

In this example, maintaining the pond at a constant temperature would require adding heat energy to the pond at the rate of 5653.1 J/s. For this and many other situations in which heat is transferred from one medium to another and the possibility of fluid movement is significant, heat transfer via this mechanism must be accounted for to realistically represent heat transfer processes.

HEAT TRANSFER BY RADIATION

In the ideal case, heat transfer via radiation can be represented by considering a so-called ideal black body. In this instance, heat transfer occurs by emission of heat, which can be described by

$$Q_{rd} = \sigma \times A \times T^4 \qquad (12.3)$$

where:

Q_{rd} is the emitted or radiated energy

σ is the Stefan–Boltzmann constant, which equals 5.669×10^{-8} W/m²K⁴

An ideal black body emits radiation that is dependent only on temperature. As a result, the wavelength of the emitted radiation is strictly inversely proportional to the temperature. At room temperature, for example, an ideal black body would emit primarily infrared radiation, whereas at very high temperatures, the radiation would be primarily ultraviolet. Real materials emit radiation in more complex ways that depend upon both the surface properties of an object and the physical characteristics of the material of which the object is composed.

In addition, when considering heat transfer from one object to another via radiation, the viewing aspect of the heat source, as seen by the object receiving radiation, must also be taken into account. A spherical object of a given surface area, for example, will impart via radiation a relatively small amount of heat to an object some distance away, compared to a flat plate with the same surface area and radiation state if the flat plate directly faced the receiving object. Taking these effects into account Equation 12.3 is generally written as

$$Q_{rd} = \varepsilon \times \zeta \times \sigma \times A \times (T_1^4 - T_2^4) \qquad (12.4)$$

where:

ε is the emissivity of the material (ε equals 1.0 for a perfect black body)

ζ is a function that accounts for the geometrical effects influencing heat transfer

T_1 and T_2 are the respective temperatures of the bodies

For most considerations involving radiative heat transfer in direct-use applications, interfaces are commonly flat plates or enclosed bodies in a fluid, thus rendering the geometrical factor of minimal importance, and Equation 12.4 simplifies to

$$Q_{rd} = \varepsilon \times \sigma \times A \times (T_1^4 - T_2^4) \qquad (12.5)$$

where:

ε is the emissivity of the radiating body at temperature T_1

A is its effective surface area

HEAT TRANSFER BY EVAPORATION

Heat transfer via evaporation can be an efficient energy transport mechanism. The factors that influence evaporation rate are temperature and pressure of the vapor that is overlying the evaporating fluid, the exposed area, the temperature of the fluid, the equilibrium vapor pressure, and the velocity of the vapor. Although these properties are relatively simple to formulate individually, the process is affected by factors similar to those that influence convective heat transfer and thus become quite complex.

One complicating factor is that the boundary layer behavior with respect to the partial pressure varies both vertically away from the interface and along the interface due to mixing via turbulent flow. As a result, the ambient vapor pressure is not easily represented rigorously.

In addition, the rate of evaporation is affected by the temperature gradient above the interface, which is affected by the properties of the boundary layer, and which, in turn, influence the equilibrium vapor pressure. Because the temperature gradient is the driving force for diffusional processes, the rate at which diffusion will transport water vapor from the surface will be affected by the local temperature conditions as well as the fluid velocity.

These complications have led to an empirical approach for establishing evaporation rates in which various functional forms are fit to datasets that span a range of conditions. A summary of various results using this method is presented by Al-Shammiri (2002). The form below is from Pauken (1999)

$$E_{ev} = a \times (P_w - P_a)^b \qquad (12.6)$$

where:

E_{ev} is the evaporation rate (g/m^2-h)

P_w and P_a are the water vapor saturation pressures (kPa) at the temperature of the water and air, respectively

$$a = 74.0 + (97.97 \times v) + (24.91 \times v^2)$$

and

$$b = 1.22 - (0.19 \times v) + (0.038 \times v^2)$$

where v is the velocity of the fluid moving over the interface (m/s).

The heat loss is

$$Q_{ev} = \frac{[a \times (P_w - P_a)^b \times A \times H_w]}{(2.778e - 7)} \quad (12.7)$$

ESTABLISHING THE FEASIBILITY OF A DIRECT-USE APPLICATION

The heat transfer mechanisms as discussed in the Section "Heat Transfer by Evaporation" indicates that the total heat loss (Q_{TL}) that must be accounted for when considering an application is the sum of all relevant heat loss mechanisms

$$Q_{TL} = Q_{cd} + Q_{cv} + Q_{rd} + Q_{ev} \quad (12.8)$$

This value represents the heat loss that must be assumed for the operating conditions of the application and does not take into account the actual heat required to do the work that the application was designed to perform. The amount of heat (Q_L) required to perform the function of the designed installation will depend upon the specific process and the size of the operation. Assuming that Q_L is constant over time, then the geothermal resource must be of sufficient temperature and flow rate to satisfy

$$Q_{Geo} > Q_{TL} + Q_L \quad (12.9)$$

For most applications, it is likely that seasonal variability will influence the value of Q_{Tot} through changes in air temperature, humidity, and other seasonal variables. For this reason, the concept of a *design load* was developed. The design load is the most severe set of conditions a facility is likely to experience. For our considerations, the design load becomes the most severe set of conditions likely to be encountered that will maximize heat loss. Hence, when evaluating the feasibility of a potential direct-use geothermal project, it is important to establish whether the resource is sufficient to meet the most demanding conditions that are likely to be encountered.

Strategies for conducting such an analysis are varied. In some instances, where an abundant resource is available, it may be suitable to size the facility in such a way that the resource meets all probable demands. In other instances, possibly for reasons of economics, it may turn out to be sufficient to design the facility such that the geothermal resource will meet the demand of some maximum percentage of probable events. The remainder, lower probability conditions can then be addressed with supplemental energy sources. In the remainder of this chapter, we will discuss these considerations in the context of specific examples. The examples chosen reflect the diversity of the types of applications that can use warm geothermal fluids and the range of issues they must address.

DISTRICT HEATING

In 2010, space heating accounted for 62,984 TJ/yr of the total 438,071 TJ/yr of energy consumed through direct-use applications (Lund et al. 2011). That is the third largest user of geothermal fluids for direct use, worldwide (Figure 12.5). The vast majority of these heating systems involve district heating systems in which multiple users are linked into a network that distributes heat to users.

EVALUATION AND OPERATION

The basic requirements for a district heating system are a source of warm geothermal fluid, a network of pipe to distribute the fluid, a control system, and a disposal or reinjection system. The design

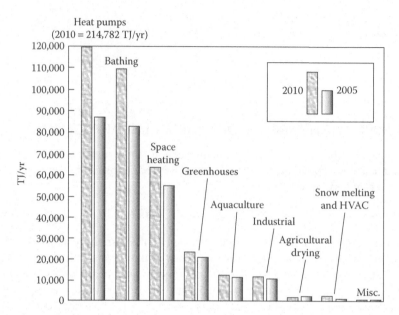

FIGURE 12.5 Energy utilization in direct-use applications, as of 2010. (Modified from Lund, J.W. et al. Direct utilization of geothermal energy: 2010 worldwide review. *Geothermics*, 40, 159–180, 2011.)

of such systems requires matching the size of the distribution network to the size of the available resource. The resource attributes that must be established are the sustainable flow rate (usually this is required to be between 30 and 200 kg/s, depending on the size of the district heating system and the temperature of the resource) and the temperature of the resource. The geothermal power (P_G) that must be provided from a resource is

$$Q_{Geo} \geq P_G = m \times C_p \times (T_G - T_R) \tag{12.10}$$

where:

  m is the mass flow rate (kg/s)
  C_p is the constant pressure heat capacity of the fluid (J/kg-K)
  T_G and T_R are the temperature (K) of the water from the geothermal source and temperature of the return water after it has been through the network, respectively

The heat demand (Q_L) that will be imposed on the system is a complex function of time. During the day, the load can vary by up to a factor of 3 and will be affected by the seasonal climatic variability. From Equation 12.10, it is evident that the only variable that can be controlled that will affect P_G is the return temperature (T_R), since the other variables are established by the properties of the natural system. Maximizing the temperature drop across the network thus becomes a means to increase the power output of the system. However, how this is addressed depends upon the operating mode that can be employed for the system (Figure 12.6).

The simplest systems obtain the required hot water from a surface hot water spring or shallow well and feed the produced fluid to a small network of user sites with direct disposal of the fluid, as in Ranga, Iceland (Harrison 1994). Such systems generally have more heat available than can be used by the district, that is, $Q_{Geo} \gg P_G$. Such systems commonly have high temperature fluids (>65°C). Because management of P_G is not necessary, the magnitude of T_R is unimportant. Nevertheless, development of such systems must be done in a way that minimizes heat losses from the piping that conveys the hot water to the locations in the network where heat is needed. Addressing this factor is important to assure that the system is operated efficiently and in a way that does not compromise potential future use of the resource.

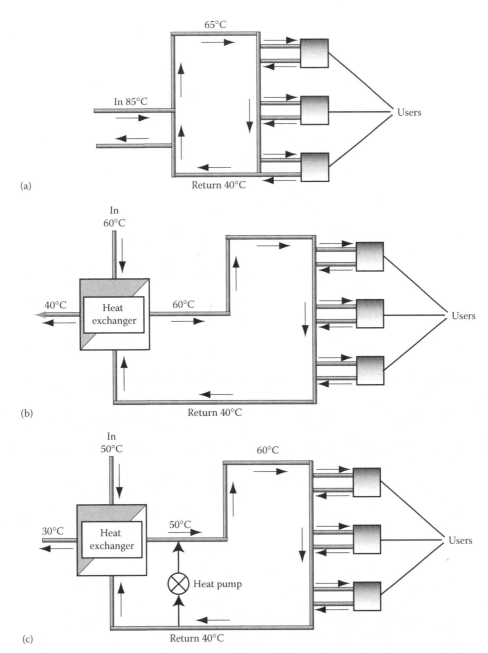

FIGURE 12.6 Schematic representation of three operating modes for district heating systems. (a) Geothermal resource with high temperature fluid (85°C) and high flow rate. Fluid is directly piped to user sites, after mixing with return fluid. (b) Moderate temperature (60°C) resource is piped through a heat exchange unit to transfer heat to a closed-loop system. The closed-loop distributes the heat to user sites. The temperatures along the flow path are meant only to indicate approximate and relative changes. (c) A low temperature (50°C) resource is piped through a heat exchange unit to transfer heat to a closed-loop system. The heat is supplemented by heat extracted from the return fluid by a heat pump. As with (b), the temperatures along the flow path are meant only to indicate approximate and relative changes.

Sites with moderate temperature fluids (50°C–65°C; Figure 12.6b) are capable of operating in a mode in which the geothermal water passes through a heat exchanger that efficiently allows the heat to be transferred to a closed-loop heating network. The fluid in the heating network is pumped to the user sites at which heat is extracted.

Sites where lower temperature fluids are available, down to about 40°C, have been developed that use heat pumps to increase the temperature of the circulating fluid (Figure 12.6c). These systems require input of some additional energy to supplement that of the geothermal fluid. Some of that additional heat can be extracted from the fluid that normally would be the fluid exiting the system.

Finally, district heating systems can be developed that essentially rely on large-scale ground source heat pump systems. These systems can utilize fluids at temperatures less than 40°C, as discussed in Chapter 11. We will not consider these further in this chapter.

MANAGING RETURN TEMPERATURE

In all operating modes, it is crucial to establish a method for controlling T_R. There are several ways to do this. In all cases, user sites employ some form of heat exchanger to extract heat from the circulating fluid. Radiators, or radiant floor, walls, and ceiling installations can accomplish this. In each case, the more heat that is extracted by the user, the lower will be the value of T_R. There are several ways to accomplish this effect. Consider, again, the relationship in Equation 12.1,

$$Q_{cd} = k \times A \times \frac{dT}{dx}$$

that describes conductive heat transfer. The primary mode for heat to be extracted from the circulating fluid in the district network is by conduction through the radiator walls. In Equation 12.1, there are two means for increasing Q_{cd} and thus reducing T_R. The parameter k is in units of W/m-K. Because a watt is equivalent to a J/s, the energy extracted from the fluid will depend upon the contact time of the fluid with the radiator. Hence, the slower the flow through the radiator, the greater will be the heat extraction. Another means for accomplishing the same thing is to increase the area (A) of contact. Hence, putting in place oversized radiators will also increase Q_{cd} and decrease T_R. Managing such networks can then be accomplished by encouraging efficiency by either charging for water usage, which would favor slower flow rates through a radiator, or charging by the temperature drop across the site, which could be done by oversizing the heat extraction units or decreasing flow rates or both.

PIPING AND HEAT LOSS

Efficient and sustainable management requires that heat losses unrelated to the primary purpose of the system be minimized. For a district heating system, this ultimately requires minimal loss of heat in the piping system between the heat source and the users sites. Insulated pipe is commercially available that can allow transmission of hot water over tens of kilometers with heat losses of 10%–15%, depending upon the flow rate. Sizing the pipe to allow sufficient flow, minimal heat loss, and minimal cost requires careful and thorough analysis of the existing demand, including the likely maximum daily and seasonal loads. In addition, it is also important to determine whether growth of the system is to be accommodated by the design or if the system is intended to support a fixed market size. Either case will dictate different approaches to sizing the system.

COMPATIBILITY OF MATERIALS AND FLUID CHEMISTRY

Finally, compatibility of materials is an important aspect of a district heating system. If a new system is being constructed that will provide service to a new community, it can be stipulated that only compatible materials be used in the piping system. However, it is often the case that a district heating system will be constructed that will serve both new and old structures. Plumbing materials are likely to be diverse, from copper and steel pipe to various types of plastic pipe. In such a case, it is important to assess the chemical aggressiveness of the geothermal fluid and the fluid used in the closed-loop, if the latter is part of the system. In some instances, oxygen diffusion through polypropylene or polybutylene pipe may affect corrosion rates in metal plumbing, and the risk of that should be evaluated (Elíasson et al. 2006).

Geothermal fluids generally contain some component of dissolved gases. Although the concentrations can be quite small, over a long period of operation, accumulation of exsolved gases can significantly impact heat transfer processes and fluid flow. It is important to consider the geochemistry of the resource fluid and evaluate the extent to which degassing may occur. If the potential for degassing is at all significant, the ability to degas the system should be an integral element of the system design. Commercially available degassing tanks and facilities can be readily designed into the network to prevent potential problems.

AQUACULTURE

One of the simplest direct-use applications for geothermal fluids is aquaculture. Geothermal fluids have been used to raise carp, catfish, bass, mullet, eels, sturgeon, tilapia, salmon, trout, tropical fish, lobsters, crayfish, crabs, alligators, algae, prawns, shrimp, mussels, scallops, clams, oysters, and abalone (Figure 12.7; Dickson and Fanelli 2006).

Geothermal fluids are used to optimize the temperature for breeding, growth, and health of the species of interest. The impact can be dramatic. For example, according to Dickson and Fanelli (2006), alligators bred under natural conditions reach a length of about 1.2 m in three years. However, if a constant temperature of about 30°C is maintained, alligators will grow to lengths of about 2 m in the same period of time. Shown in Figure 12.8 are growth curves for other animals as a function of temperature. It is evident that, at least in terms of controlling growth rates, optimal temperatures exist that geothermal fluids can maintain.

The design of an aquaculture facility must take into account heat loss from a variety of processes. If the facility uses open pools, heat losses due to conduction from the pool walls must be taken into

FIGURE 12.7 Fish farm in Idaho where geothermal waters as warm as 35°C (95°F) are used to grow tilapia, catfish, and alligators. (Photograph of Ray's Fishbreeders of Idaho, Inc., from *Geothermal Resources Council Bulletin*, November/December, 2001.)

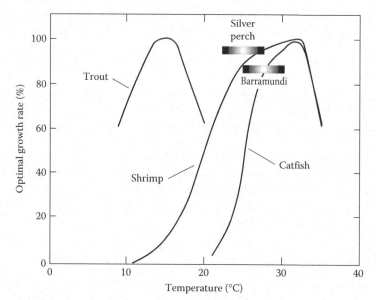

FIGURE 12.8 Rate of growth of fish as a function of environmental temperature. The optimal temperatures for silver perch and barramundi are indicated by the shaded bars. The dependence of growth rate on temperature for trout, shrimp, and catfish are shown as curves over their growth range. Note the steep drop-off of growth rate at the high temperature portion of the growth ranges for shrimp, and catfish. This phenomenon is common among aquatic animals. (Data from Beall, S.E. and Samuels, G., The use of warm water for heating and cooling plants and animal enclosures, Oak Ridge National Laboratory Report ORNL+TM+3381, 1971; Mosig, J. and Fallu, R., *Australian Fish Farmer: A Practical Guide*, Landlinks Press, Collingwood, Australia, 444, 2004.)

account, as well as heat loss from the surface by convection, radiation, and evaporation. The magnitude of the heat losses can be determined by considering an example fish pond. We will consider a facility that measures 10×15 m and is 1.5 m deep and is constructed with concrete walls 10 cm thick in the ground. We will assume that winter air blows over it at a rate of 1.0 m/s at a winter low temperature of 10°C. We will assume the temperature of the pond is maintained at 27°C. The total volume of the facility is 225 m³, which is 225,000 liters.

The heat loss by conduction is

$$Q_{cd} = k \times A \times \frac{dT}{dx}$$

$$Q_{cd} = 1.4 \text{ W/m-K} \times 225 \text{ m}^2 \times \left(\frac{12 \text{ K}}{0.1 \text{ m}} \right)$$

$$Q_{cd} = 37,800 \text{ J/s}$$

Heat loss via convection is

$$Q_{cv} = (9.045 \times v) \times A \times dT$$

$$Q_{cv} = (9.045 \times 1.0 \text{ m/s}) \times 150 \text{ m}^2 \times 17 \text{ K}$$

$$Q_{cv} = 6407 \text{ J/s}$$

Heat loss via radiation is

$$Q_{rd} = \varepsilon \times \sigma \times A \times (T_1^4 - T_2^4)$$

$$Q_{rd} = 0.99 \times 5.669 \times 10^{-8} \, \text{W/m}^2 \text{K}^4 \times 150 \, \text{m}^2 \times (300 \, \text{K}^4 - 283 \, \text{K}^4)$$

$$Q_{rd} = 14,191 \, \text{J/s}$$

Heat loss via evaporation is

$$Q_{ev} = a \times (P_w - P_a)^b \times H_w$$

$$Q_{ev} = 196.88 \times (3.7 - 1.23 \, \text{kPa})^{1.068}$$

$$Q_{ev} = 52,575 \, \text{J/s}$$

The total heat loss (Q_{TL}) is thus

$$Q_{TL} = 37,800 + 6,407 + 14,191 + 52,575 \, \text{J/s} = 110,973 \, \text{J/s}$$

This heat loss value overestimates the average load on the facility for several reasons. The heat lost by conduction will decrease over time because the thermal gradient driving conduction will diminish as the ground around the pond heats up. This term will eventually become negligible and can be ignored for long-term considerations, but must be considered during the start-up phase of the operation. In addition, the calculation was made for winter conditions and a modest breeze. Neither of these conditions is likely to be the average condition for the pond. Under average conditions, it can thus be assumed that the actual heat loss will be significantly less than computed.

Managing the flow of water into the pond requires balancing the rate at which freshwater needs to be added to the facility to keep the stock healthy, and the addition of heat to maintain temperature. Using the heat loss computed above as the actual rate at which makeup heat must be added to keep the pond at 27°C, we can compute the inflow rate as a function of the resource temperature from

$$F_{in} = \frac{Q_L}{C_p \times (T_G - T_P)} \tag{12.11}$$

where:
   F_{in} is the rate at which fluid must be added (l/s)
   Q_L is the total heat loss (J/s)
   C_p is the constant pressure heat capacity of water (J/l-K)
   T_G is the temperature of the geothermal fluid (K)
   T_P is the temperature of the pond (K)

Plotted in Figure 12.9 is the rate at which fluid must be added to keep the pond at 27°C, as a function of the temperature of the incoming fluid. The curve in the figure indicates that the amount of geothermal water that needs to be added to the pond is very high if the geothermal water temperature is within about 15°–20° of the pond temperature. Beyond that point, the amount of makeup fluid that must be added changes at a relatively small rate over a broad range of geothermal water temperatures. This is often considered a rule of thumb for applications such as this (Rafferty 2004), where makeup fluid is added to replenish heat lost through the natural processes we have been considering.

This type of direct-use application must also take into account the chemical effects of the water that is being used. Sensitivity of the animals and plants to nutrient and trace metal concentrations, pH, and dissolved gases must be evaluated in terms of the local water chemistry before direct injection of geothermal fluids into the pond can take place. In instances where the chemical composition of the geothermal fluid is inappropriate for the animals or plants that are to be produced, heat exchangers can be easily installed to accommodate the needed heat transfer.

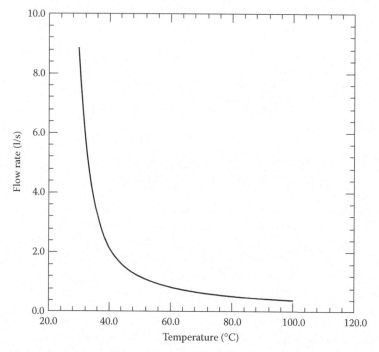

FIGURE 12.9 Flow rate (l/s) required to replenish heat lost from a pond, as a function of the inflow temperature. The pond temperature is assumed to be 27°C.

DRYING

Drying commodities of various kinds is a well-established use of geothermal heat. Installations exist around the world for drying onions, garlic, coconut, meat, fruits, lumber, potatoes, spices, sugar, concrete blocks, and a range of other products. The advantages of using this heat source are elimination of fuel costs to fire boilers and heaters and reduction of risk of fire by eliminating the need for combustion.

The intent in all of these applications is to reduce the water content of the commodity of interest. In the case of many vegetable and lumber products, the requirement is to reduce the water content from about 50%–60%, to as low as 3%. The actual end point to be attained depends on the nature of the product and the anticipated shelf life.

Drying vegetables is usually accomplished by having a geothermal fluid with temperatures between ~110°C and ~170°C pass through a hot-water-to-air heat exchanger. The heated air is then blown through drying ovens with perforated stainless steel conveyer belts or in heating cabinets. In many applications, the drying will be accomplished by having the vegetable products pass through multiple drying stages at different temperatures, which allows the most efficient use of the available heat. The end product usually has moisture contents of 3%–6%. The geothermal fluid exiting the heat exchangers is generally in the temperature range of 30°C–50°C.

Lumber drying requires lower temperature geothermal fluids, generally in the range of 93°C–116°C (Lineau 2006). This lower temperature reflects the fact that the drying process for lumber requires more time (days to weeks), slow water extraction to prevent damage to the lumber, and higher moisture contents for the product (~6%–14%, depending on species and product specifications). Drying is accomplished by placing cut, dimensioned lumber in spaced stacks in a building-sized kiln and allowing warm air to circulate around it. The air is heated by hot water finned heat exchangers around which air is circulated by a ventilation system.

Although the thermodynamic efficiencies of geothermal drying systems are relatively low because of the low temperatures and low ΔT of the process, they are economically and environmentally advantageous because they eliminate the need for a fuel cycle, have zero emissions associated with the heating process, allow easy scheduling for indefinite durations, and have a predictable and constant cost.

The heat losses and energy demands associated with drying are relatively simple to compute, compared to the aquaculture case discussed in the Section "Aquaculture". The primary parameters that must be dealt with, and which are usually well established for any given industry, are the initial moisture content, of the material, the target moisture content, and the drying rate for a given humidity and temperature. The exchange efficiencies of commercially available heat exchangers are well known and therefore do not need to be computed. The principle process of interest is convective heat transfer and the associated energy and temperature change associated with removing water from the commodity of interest. In most cases, empirical relationships and experience form the basis for the design and operation of such facilities.

An important consideration for efficient and environmentally sound management of direct-use systems is the possibility of constructing *cascaded* systems. A cascaded application is one in which the outflow of warm water from one process is used as the source of heat for a process that can function at that temperature. In Canby, California (see Case Study, below), such a cascaded system has been employed to meet multiple needs of a community.

SYNOPSIS

Geothermal resources that have temperatures in the range of 10°C–150°C are capable of providing heat energy for a variety of direct-use applications. Such resources are currently employed around the world in a broad range of industries. Total installed capacity for direct-use applications is in excess of 122 TWh/yr. Direct-use applications are based on the fundamental processes that control heat transfer: conduction, convection, radiation, and evaporation. The extent to which each of these processes affect the performance of a particular direct-use application depends upon the specific needs and design of the application. As a result, each system must be rigorously evaluated for heat losses, demand load, and the magnitude of the potential geothermal heat supply. Efficient use of these low- to moderate-temperature resources can be enhanced by coupling together several applications that cascade from one to the other. Such systems allow the maximum amount of heat to be used for useful work. Because geothermal direct-use applications reduce or eliminate the need for a fuel cycle, have a high capacity factor, and reduce fire risk by eliminating the need for combustion, they can provide significant benefits over traditional technologies. By reducing or eliminating the fuel cycle, as well as implicitly reducing the need for electricity generation, they significantly reduce greenhouse gas emissions.

CASE STUDY: CANBY CASCADED SYSTEM

Cascaded systems allow multiple applications to be linked to a single geothermal fluid source. The basic design takes advantage of the broad temperature range that different applications require. By staging applications in a sequence in which successive applications require lower temperatures, it is possible to cascade a series of applications into a single system. A few examples of such systems that are either installed or underway include the following:

- Klamath Falls, Oregon—power generation cascaded to district heating, greenhouses, brewery, and snow melting
- Cotton City, New Mexico—power generation cascaded to a greenhouse complex
- Geinberg, Austria—power generation cascaded to district heating, spas, swimming, and greenhouses
- Hungary—power generation cascaded to greenhouses
- Podhale, Poland—power generation cascaded to timber drying, greenhouses, fish farming, and space and hot water heating
- Lendova, Slovenia—power generation cascaded to district heating and cooling, aquaculture, greenhouses, and spas

The case study presented here represents a low-cost effort that efficiently utilizes a small resource. The experiences from this development provide insight regarding how to deal with potentially adverse chemical characteristics of geothermal waters and the types of applications limited resources can support.

Many locations have the potential to develop cascaded geothermal applications. In 2008, the Geo-Heat Center of the Oregon Institute of Technology published a survey of western states in the United States to determine the number of communities located within a few kilometers of geothermal resources. The study documented that 404 communities have the potential to utilize geothermal resources (Boyd 2008; Figure 12.10). The resources available to these communities are sufficient to allow the development of cascaded applications. Although very few of these communities have undertaken such efforts, one community in Northern California has been remarkably successful in developing a small resource.

The town of Canby is located in northeastern California in a region that has hot springs and warm water at relatively shallow depths. Tests in water wells had suggested that the local geothermal gradient was such that 60°C–70°C water would be reached at depths of about 487 m. Local hydrology implied the aquifer at that depth and location would support a flow rate of about 9–13 l/s. Such a resource would be very suitable for a small district heating system, with sufficient unused capacity to allow for cascaded uses.

FIGURE 12.10 Map of the western United States showing locations where geothermal resources with temperatures greater than 50°C are located within 8 km of a community. (Reproduced with permission from The Geo-Heat Center, http://geoheat.oit.edu/.)

FIGURE 12.11   Drill rig used for the Canby geothermal project. The drilling rig was a rotary platform system with a tower height of about 18.5 m. (Photograph courtesy of Dale Merrick.)

In 2000, a geothermal well was drilled (Figure 12.11) in Canby, Oregon. When the target depth was reached, it was found that there was inadequate flow and the rock at that depth was unstable. Drilling was continued, at increased cost, ultimately reaching a depth of about 640 m, at which point the temperature was found to be about 85°C. The measured flow rate was 2.33 l/s (37 gallons per minute). Although the flow rate was significantly lower than expected, the higher-than-expected temperature of the resource compensated for the low flow rate and established that the resource was adequate for direct-use purposes.

Chemical analyses of the water, however, showed mercury concentrations at about 282 ng/l, which were well above allowable limits. After considerable effort to find remediation techniques that would resolve the problem, it was found that a granular activated carbon (GAC) filtration system cleaned the water to less than 1 ng/l, which was well below required standards, allowing discharge of the return water from the district heating system to be made to a local river. The resulting discharge had contaminant levels below those required for surface release to streams and below that of the river into which discharge took place.

The design heating load for the district heating system was calculated based on the number of buildings (34) to be included in the system, building size (nearly 5000 m² of total floor space) and construction type (insulated and uninsulated buildings). The peak heating load was based on American Society of Heating, Refrigerating and Air-Conditioning Engineers (ASHRAE) heating design standards for 99.6% of the climate conditions in the area, which was 18°C inside temperature and −15°C outside temperature. This combination of climate conditions and building population resulted in a computed heating load of 389,784 J/s.

Analysis of the hydrological state of the aquifer suggested that long-term production capacity was about 2.3 l/s with a potential maximum draw down of the water table of about 75 m. In order to assure that the system would be operated in a sustainable way, a computer control system was installed to monitor and control flow rates so that the actual load required by the thermostatic controls in the buildings was matched by the flow rate in the system. This necessitated emplacing a pump in the well, which was seated at a depth of about 73 m, which allows precise control of flow rates.

Such a control capability is an important element in a system in which a limited resource must be carefully managed. By monitoring demand, the system can follow the load, thus eliminating or reducing production when there is no or little demand. Such an approach minimizes draw down and allows recovery of the system when demand is low.

The district heating system (Figure 12.12) required approximately 2050 m of pre-insulated copper pipe for the main distribution lines. Approximately 550 m of 2.5 cm diameter pre-insulated cross-linked polyethylene pipe was used for connecting supply and return pipes to each building. Propane-fired furnaces that were in most of the buildings were retrofitted with hot water-to-air heat exchange systems and blowers. In-floor radiant heating was installed in one building. It was realized that the hot water in the district heating system could also be used for domestic hot water heating,

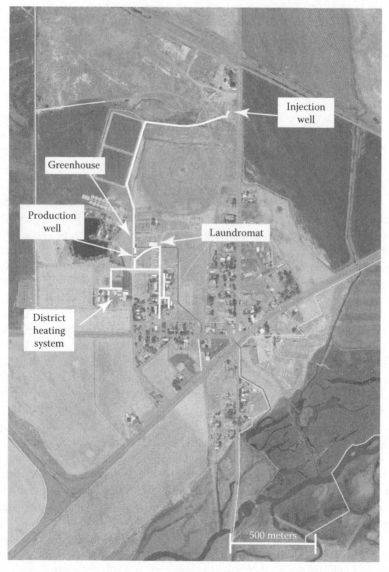

FIGURE 12.12   Air photograph of the town of Canby, showing the district heating layout, production well location and discharge line. The location of the production well, which is colocated with the control facility and GAC filter system, is indicated. (Modified from diagram provided by Dale Merrick and used with permission.)

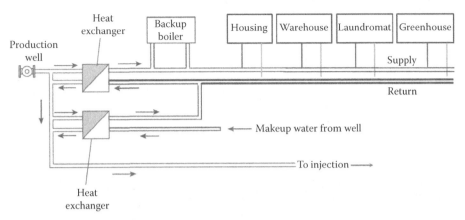

FIGURE 12.13 Schematic of the production well, distribution system, and disposal system for the Canby district heating system. (Modified from an original diagram by Dale Merrick and used with permission.)

so such a system was also installed. Approximately 1550 m of PVC pipe was used for the discharge line (Figure 12.12).

Figure 12.13 shows the component layout of the control and production well facility. Production from the well flows to a heat exchanger capable of transferring 433,745 J/s to the heating fluid in the closed-loop district heating system, which is more than sufficient to meet the design load of 2.3 l/s at a temperature of 88°C. When the heat exchanger operates optimally, it is capable of heating 38°C return water from the district heating system to 66°C. The geothermal fluid that leaves the heat exchanger at 43°C flows to a secondary heat exchanger which preheats return water, thus maximizing heat transfer and use.

It was quickly found that optimal performance for the heat exchanger was not achieved because the water has an iron content of about 3 mg/l. The iron in the water forms deposits on the surfaces of the heat exchanger plates, reducing the thermal conductivity of the plates. Within a period of about three months, the heat transfer efficiency of the primary heat exchanger drops by about 15%. As a result, in order to maintain the needed level of heat transfer on a consistent basis, a maintenance program was incorporated into the operational schedule for the system, such that the heat exchange plates are replaced and/or cleaned every three months.

Experience has shown that, on average, the district heating system load is about half that of the design load. Because the computer monitoring and control system functions on a load following basis, this has allowed the resource to be minimally affected. Experience has shown that it is only for brief winter periods that the design load must be met. On a few occasions, flow rates as high as 2.5 l/s have been required to satisfy the peak demand. However, such periods are short-lived and recovery from these demands is rapid.

Figure 12.14 shows how much energy per second can be obtained from geothermal fluids for a given flow rate and temperature drop across the system. Also shown is the range of operating conditions that the Canby district heating system has encountered. The dashed line is the design load requirements. The fact that operation of the system falls below the design load requirements for all but a small fraction of the time has allowed additional, cascaded direct-use applications to be added.

A clothes washing and drying facility with five high-efficiency dryers and washing machines were added to the system. The dryers were retrofitted with increased airflow passages and hot water-to-air heat exchangers. The hot water supply is provided by a third heat exchange unit through which the geothermal fluids heat the system water. The air for the dryers is heated to between 62°C and 71°C. The variable temperature range reflects the fact that the ambient outside air that passes through the water-to-air heat exchanger can vary significantly with the seasons.

The geothermal hot water cascades from the laundromat facility to a snow melt system installed in attached paved areas keeps open areas free of snow during winter months.

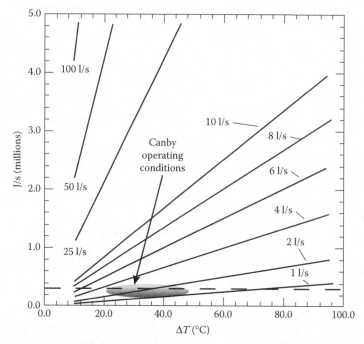

FIGURE 12.14 The amount of useful heat extracted from geothermal fluids as a function of the temperature difference between the supply and return fluids, contoured in terms of the flow rate, in l/s. The range of operating conditions for the Canby direct-use cascaded system are enclosed within the shaded area. The dashed line indicates the design load for the district heating system.

Until 2013, discharge of the used geothermal fluid was to the local river. Environmental requirements stipulated that the discharge temperature could not be higher than 27°C. To remove sufficient heat from the geothermal water in order to meet this temperature goal, the fluid was cascaded to a 270 m² greenhouse facility. The function of the greenhouse, in addition to cooling the geothermal water, is to provide the local community with fresh vegetables and a potential commercial product. The entering geothermal fluid temperature fluctuates between 26.6°C and 54.4°C, depending on the load from the district heating system and other facilities in the system. This fluid passes through a high-efficiency heat exchanger, from which the heat is transferred to the soil in the greenhouse and to the air space to maintain controlled warm air conditions in the greenhouse year round. The fluid leaves the heat exchanger at temperatures low enough to assure it is discharged to the river well below the 27°C constraint.

In 2013, a new geothermal well was drilled (Figure 12.12). The well is now used as an injection well for the used geothermal fluid that has circulated through the system. This well eliminated the need to discharge the water to the river. By careful monitoring of fluid flow and temperatures into and out of each facility in the cascaded system, energy extraction can be maximized and the temperature of the injected fluid better controlled.

The entire system is backed up by a propane-fired boiler, in case of disruption to the heat supply or if heat from the geothermal supply was inadequate. To date, the backup system has been used once, when the downhole pump failed and had to be replaced. Otherwise, the backup system has not been needed because the geothermal supply has been adequate to meet all load demands.

The economics of this direct-use complex are positive. In 2005, the savings in propane and electrical costs for the district heating system alone was $43,355. Given the then-current propane costs, it was determined that the avoided costs of propane would allow a simple payback time of just over eight years. The laundry facility replaced an older facility that used more than 26,700 l of propane per year. With the use of geothermal heat instead of the propane for clothes washing and drying, the avoided cost is sufficient to allow a simple payback period of about three years. In 2013, the cost savings for the community using this cascaded system was approximately $100,000.

PROBLEMS

12.1. For a direct-use fish pond application, what are the two most significant sources of heat loss?

12.2. What methods or strategies could be employed to reduce the heat losses in problem 12.1?

12.3. Plot the convective heat loss, as a function of wind velocity from 0 to 10 m/s, for a 5 m by 5 m pool. Assume that the air temperature is 5°C and the pool temperature is 35°C. Do the same calculation for an air temperature of 10°C.

12.4. Plot how long it would take for the pool in problem 12.3 to drop by 5 degrees, as a function of wind speed, if its depth was 1 m?

12.5. Assume a geothermal water resource of 40°C was available. Plot the flow rate necessary to prevent the pool temperature from dropping more than 5°C.

12.6. Consider the effects of radiation and evaporation in the above calculations and replot the needed input flow rate determined in problem 12.5.

12.7. Figure 12.1 indicates that curing concrete blocks requires temperatures between about 65°C and 80°C. If a geothermal resource were available with water temperatures of 50°C, is there any method that could be employed to use it for the curing process?

12.8. Using Figure 12.1, propose a cascaded system consisting of four applications that could be developed using a geothermal resource with a temperature of 93°C.

12.9. If the Canby geothermal resource were functioning at a flow rate of 5 l/s, what changes might be possible to increase the use of the available energy.

12.10. Discuss three environmental considerations that must be addressed when developing a direct-use application.

12.11. Describe the considerations that would have to be addressed if a district heating system were to be developed using ground source heat pumps.

REFERENCES

Al-Shammiri, M., 2002. Evaporation rate as a function of water salinity. *Desalination*, 150, 189–203.

Beall, S.E. and Samuels, G., 1971. The use of warm water for heating and cooling plants and animal enclosures. Oak Ridge National Laboratory Report ORNL-TM-3381.

Boyd, T., 2008. Communities with geothermal resource development potential. *Geo-Heat Center,* Oregon Institute of Technology Report, Klamath Falls, Oregon, 72 pp.

Dickson, M.H. and Fanelli, M., 2006. Geothermal background. In *Geothermal Energy: Utilization and Technology*, eds. M.H. Dickson and M. Fanelli. London: Earthscan, pp. 1–27.

Elíasson, E.T., Ármannsson, H., Thórhallsson, S., Gunnarsdóttir, M.J., Björnsson, O.B., and Karlsson, T., 2006. Space and district heating. In *Geothermal Energy: Utilization and Technology*, eds. M.H. Dickson and M. Fanelli. London: Earthscan, pp. 53–73.

Energy Information Administration, 2009. International Energy Statistics. http://tonto.eia.doe.gov/cfapps/ ipdbproject/iedindex3.cfm?tid = 2&pid = 2&aid = 2&cid = ww,&syid = 2003&eyid = 2007&unit = BKWH.

Fridleifsson, I.B., Bertani, R., Huenges, E., Lund, J.W., Ragnarsson, A., and Rybach, L., 2008. The possible role and contribution of geothermal energy to the mitigation of climate change. *Proceedings of the IPCC Scoping Meeting on Renewable Energy Sources*, eds. O. Hohmeyer and T. Trittin. Luebeck, Germany, January 20–25, pp. 59–80.

Harrison, R., 1994. The design and economics of European geothermal heating installations. *Geothermics*, 23, 61–71.

Holman, J.P., 1990. *Heat Transfer*. 7th Edition. New York: McGraw-Hill.

Lienau, P.J., 2006. Industrial applications. In *Geothermal energy: Utilization and technology*. eds. M.H. Dickson and M. Fanelli. Earthscan, London. pp. 129–154.

Lund, J.W., 2007. Characteristics, development and utilization of geothermal resources. *Geo-Heat Center Bulletin*, June, 1–9.

Lund, J.W., Freeston, D.H., and Boyd, T.L., 2005. Direct application of geothermal energy: 2005 Worldwide review. *Geothermics*, 34, 691–727.

Lund, J.W., Freeston, D.H., and Boyd, T.L., 2011: Direct utilization of geothermal energy: 2010 Worldwide review. *Geothermics*, 40, 159–180.

Mosig, J. and Fallu, R., 2004. *Australian Fish Farmer: A Practical Guide*. Collingwood, Australia: Landlinks Press, 444 pp.

Pauken, M.T., 1999. An experimental investigation of combined turbulent free and forced evaporation. *Experimental Thermal and Fluid Science*, 18, 334–340.

Rafferty, K.D., 2004. Direct-use temperature requirements: A few rules of thumb. *Geo-Heat Center Bulletin*, June, pp. 1–3.

Rafferty, K.D., 2006. Aquaculture technology. In *Geothermal Energy: Utilization and Technology*, eds. M.H. Dickson and M. Fanelli. London: Earthscan, pp. 121–128.

Wolf, H., 1983. *Heat Transfer*. New York: Harper & Row.

FURTHER INFORMATION

American Society of Heating, Refrigerating and Air-Conditioning Engineers (ASHRAE; http://www.ashrae.org/).
    ASHRAE provides data, instruction, and standards for most applications involving HVAC activities. They provide information that is needed to develop design concepts and calculations for many direct-use applications.

Cataldi, R., Hodgson, S.F., and Lund, J.W., 1999. *Stories from a Heated Earth*. Sacramento, CA: Geothermal Resources Council and International Geothermal Association, 569 pp.
    This book provides an excellent description of many applications of direct-use systems from around the world. It also provides numerous historical references that are useful for gaining perspective on the development of geothermal applications.

Dickson, M.H. and Fanelli, M., 2006. *Geothermal Energy: Utilization and Technology*. London: Earthscan.
    This book provides descriptions of numerous direct-use applications, some of which are quite detailed. It provides a useful introduction to the breadth of direct-use applications.

Geo-Heat Center, Oregon Institute of Technology (GHC; http://geoheat.oit.edu/).
    GHC is one of the world's premier institutions for the study and analysis of direct-use applications. They maintain a database of applications, including international information regarding the most recent developments in technology development and deployment. They also provide software useful for many direct-use applications.

13 Enhanced Geothermal Systems

Throughout this book we have noted that the Earth is a tectonically active, heat-driven planet with a vast thermal resource in the subsurface. Access to this resource has, until recently, focused exclusively on resources that are within a few kilometers of the surface, reflecting the reality of technological limitations and economic challenges when attempting to access deeper resources. Even so, interest in the deep thermal resource has continued to grow as technological capabilities have improved. Several projects currently underway demonstrate the likelihood that development of these deeper geothermal resources will lead to a greater role for these resources in the power generation market. This chapter outlines the history of the concept of *enhanced geothermal systems* (EGS), their magnitude and characteristics, and the state of development of the technology for accessing it.

CONCEPT OF EGS

To be viable, use of a geothermal resource for power generation depends on satisfying the four following conditions:

1. *The existence of sufficient heat.* Usually, for the current state-of-the-art power generation equipment, the minimum temperature is about 95°C. However, the actual minimum temperature at any particular site depends upon the temperature at the back-end of the power generation cycle, that is, the ΔT. For most locations, the minimum temperature will be tens of degrees greater than the minimum that is currently possible.
2. *The availability of sufficient fluid to transfer heat from the subsurface to the power generation equipment.* Although water is commonly used as the heat transfer fluid, other fluids, for example, brine, seawater, and CO_2, may be used (see discussion in Chapter 16 regarding future technological advances).
3. *Adequate permeability to allow a sufficient mass of fluid to circulate through the geothermal reservoir at a rate high enough to achieve reasonable power outputs.* This requires that interconnected porosity be adequate to allow mass flow rates of approximately 10 kg/s or higher. The actual mass flow rate, of course, will depend on the power output required for the designed facility.
4. *Sufficient stability to allow sustainable power generation.* To be environmentally sound and economically attractive, a resource should be managed such that the minimum lifetime of a power generation facility is 20 years. However, longer lifetimes should be used in the design and operation of a facility—experience has shown that a geothermal resource can be operated in a manner that allows it to provide power for many decades.

Development of conventional dry steam and hydrothermal systems has relied on demonstrating that these four requirements are met at a particular site. Much of the expense of developing a geothermal resource, in fact, is incurred because of the necessity to prove that adequate heat exists in reservoir, that permeability can be accessed, that there is either sufficient naturally occurring fluid in the subsurface to transfer heat to the generation facility or that there is sufficient water available to develop a reinjection system, and that the system is sustainable. Historically, satisfying these conditions could only be achieved in a few settings where heat was available within a few kilometers of the surface, reflecting the limitations of existing technological capabilities. Even so, it has long been recognized that the deeper subsurface was, for practical purposes, something close to a near-infinite energy reservoir, as noted in Chapter 2.

Over the last decade, sufficient technological progress has been made so that previously unusable thermal reservoirs can begin to be considered useful for power generation. The key advances that have allowed this change are the result of improved methods that address the latter two conditions listed above. Specifically, new methods have been developed for improving permeability in regions where the natural rock formation may have very low permeability and establishing means for assuring that permeability and high mass flow can be maintained for the long term, both of which are discussed in more detail in sections that follow in this Chapter.

Both of these technological advances reflect improved capabilities to actively engineer subsurface properties of geological systems. These capabilities include the technology to activate existing fractures without inducing new fractures (*hydro-shearing*), the ability to induce new fractures in impermeable rock in order to generate fracture permeability (*hydro-fracturing*), and the use of naturally occurring minerals that are in chemical equilibrium with the reservoir rock at the pressure and temperature of the geothermal reservoir in order to keep fractures open so that permeability enhancement is sustainable. These technological developments, as well as other relevant issues, are discussed in this chapter. These engineered, modified physical conditions are the *enhanced* component of EGS.

Note that these enhancements are independent of the depth of the geothermal reservoir. Although many discussions regarding EGS are commonly referenced to relatively deep environments (~>6 km), enhancement of geothermal systems using engineering methods can be accomplished under any set of conditions. Hence, the phrase "enhanced geothermal systems" should not be restricted to only deep geothermal systems—any system that has been engineered by technological means in such a way that improved permeability or fluid mass flow can be considered an enhanced geothermal system. Although most of the remaining discussion in this chapter will focus on applications involving deep resources, it should be accepted that these same considerations and methods can be relevant for shallow geothermal reservoirs as well.

MAGNITUDE OF EGS

The geothermal systems we have discussed in Chapter 10 for generating electricity have been developed in regions where temperatures greater than about 130°C–150°C can be found within a few kilometers of the earth's surface. Such resources are not common, although they are more abundant than the existing distribution of geothermal generating facilities would suggest. However, as was pointed out in Chapter 2, it has long been known that any location on the planet could, in principle, support construction of geothermal power production facilities since temperatures in excess of 130°C–150°C can be found everywhere—the only question is at what depth?

The magnitude of this resource can be understood by considering the amount of energy available in 1 km^3 of rock that is at elevated temperature. Note that temperatures in excess of 150°C occur at depths between 5 and 10 km over more than 90% of the United States. Over half that area has temperatures in excess of 250°C in that depth range (see Figure 13.1 for temperatures at a depth of 6 km in the United States). The amount of energy that can be extracted (Q_{ex}) is a function of the heat capacity (C_p) of the rock (J/m^3-K), the number of degrees by which the temperature is decreased in the power production cycle (ΔT) the density of the rock (ρ) and the rock volume (V)

$$Q_{ex} = V \times \rho \times C_p \times \Delta T \tag{13.1}$$

For 1 km^3 of rock with a density of 2550 kg/m^3 and a heat capacity of 1000 J/kg-K, the extracted energy as a function of the ΔT is shown by the upper curve in Figure 13.2. Also shown by the lower curve is the amount of thermal energy that could be extracted if only 1% of the rock volume were available for energy production. The amount of energy that could be extracted assuming the temperature of the rock volume is 150°C is delineated by the bracketed region of the line. The overall length of the curves represents the energy extractable if the temperature were 250°C.

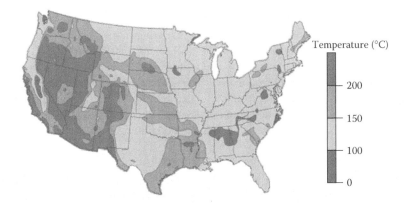

FIGURE 13.1 (**See color insert.**) Subsurface temperatures at a depth of 6 km in the continental United States. (From US Department of Energy, Energy Efficiency and Renewable Energy Office, http://www1.eere. energy.gov/geothermal/geomap.html.)

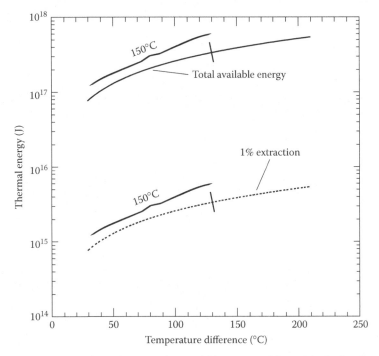

FIGURE 13.2 The amount of thermal energy that could be extracted from 1 km³ of rock, as a function of the temperature difference between initial and final states. The top curve defines the total thermal energy that would be extracted from the entire rock volume and the lower curve represents the extracted energy from 1% of the rock volume. These calculations assume the same C_p at 150°C and 250°C and that the minimum final temperature is 15°C.

Integrating all of the available energy as a function of depth is difficult because of the irregular heat distribution (Figure 13.1) and the absence of subsurface data in many parts of the United States. Figure 13.1 is developed on the basis of scattered measurements of geothermal gradients, surface heat flow, thermal conductivities, and the thickness of sedimentary rocks overlying crystalline basement. The irregular temperature distribution that is documented by these data reflects the effects of geological processes that have shaped the continent. As an example of the impact of geological processes on heat distribution, the higher temperatures in the western United States, and the Basin and Range Province in particular, reflect the effects of rifting in the western United States.

Taking into account the obvious variability, it has been estimated (Tester et al. 2006) that down to a depth of 10 km in the subsurface of the United States, there is a thermal energy resource base in excess of 13 million exajoules (1 exajoule = 10^{18} J). It has also been estimated that about 1.5% of that resource could be extracted using technology that is either currently available or is in development (Tester et al. 2006). Although that amount of extracted energy (~200,000 exajoules) is a small fraction of what is available, it represents more than 2000 times the amount of energy consumed by the United States.

Figure 13.3 shows the amount of energy that could be extracted, as a function of the percent of the total thermal resource that is available within 10 km of the surface. The shaded box represents that portion of the resource for which there is reasonable expectation that the technology may be developed in the foreseeable future to extract that energy. Clearly, the available resource is more than sufficient to satisfy the energy needs of the United States.

From a global perspective, similar arguments can be made for every continental landmass. Figure 13.4 shows how temperature varies with depth and in response to the type of geological setting. Rift settings, where segments of crust are diverging from each other and mantle up-flow occurs in response, are generally the regions where the shallowest high temperature environments occur, as discussed in Chapter 6. Stable continental settings with low heat flow are the regions with the coolest temperatures, as a function of depth. These gradients demonstrate, however, that temperatures of 200°C can be found in all settings at depths of less than ~18 km and less than 10 km for most settings, as indicated by the gray bar in the figure. The clear implication of this fact is that, in principle, all heat required for power generation can be obtained from geothermal resources available within 10–15 km of the ground surface. Similarly, all energy needed for heating and cooling building spaces and most industrial applications could be obtained from that same resource. In other words, the heat energy within the earth is more than sufficient to provide all of the power humanity requires.

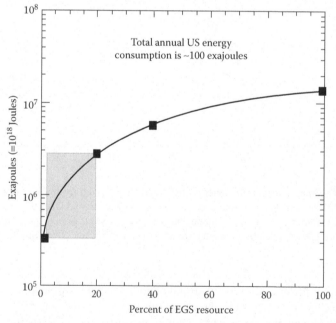

FIGURE 13.3   The amount of energy that would be recovered from the EGS resource base of the continental United States, as a function of the percentage of the EGS resource that is extracted. The shaded box encloses the region that represents 1%–20% recovery. For reference, the amount of energy consumed by the United States is about 100 exajoules.

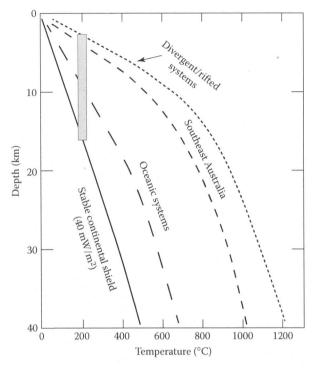

FIGURE 13.4 The variation of temperature with depth as a function of the type of crust. The different crustal types bound the range of temperature gradients for most of the earth's crustal provinces. Note, too, that Southeast Australia is a stable continental domain, but has high heat flow due to the radiogenic nature of rocks in that region, thus providing an indication of the variability to be expected within nondivergent continental crust. The shaded region bounds those conditions at which the temperature is 200°C ± 10°C. (Modified from O'Reilly, S.Y. and Griffin, W.L., 2011. 4D Lithosphere Mapping: A methodology and philosophy for tracing the architecture and composition of the lithosphere through time. http://gemoc.mq.edu.au/Participants/AcademManag/SueResources/4DMapping.html.)

CHARACTERISTICS OF EGS

As discussed in detail in Chapter 10, the principle means through which heat is extracted from the subsurface and used to generate electricity is by bringing to the surface water that has equilibrated with a geothermal reservoir. However, in most deeper regions in the subsurface where temperatures are in excess of 150°C, there is insufficient natural porosity and permeability to accommodate significant mass flow of fluid. As noted above, to overcome this problem, the strategy has been developed to enhance the porosity and permeability of a volume of rock by *stimulating* a geothermal system through hydro-shearing or hydro-fracturing. Once a volume of rock has been *stimulated*, other wells can be drilled into the stimulated zone of fractured rock (Figure 13.5). The resulting stimulated zone provides a complete flow pathway in which an injection well or wells can be used to pump fluid into the subsurface in the region where the rock has been stimulated. The fluid is then induced to flow, via pressure gradients, through the stimulated zone where it picks up heat from the enclosing hot rock and then is pumped to the surface at the production wells. The hot fluid can then be either piped through a heat exchanger of a binary plant or introduced into the turbine of a flash steam plant.

Commercially viable generating facilities have design lifetimes of about 20–30 years. That time period places an important constraint on an enhanced geothermal system. The reservoir must be managed such that it can provide the required heat to power the facility over its design lifetime. That requirement places constraints on the rate at which heat can be extracted from the reservoir; that is,

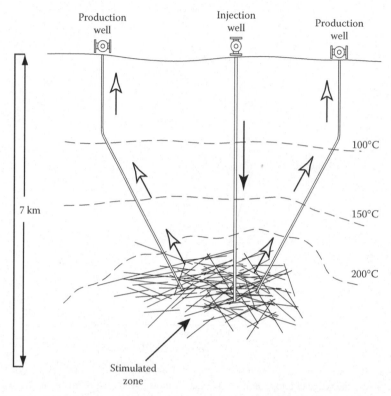

FIGURE 13.5 Schematic diagram of an enhanced geothermal system. The injection well is initially used to stimulate a zone in the rock that is at the target temperature which, in this case, is 200°C at a depth of about 6.5 km. That same well is then used to pump fluid into the stimulated zone. Production wells that have been drilled into the stimulated zone then recover the heated fluid and transfer it to power generating facilities.

it is important to maximize the exposed fracture surface area over which a given amount of heat is extracted. This approach allows the temperature drop in the system to be minimized, per unit area of fracture surface. As shown in Chapter 10, the rate of fluid flow that is required to maintain a 5 MW power plant where the resource temperature is in the range of 200°C–250°C is on the order of 50 kg/s—maximizing the stimulated zone properties so that this flow rate can be maintained while minimizing the thermal draw down becomes an important reservoir management issue.

IDENTIFYING CANDIDATE ZONES FOR STIMULATION

Stimulation, as noted above, is accomplished by increasing the permeability of rocks in the geothermal reservoir. Generally, this is done by pumping fluids into the borehole at pressures that are high enough to either activate existing fractures so that they slip (i.e., hydro-shearing) or fracture intact rock or sealed fractures (i.e., hydro-fracturing). In either case, it is important to target for stimulation only those zones that will provide access to the best thermal regimes. This goal is important for several reasons. Most obvious is that increasing the permeability in the highest temperature zones will provide the greatest efficiency and highest power output, for a given mass flow rate. Another reason is that stimulating zones that are cool will result in the dilution of any high-temperature fluids, thus reducing the thermal energy provided to the generating equipment. Third, it is impractical and expensive to stimulate the entire length of a borehole. The strategy that is most commonly used, therefore, is to isolate the zones that are the target stimulation zones and pressurize fluids only in those borehole intervals.

To accomplish this approach, however, requires knowledge of where the appropriate zones are in the borehole. Several methods are used to identify target stimulation zones. One method that is used is to note the depth intervals at which circulation of drilling muds is lost during drilling. By mapping these zones of lost circulation during drilling, it is possible to locate those regions of modest to high permeability that naturally exist in the geological system. If shallow zones of high permeability occur, they are target depth intervals for sealing or isolation because, if they are not blocked, either they could provide the means for cold fluids to enter the flow stream and cool the hot geothermal waters or they could be zones that would allow the hot fluids to leak from the borehole. Deeper zones associated with high temperature regions are targets for stimulation because they would be the likely path through which water pumped into the borehole at high pressure during the stimulation process would activate existing fractures and/or allow access of high-pressure fluids to a large volume of the reservoir.

The limiting aspect of using lost circulation zones for targeting regions for stimulation is that they do not provide any useful information regarding the existing flow regime in the natural geological system. A zone of lost circulation may indicate the presence of high porosity, fluid-absent rock that does not extend or connect with any extensive, interconnected flow path, but which may be of significant areal extent. It may also not be in a region where the temperature is sufficiently high to be of use for geothermal purposes. In addition, even at very deep intervals, it is possible for water to exist and migrate through rocks—naturally occurring waters have been observed to exist at depths in excess of 15 km in deep oilfield systems. However, such fluids can be isolated from high-permeability zones by impermeable rock units. In contrast, some high-permeability regions may exist in which fluid flow does occur naturally, and lost circulation in these regions would be potentially important indicators. Being able to distinguish between these situations can provide valuable information about the likelihood of being able to develop fluid flow networks that can be used for injection and production in EGS.

A method for detecting where flow may occur is to map temperature perturbations in boreholes when a well is drilled and monitored. One methodology for doing this is called *distributed temperature sensing* (DTS; Erbas et al. 1999; Henninges et al. 2003). In a DTS system, fiber-optic cables are enclosed within a high-grade steel tube a few millimeters in diameter. The cable is then suspended the entire length of the borehole. A laser tuned to a specific frequency is then connected to the optical fiber. The laser light is transmitted down the optical fiber. However, at every point along the travel path of the light, a small fraction of the light is scattered and reflected through interaction with the atoms making up the fiber back up the fiber to the point of origin. That interaction includes Raman scattering, which involves very subtle shifts in the wavelength of the reflected light. This interaction is temperature sensitive, leading to measureable temperature-dependent wavelength shifts. By using sophisticated detection equipment and processing algorithms, the temperature of the DTS cable can be determined everywhere along its length in virtual real time. An example of a DTS record from the Newberry EGS site (Cladouhos et al. 2009, 2011a, 2011b, 2013; Petty et al. 2013) is shown in Figure 13.6. The figure shows the measured initial temperature distribution ("Static") over the depth interval of 1900–3000 m. Note the nearly linear, smooth temperature profile. Two other profiles ("October 24, 2010" and "November 1, 2010") are shown, both of which were obtained after pumping in the well during stimulation and flow testing. Pumping forces fluid into existing zones in which there is some natural permeability. If the porosity volume is sufficient, flow-back will occur after the pressure is lowered. Note that both of the profiles show, in addition to substantial temperature reductions, noticeable deflections in the temperature profiles at exactly the same depth. These deflections mark zones where fluid entered the rock during pumping and is flowing back into the borehole after pumping was halted. Both of these zones indicate regions of high temperature that are candidates for stimulation.

Note, too, that these zones do not directly correspond to the regions that were logged during drilling as depth intervals at which circulation of the drilling fluid was affected by fluid loss. Although these "mud loss" zones are sometimes used to map candidate stimulation horizons, the

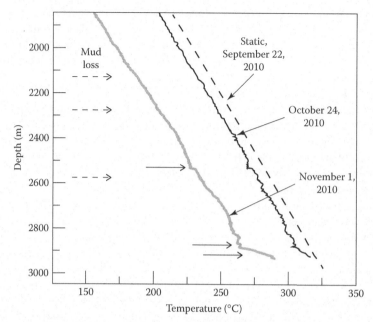

FIGURE 13.6 Temperature as a function of depth at the Newberry EGS demonstration site, as measured using a DTS system. The temperature variation is shown at three different dates. The dashed line labeled "Static" represents the equilibrium temperature conditions in the borehole prior to pumping activities. The two solid lines with later dates show the temperature distribution after pumping activity. Note the perturbation to the smooth temperature curves at the points indicated by the solid arrows, which indicate locations where fluid flow into and out of the enclosing rock has occurred due to pumping activity. These locations are potential candidate zones for stimulation. Note also that the regions where mud loss was recorded during drilling (dashed arrows) do not provide a consistent match with zones that may be good candidates for stimulation.

DTS measurements show that only the deeper zone is the best candidate for enhancing permeability. It may be that the other zones where mud loss occurred could also be stimulated, but the available data are not adequate to demonstrate that.

The primary limitation of DTS systems is their survivability at high temperature. Although the ability to accurately measure temperatures in excess of 350°C is possible with these systems, the components used to construct the cable are not stable above about 200°C. Above that temperature the components of the cable degrade.

IMPROVING PERMEABILITY: REQUIRED ROCK VOLUME

Obtaining heat from rock and transporting it to the surface in sufficient quantity to allow useful power generation is the overall goal of an enhanced geothermal system. Because enhancement is a process of generating pathways through which water will flow, a key challenge for an enhanced geothermal system is how large a rock volume can be stimulated in order to obtain adequate heat.

In order to develop a useful reservoir that has sufficient lifetime to justify the investment necessary for a geothermal power generation facility, several issues must be addressed. One issue that must be considered is the need to keep the temperature reduction along the flow path to a minimum so that the resource can be sustained for many years. Another issue that must be addressed is assuring that an adequate ΔT can be achieved in order to support efficient power generation. Finally, an adequate mass flow must be obtained such that the rate at which heat is transferred to the power generating facility is high enough to allow megawatts of power to be generated. As noted above, the

TABLE 13.1
Surface Area of Fractures (m²) for the Indicated Dimensions

Length (m)	Distance from Injection Well (m)			
	50 m	100 m	1,000 m	5,000 m
2	200	400	4,000	20,000
4	400	800	8,000	40,000
6	600	1,200	12,000	60,000
8	800	1,600	16,000	80,000
10	1,000	2,000	20,000	100,000
20	2,000	4,000	40,000	200,000
50	5,000	10,000	100,000	500,000
100	10,000	20,000	200,000	1,000,000

rate of fluid flow that is required to maintain a 5 MW power plant where the resource temperature is in the range of 200°C–250°C is on the order of 50 kg/s.

Armstead and Tester (1987) have shown that surface areas on the order of 100,000 m² per fracture are required at these flow rates in order to keep the temperature drop along the fracture surface small enough to assure adequate reservoir heat for the required lifetime. Table 13.1 lists the exposed surface areas of rectangular fractures that have propagated the indicated distance from a well and which have the stipulated length perpendicular to that direction. Clearly, this stipulation requires that the ideal situation to maintain a long-lived thermal reservoir is that stimulation result in a few fractures that propagate over significant distances. Such networks must have fracture permeabilities on the order of 10–50 mD in order to sustain adequate flow for power generation.

The minimum volume of fractured rock that can be generated that will satisfy these criteria while also being within the range of existing technologies (see below) is about 2 km³ (see results presented in Baria et al. 2006). Because preexisting fracture sets formed in response to natural stress fields that had specific orientations for the maximum, minimum, and intermediate principal stresses at the time the fractures formed, the enhanced permeability that results from the stimulation process will generally not be distributed in random orientations. Instead, it is likely that a preferred orientation will exist, resulting in an asymmetric reservoir in which flow pathways occur in a limited number of directions.

PHYSICAL PRINCIPLES OF THE STIMULATION PROCESS

As noted previously, hydro-shearing or hydro-fracturing are techniques that can be employed to enhance permeability. Both techniques are based on the same physical principles, but they rely on different methods for applying them.

Recall that rocks in the subsurface exist within a stress field (see discussion of stress and strain in Sidebar 4.1). There are three principal stresses, each oriented perpendicular to the other. These result in maximum, intermediate, and minimum stress values that the rock experiences. The orientation in space of these three principal stress directions depends upon whether tectonic forces are acting on the rock body. If any plane in the rock (either an imagined plane through intact rock or an idealized fracture) is considered abstractly, these principal stresses resolve into normal and shear stresses, the values of which depend upon the orientation of the plane to the principal stress directions,

$$\sigma_n = 0.5 \times (\sigma_1 + \sigma_3) + 0.5 \times (\sigma_1 - \sigma_3) \times \text{cosine}(2 \times \theta) \tag{13.2}$$

$$\sigma_s = 0.5 \times (\sigma_1 - \sigma_3) \times \text{sine}(2 \times \theta) \tag{13.3}$$

Compressive stresses are positive and tensional stresses are negative. If a graph is drawn with two orthogonal axes, the vertical axis being shear stress and the horizontal axis being normal stress, it is apparent through inspection of Equations 13.2 and 13.3 that a plot for the normal and shear stresses acting on the plane, for all values of the angle theta, a circle will be generated (Figure 13.7). This circle defines all combinations of normal and shear stress values that the plane will experience, for any orientation it may have relative to the stress acting on it. Because the stress state is computed using the 2θ value of the orientation, rather than the θ value, the plot demonstrates that the maximum shear stress the plane will experience occurs when the angle between the plane and the applied normal stress is 45°, which is intuitively apparent.

There is a combination of normal and shear stress values at which a rock will fail. This combination of values resolves into a locus of points that define a *Coulomb failure envelope* (also known as a *Mohr–Coulomb failure envelope*). When plotted on a stress–strain diagram, such as that drawn in Figure 13.7, the region enclosed by the Coulomb failure envelope encompasses those stress–strain values for which the rock will maintain its integrity. Stress values that fall outside of that Coulomb failure envelope will result in the rock failing by brittle fracture (the field labeled *FAILURE* in Figure 13.7). The values that define the Coulomb failure envelope are determined experimentally for each rock of interest, because the grain size, grain orientation, and mineralogy, among other things, are the parameters that control rock strength and are unique to each rock unit. For the set of stress values defined by the right-hand circle in Figure 13.7, all conditions that fall within the Coulomb failure envelope and the rock is therefore in a stable state and will not fail.

Two hypothetical Coulomb failure envelopes are shown in Figure 13.7: the one labeled "Intact rock" schematically represents the stress conditions under which some hypothetical rock would

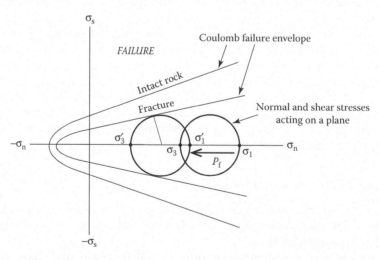

FIGURE 13.7   Two-dimensional representation of the state of stress acting on a plane. Normal (σ_n; horizontal axis) and shear (σ_s; vertical axis) stresses acting on a plane define a circle (Mohr's circle) as the orientation of the principal stress direction is rotated through 180°. For the schematic representation in this example, σ_1 and σ_3 are the maximum and minimum normal stresses acting on the hypothetical plane. The lines labeled "Fracture" and "Intact rock" schematically delineate the combinations of σ_n and σ_s that will result in brittle failure of the respective materials (the Coulomb failure envelope). For the original stress state defined by σ_1 and σ_3, the Mohr's circle falls within conditions that are not sufficiently severe to result in failure. If fluid pressure (P_f) is increased, the effective stresses (σ_n' and σ_s') acting on the rock shift toward the origin until the Mohr's circle intersects one of the Coulomb failure envelopes, at which point brittle failure will occur. In the case depicted here, failure will occur along the fractures in the rock that are oriented at an angle indicated by the line drawn from the intersection of the Mohr's circle and the Coulomb failure envelope for fractures to the center of the shifted circle.

fail through brittle fracture; the other labeled "Fracture" delineates the conditions for failure for a healed fracture in that same rock. The relationship between these Coulomb failure envelopes illustrates important points that affect techniques for permeability enhancement.

First, healed fractures commonly (but not always) are weaker than the rocks they occur within. This reflects the fact that fracture healing is associated with rock alteration in which metamorphic minerals grow as fluid migrates through the fractures and interacts with the enclosing rocks. The minerals that grow are typically weaker than the minerals in the host rock because the former are usually relatively low metamorphic grade minerals, which are likely to be hydrous, weak, and/or easily cleaved. Second, the fractures commonly are approximately planar in form, whereas the enclosing rock often has a much more intergrown, complex structure, which gives it a higher degree of intrinsic strength. The consequences of these relationships can be appreciated when considering the techniques that are commonly used to enhance rock permeability.

The primary method to enhance rock permeability involves pumping water into the well until failure occurs. Increasing fluid pressure counteracts the compressive stress the rock experiences under natural loading conditions when it is buried and confined. The magnitude of the effective normal and shear stresses at elevated fluid pressures can be computed as follows:

$$\sigma_1' = \sigma_1 - P_f \tag{13.4}$$

$$\sigma_3' = \sigma_3 - P_f \tag{13.5}$$

where:
σ_1' and σ_3' are the effective maximum and minimum normal stresses acting on the plane
P_f is the applied fluid pressure

Note that, as discussed in Chapter 4, these variables all have the dimension of force per unit area.

The consequence of this elevated fluid pressure on the mechanical properties of the rock is to shift the relevant Mohr's circle to the left, as indicated by the left-hand circle in Figure 13.7. As the pressure is increased and the circle migrates, it will eventually reach conditions at which the Coulomb failure envelope becomes tangent to the circle. This point of tangency defines the normal and shear stresses at which brittle failure will occur. In addition, the line drawn from the point of tangency to the center of the relevant circle defines the angle at which brittle failure will occur in the rock. For rocks that have a population of fractures, and the fracture orientations are approximately that indicated in the diagram, hydro-shearing will result. That is, the healed fractures will be reactivated and slip. In this case, the unfractured rock that hosts the healed fractures will not fail but will remain intact (Cladouhos et al. 2009). The amount of slip an activated fracture will undergo will depend upon the characteristics of the fracture, such as its length and planarity, roughness, and other properties.

Because of the limited amount of slip that a preexisting fracture can experience, increasing the fluid pressure beyond that required for hydro-shearing is possible. In this case, the Mohr's circle will migrate further to the left in the diagram, beyond the Coulomb failure envelope for the healed fractures, until either it reaches the Coulomb failure envelope for the intact rock σ_3' intersects the shear stress axis (which also means the normal stress has been reduced to zero). At that point, failure of the intact rock by brittle fracturing will occur, and hydro-fracturing will be initiated.

The orientations, distribution, and lengths of fractures that will develop, and the volume of rock they will access, will depend upon several factors. One important factor is the orientation of the imposed stress field that the rock is experiencing. Tectonic forces, loading, and other geological factors will determine the overall state of stress the rock and the orientation of the imposed stress field. Also important is the extent to which a preexisting fabric may have developed in the rock that results from preferred mineral orientations and compositional heterogeneity. Finally, if the rock consists of a variety of minerals with diverse mechanical properties, their orientations, relative abundances, and distribution throughout the rock will affect the characteristics of the fractures that develop.

MANAGING THE STIMULATION PROCESS

Permeability enhancement is accomplished by developing a network of fractures that provide access for fluids to extract heat from a sufficiently large rock volume. The most effective way to accomplish this is by injecting fluids at pressures that are high enough to induce brittle failure of existing fractures or intact rock. However, because boreholes may extend over distances of several meters to tens of kilometers, pressurizing an entire borehole would not be appropriate. Indiscriminate pressurization would result in developing fractures in regions that would not be the most suitable for the purposes of power generation and cold zones may be stimulated. As a result, methods for isolating selected regions for permeability enhancement have been developed.

In the oil and gas industries, "packers" are commonly used. These can be mechanical or pneumatic devices that are used to seal the borehole above and below the region that is being stimulated (Figure 13.8). Packers can be made of expandable, accordion-like segments of flexible metal that are pressurized to expand and seal a borehole, or of flexible, elastomeric materials that can be inflated. Additionally, sections of a borehole can be cemented closed over certain intervals to allow for a closed off section of the well. In all of these cases, a pipe or some other access port is required that passes through the packer so that fluid pressure can be built up within the sealed-off zone. Such systems usually require a drilling rig to be on site to maintain the piping system and access to the packers for modification, if needed, and removal. Once stimulation is completed, the packer is removed either by deflating it

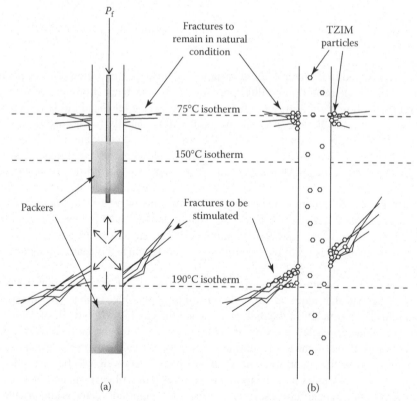

FIGURE 13.8 Schematic representations of two methods to stimulate fractures. (a) Use of packers to isolate a zone for stimulation. P_f represents a method to introduce a high-pressure fluid into the packed-off interval. (b) Use of thermally-degradable zonal isolation materials (TZIM) to close off fractures that are not to be stimulated (upper fracture set). The TZIM is pumped into the borehole at high pressure until it has plugged up all of the fractures and permeability. The TZIM is chosen so that it degrades slowly at the temperature of the upper fracture set, but degrades quickly at the temperature of the lower fracture set, allowing the lower zone to be stimulated shortly after TZIM emplacement while keeping the upper zone blocked so fluid pressure in those fractures cannot increase.

or drilling it out.thermally-degradableAlthough packers have been employed for work in geothermal systems, they are not reliable at the higher temperatures likely to be encountered in EGS, that is, at temperatures exceeding 200°C. Generally, the materials that are critical for operation of a packer, such as elastomeric materials, control elements, and various combinations of metals with different expansivities, tend to fail under the extreme temperatures and pressures of geothermal systems. For this reason, other approaches are being considered and developed for isolating the regions where stimulation and permeability enhancement will be undertaken.

A newer method that has recently been developed for isolating zones for stimulation avoids the use of mechanical or physical devices to block off segments of the borehole and instead uses materials that are unstable to plug existing fracture networks or permeability zones. These materials are called thermally-degradable zonal isolation materials (TZIM; Petty et al., 2013). These compounds are designed to chemically or thermally breakdown over time to benign products, such as H_2O, CO_2, or other nonreactive and nontoxic molecules. The rate at which they decay will depend upon the temperature or the chemical properties of the environment. By measuring the breakdown rates in the laboratory over a range of conditions, a specific compound can be selected to be most useful in a specific environment (Petty et al. 2013). Figure 13.9 shows the degradation curves for two hypothetical compounds that are temperature sensitive.

The approach that would be taken to employ these compounds is to inject into a borehole finely ground TZIM material that will rapidly breakdown under the conditions that are observed in the region that is the stimulation target, but will persist for longer times to block those regions for which stimulation is not desired. Referring to Figure 13.8, the deep zone that is the stimulation target region is at temperatures significantly above 150°C, whereas the upper zone that should remain unaffected by the stimulation process is at temperatures significantly below 150°C. Hence, a TZIM that would rapidly breakdown above 150°C but persist for weeks at lower temperatures is desirable. In Figure 13.9, hypothetical compound A would be a good candidate compound, whereas compound B would not be suitable because it would persist for long periods under all conditions in the borehole.

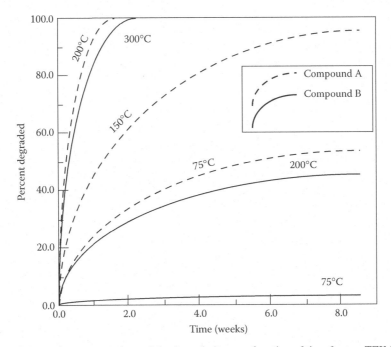

FIGURE 13.9 Schematic representations of the degradation as a function of time for two TZIM compounds. Compound A degrades relatively quickly at moderate temperatures, thus being suitable for stimulation of systems at low to moderate temperatures. Compound B is a more stable TZIM, appropriate for higher temperature applications.

The approach that would be taken to stimulate this well would be to first establish the approximate aperture of the fractures in the various zones through some geophysical or other remote sensing technique. Once known, a suite of grain sizes of compound A would be mixed, the size distribution being such that the compound would be able to easily flow into the fractures. It would then be injected into the well at moderately high pressure to assure it enters the fracture system. The system would then be allowed to equilibrate, allowing the higher temperature zone a few days to sufficiently degrade the injected TZIM compound. A general rule of thumb is that about 50% degradation is sufficient to allow stimulation to proceed. In the case we are considering, that degree of degradation would occur in the higher temperature fracture zone within a few days. The amount of degradation in the shallower, cooler region would be less than about 15%. Hence, after a few days stimulation of the deeper zone using high-pressure water injection can be initiated without initiating hydro-shearing in the cooler fractures. Once stimulation is completed, allowing the system to approach steady-state conditions for a period of approximately eight weeks would allow the partial degradation of the shallower zone, at which point a more stable TZIM could be injected into that zone if it was desired to seal it indefinitely.

Enhancing the permeability of a rock volume is a process that perturbs the existing steady-state condition of the rock mass. Without some means to prop open the fractures that dilated during the hydro-shearing process, it would be anticipated that the fractures would close back up once the high pressure was reduced. However, the hydro-shearing process encourages rock failure to occur by shear along pre-existing fractures. Because fracture surfaces tend to be irregular and rough, shearing of one fracture surface past another surface will usually cause rock fragments to shear off the fracture wall and remain as loose particles within the fracture opening. Once the fluid pressure is removed when the stimulation process is completed, these rock fragments will tend to prop open the fracture, maintaining the enhanced porosity. Within the oil and gas industries, it has become common practice to introduce into the hydro-fracturing fluid artificial proppants, such as sand or other materials of selected sizes and hardness, to maintain fracture apertures once a system has been stimulated. However, such compounds are not necessarily thermodynamically stable in the environment of the natural system and will tend to degrade over time, allowing fractures to close-up. The hydro-shearing process, however uses as proppants the natural minerals that have developed as the thermodynamically stable compounds in that environment. Hence, these natural proppants have a much greater probability of maintaining the enhanced permeability for the lifetime of the enhanced geothermal system that has been engineered.

MONITORING THE STIMULATION PROCESS

When slip due to brittle failure occurs along fractures, seismic energy is released. This will be the case regardless of the amount of slip. If seismometers are placed in a variety of surface locations around the stimulation site, as well as in boreholes specifically intended for seismic monitoring, it is possible to detect the small seismic events associated with slip along fractures during hydro-shearing and/or hydro-fracturing. Monitoring this microseismic activity allows the shape and location of the stimulated zone to be mapped. Using that information, the stimulation process can be managed effectively so that the specific zones of interest are actually stimulated, unwanted zones of stimulation can be detected and remedied, and any unexpected seismic response can be detected.

Mapping the stimulated zones is an effective way to establish the drilling targets for production wells. The overall shape and depth of the stimulated zones can be mapped, and strategies for placement of the wells in the system can be developed and efficiently implemented. This allows drilling to be undertaken to intersect the most likely regions of enhanced permeability.

An additional benefit of collecting seismic data is the ability to map the mechanism by which brittle failure occurs. When brittle failure occurs along a fracture, motion can be pure dilation (Figure 13.10), pure strike slip (motion purely in the horizontal direction), pure dip slip, pure reverse slip, or some combination that results in oblique slip. When the seismic energy is released upon the first motion during failure, the compressional and shear waves that propagate away from the point of failure have an orientation and polarity that can be detected with sufficiently sensitive seismometers.

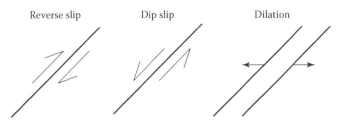

FIGURE 13.10   Schematic representation of the sense of motion along a fracture. This figure is drawn as a vertical slice through a fracture showing how the two sides of the fracture would move relative to each other for reverse slip, dip slip, and dilation. Pure strike slip motion would occur with the sides of the fracture moving in opposite directions into and out of the vertical plane represented in the figure.

If there are a sufficient number of seismometers deployed around the site, detailed analysis of this "first motion" from all of the responding seismometers allows the type and direction of slip to be determined. By mapping the distribution of the various failure mechanisms of a population of events within the stimulation zone, the state of stress in the rock and the orientation of slip planes can be inferred. This can be valuable information useful for planning and controlling stimulation efforts. Such information can greatly reduce the overall cost of the stimulation process by providing better control. In addition, reducing the risk of large seismic events can be enhanced by access to information on the state of the stress regime and its orientation (see discussion in Chapter 15).

HISTORY OF ENHANCED GEOTHERMAL SYSTEM DEVELOPMENT

Attempts to develop EGS began in the late 1970s and early 1980s with the Fenton Hill "Hot Dry Rock" project that was carried out at Los Alamos National Laboratory, supported by the Department of Energy (Smith 1983). This project was the first to undertake deep drilling, stimulation, and fluid injection in order to produce power. Although the project was a technical success, in the sense that it proved that the concept of stimulating hot rock to allow circulation and heat exchange was sound, a number of technical challenges prevented it from being a successful demonstration of an economically viable concept. Since that time, EGS projects have been pursued in Australia, France, Germany, Japan, Sweden, Switzerland, the United Kingdom, and the United States (Table 13.2).

TABLE 13.2
Locations and Characteristics of EGS Efforts Worldwide

Location	Years	Depth (m)	Temperature of Producing Fluid (°C)	Flow Rate (l/s)	Power Facility (MW)
Fenton Hill, New Mexico	1972–1996	3,600	191	13	–
Rosemanowes, United Kingdom	1984–1991	2,200	70	16	–
La Mayet, France	1984–1994	800	22	5.2	–
Hijiori, Japan	1985–2002	2,200	180	2.8	–
Soultz-sous-Forêts (I), France	1987–1995	3,800	135	21	–
Soultz-sous-Forêts (II), France	1996–present	5,000	155	25	1.0
Landau, Germany	2005–present	2,600	160	76	1.5
Habanero, Australia	2003–present	4,250	212	30	1.5
Newberry, Oregon	2009–present	3,075	290	25	–
Desert Peak, Nevada	2002–present	1,800	200	32–101	–

Source: Wyborn, D., Hydraulic Stimulation of the Habanero Enhanced Geothermal System, South Australia, 5th British Columbia Unconventional Gas Technical Forum, Victoria, Canada, April 5–6, 2011.

Although only one of these projects is currently functioning as a commercial enterprise as an integral part of a power grid (Landau, Germany), pilot projects and initial test operations have been successful in several other locations. Successful pilot project completion was achieved at the Habanero enhanced geothermal system power plant, located ~10 km south of Innamincka in South Australia. In late 2013, this plant completed an operational test program on a scaled down 1 MW facility. Also in late 2013, the Desert Peak, Nevada Enhanced Geothermal System project completed its initial development and test phase, as did a project in Oregon at the Newberry Volcano. The Newberry Volcano EGS effort is discussed in more detail under the Section "Case Study" later in this Chapter.

Completion of these projects, and those that preceded them, has provided important lessons that have refined the research needs to advance EGS to commercial status. Some of the key challenges and technological achievements of these efforts are summarized in Sections "Drilling and Downhole Equipment" and "Drilling Fluids."

DRILLING AND DOWNHOLE EQUIPMENT

The requirements to successfully complete a well have been discussed in Chapter 8 and the methods for stimulating a well have been discussed above in sections relating to the stimulation process. Accomplishing a completed, stimulated well is particularly difficult at the high temperature and pressures encountered in EGS projects. Equipment failure of downhole-sensing equipment, packers, and other electronic devices is common. This is an important challenge. Downhole cameras and other optical devices are often used to map fractures and before the stimulation process occurs. This information is useful for planning and designing the stimulation program. Devices for measuring the borehole shape and how it changes over time can be useful for establishing the stress state in the rock at depth. If directional drilling is to be employed, electronic or mechanical equipment is needed to establish the orientation and location of the drill string and to allow steering of the bit. In addition, drilling bits and pipe that are usually used for EGS holes often are identical to equipments used in oil and gas drilling, which does not encounter such severe conditions. All of these issues emphasize the fact that much research needs to be done to overcome the technological challenges of drilling at the kinds of temperatures most EGS efforts are likely to encounter.

DRILLING FLUIDS

It is relatively common to encounter zones in the subsurface where high permeability allows drilling fluids to escape to the surrounding rock, thus short-circuiting the fluid that would cool the bit and drill string and remove the cuttings. Materials normally used to seal such zones often have limited ability to perform at elevated temperatures, thus making it difficult to adequately and permanently seal off such zones.

In addition, drilling muds normally used in deep wells often begin to breakdown at the high-temperature conditions encountered in these wells. Clays in the muds can begin to dehydrate, as described in Chapter 8 in the discussion of drilling muds. This was dramatically demonstrated in the Kakkonda case study, detailed in Chapter 8. In addition, chemical compounds in the natural environment may react with other compounds in the muds, modifying the viscosity and thermal properties of the materials. Finally, chemical additives in the drilling muds intended to improve their performance are usually employed based on shallower, lower temperature experience. At the elevated temperatures of EGS, the additives can become unstable. All of these effects degrade the ability of the drilling fluids to flow, lubricate, and seal.

RESERVOIR ENGINEERING

The ability to stimulate a zone using hydraulic techniques has now been well established. Controlling the fracture properties and their geometry remains, however, an important challenge. The ability to assess the orientation and properties of fractures that will open when stimulation is underway needs further improvement. Measuring the orientation and magnitude of subsurface stresses at high temperatures, as well as the properties and orientations of existing fracture sets, needs refinement. As stimulation proceeds, it is also important to be able to control the rate of fracture growth through knowledge of the likely response of the rock mass to changes in pressure, pumping rate, and fluid properties. Although it is unquestionable that efforts to date have been remarkably successful at accomplishing this, achieving these goals *with a high degree of certainty* is currently beyond available technology. Although the scientific principles involved in rock fracture mechanics are well understood, the ability to accurately and precisely control fracture development and propagation remains an area of active research and development.

At the elevated temperatures of EGS, fluids will react with fracture surfaces that are newly exposed. This *reactive transport* has the potential to modify fracture apertures, roughness, and other properties. Some of these changes can be beneficial, such as dissolving minerals along the fracture and thus increasing fracture apertures or depositing stable minerals at asperities, thus more securely propping open the fractures. However, the effects of reactive transport could also be detrimental, such as restricting fluid flow and decreasing permeability by depositing large volumes of secondary minerals along flow paths or causing unwanted fracture growth by corrosion. The ability to model these processes has become quite sophisticated (Clement et al. 1998; Xu and Pruess 1998, 2001; Parkhurst and Appelo 1999; Glassley et al. 2003; Phanikumar and McGuire 2004; Steefel et al. 2005; Rinaldi et al. 2012). It is now possible to model three-dimensional hydrological and geochemical processes at very high resolution, taking into account multiphase flow in fractured, porous media. The results can assist with forecasting the short- and long-term interactions that can be expected for injected fluids of different chemistries. Using this approach, it has been suggested that the use of CO_2, rather than water, as the heat-carrying fluid may be beneficial in reducing chemical interactions (Pruess and Azaroual 2006).

Currently, the primary limitation of this modeling approach is the inability to obtain detailed characterization of the *in situ* conditions that are the starting point for such simulations and representing mechanically realistic fractures. Nevertheless, this capability holds promise for providing the ability to develop detailed projections of the evolution of porosity, permeability, and geochemical and mineralogical properties in producing geothermal reservoirs.

RESERVOIR MANAGEMENT FOR SUSTAINABILITY

When undisturbed, geothermal reservoirs represent the natural evolution of a geological system in which heat transfer is influenced by conduction and convection processes. The balance of these two heat transfer mechanisms reflects the local geological and hydrological properties of the site. In some hydrothermal systems, convective fluid movement plays an important role in establishing the thermal regime. The geothermal regime at Long Valley Caldera, for example, (Chapter 6) is strongly affected by convective heat transfer. In other systems, particularly those that are being considered as sites for EGS development, conduction is the dominant heat transfer mechanism and convection plays a very minor role. When EGS development takes place in these settings, convection-dominated heat transfer is introduced into the reservoir when the reservoir is stimulated and fluid pumped through the fracture system. Because heat transfer by convection will occur at a rate that is many times faster than the *in situ* conduction rate, heat will be removed faster than it will be replaced. The challenge then becomes how best to manage the resource to assure that it can sustain power production over the lifetime of the power generating facility. It also becomes important to understand the time required for recovery of the system, in order to guide responsible resource management.

Realistic modeling of the behavior of these systems requires the use of sophisticated simulation tools that can account for fluid flow, chemical reactions, and heat transfer in complex, three-dimensional fracture networks. The reactive transport simulation tools referenced above are the best means for accomplishing this. However, less computationally intensive approaches can be used to approximate the behavior of these systems and gain important insight into their behavior and evolution.

One such approach is that outlined by Gringarten and Sauty (1975), who considered the time required for a temperature drop to propagate from an injection well to a production well

$$T_b = \frac{(\pi \times \gamma_t \times d^2 \times t)}{(3 \times \gamma_f \times v)} \qquad (13.6)$$

where:

T_b is the time (h)
γ_t is the heat capacity (J/m³K) of the reservoir
γ_f is the heat capacity of the fluid (J/m³K)
d is the distance between wells (m)
t is the reservoir thickness (m)
v is the flow rate (m³/h)

This approach assumes constant temperatures on either side of the reservoir body and does not account for the change in conductivity with temperature. Rigorous representation of the rate of temperature draw down also requires detailed knowledge of the flow geometry, heat capacity of the rock, and exposed surface area along the flow path. Nevertheless, this approach does provide an approximate indication of the time required to achieve a given temperature drop.

Figure 13.11 shows the results computed for a temperature change of 10°C. Note that both the well separation and the flow rate have a significant impact on the time it takes to realize a specified temperature reduction. In this particular case, if a lifetime of 30 years is assumed for the power facility, reasonable flow rates (<80 kg/s which approximately equates to 280 m³/h) can be supported, provided the well spacing is greater than 500 m. Shorter well spacings would diminish the possible flow rate.

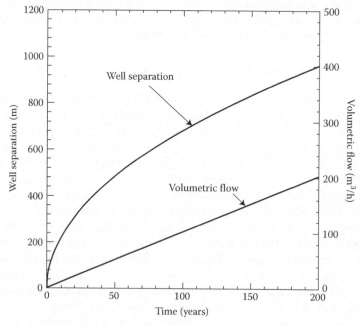

FIGURE 13.11 The time required to achieve a 10°C temperature drop, as a function of the spacing between the injection well and the production well (left axis), and the volumetric flow rate. Both curves were calculated assuming a porosity less than 0.001, a rock heat capacity of 2.7e6 J/m³K, a fluid heat capacity of 4.18e6 J/m³K, and a reservoir thickness of 25 m. The well separation distance was calculated assuming a constant volumetric rate of 88 m³/h, and the volumetric flow rate curve was calculated assuming a constant well separation of 500 m.

The importance of these considerations is emphasized by considering the time it takes to replenish the heat. The ratio of the rate of heat transfer by convection to that of conduction (see Chapter 12 for a discussion of these heat transfer mechanisms, specifically Equations 12.1 and 12.2) is

$$\frac{Q_{cv}}{Q_{cd}} = \frac{(h \times A \times dT)}{(k \times A \times dT/dx)} \tag{13.7}$$

The convective heat transfer coefficients for fluid flowing through fractures are not well characterized, but will be in the range of several hundred to 1000 J/s-m^2-K. The thermal conductivity, however, will be less than 10 J/s-m-K. Assuming midrange values for these coefficients (300 J/s-m^2-K and 5 J/s-m-K, respectively) and assuming the thermal gradients are in the immediate vicinity of fracture surfaces and act over a distance of 10 m, these assumptions result in (after reconciling the units)

$$\frac{Q_{cv}}{Q_{cd}} = \frac{(300\,\text{J/s-m})}{(5\,\text{J/s-m})} = 60$$

This result indicates that, in the immediate vicinity of a fracture, replenishing the heat removed by convection with heat in the surrounding rock mass will take significantly longer than the time it took to remove the heat. Accurately estimating the amount of time for an entire enhanced geothermal system to recover is a complex problem (Elsworth 1989, 1990) and is not possible without detailed information about geometry of flow paths, heat capacities, and thermal conductivities of the affected materials and the local heat flow. However, for reasonable assumptions about these parameters and the geometry of an EGS-stimulated region, Pritchett (1998) and Tester et al. (2006) have shown that the time it takes to replenish 90% of the heat is approximately three times the operational period.

This result suggests two operational scenarios. One scenario would be to use an EGS reservoir system for 30 years and then abandon it for about 100 years. The wells originally put in place could then be refurbished at significantly lower cost than the original drilling effort and the system could then be restarted. An alternative approach, and one which would be less disruptive to the generation effort, would be to use deviated wells to develop an area three to five times the size needed for the operation. Then, systematically cycling fluid production through the various wells on a regular but limited-time basis would allow operations to proceed in an uninterrupted fashion for a 100 years or more.

SYNOPSIS

The development of technology suitable for generating power from geothermal systems has improved significantly since the 1950s, when electrical power generation using geothermal energy resources grew to an international enterprise. The growth of the power generation industry has been steady since that time. Even so, geothermal resources of even greater magnitude are available but have yet to be developed. These are resources at 250°C and higher that occur everywhere in the earth. Although the depth to these resources is variable, it generally is between 3 and 10 km. This ubiquitous geothermal reservoir is immense—conservative estimates place the magnitude of this resource at several thousand times the total energy need of every nation on the globe. However, accessing that resource is currently a technological challenge. Because such systems generally lack significant permeability and fluid, enhancing the natural reservoir permeability is required. This has led to the concept of *Enhanced Geothermal Systems*. The necessary features of a sustainable enhanced geothermal system are that a stimulated zone volume of about a few km^3 must have a fracture network sufficient to sustain a flow rate of about 50 kg/s for about 30 years. The location and orientation of the stimulated zone must be sufficiently well characterized to allow it to be a drilling target for production wells. Flow rates must be matched to the actual permeability structure of the network in order to assure a sustainable resource. Significant strides have been made at accomplishing these characteristics, but much work remains to be done. Current research and development efforts and projects have shown the feasibility of this technology. The success rate of these efforts suggests that EGS could be deployed for power generation within the next 10–15 years on a routine basis.

CASE STUDY: NEWBERRY VOLCANO ENHANCED GEOTHERMAL SYSTEM DEMONSTRATION PROJECT

Newberry Volcano is located in central Oregon, just to the east of the Cascade Mountain Range (Figure 13.12). Unlike other volcanoes in the Cascades, which are stratovolcanoes, Newberry is a large shield volcano, covering a total area of 3100 km² (about the size of the state of Rhode Island in the United States). It first erupted about 400,000 years ago. About 75,000 years ago it went

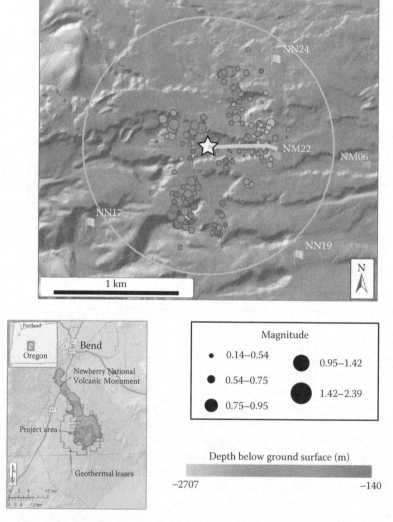

FIGURE 13.12 (See color insert.) The location of the Newberry EGS demonstration project in Oregon (lower left). The upper figure is a topographic map that shows the location of the wellhead (star), the trajectory of the well (light blue), and the locations of microseismic events (depth and magnitude are indicated by the color scale and magnitude chart below the figure). (Modified from Cladouhos, T.T. et al., *Geothermal Resources Council Transactions*, 37, 133–140, 2013.)

through a catastrophic caldera-forming eruption that resulted in collapse of the volcanic edifice and development of two crater lakes. It last erupted about 1300 years ago, thus it is classified as an active volcano (Bard et al. 2013). The presence of hotsprings and recent volcanic activity made Newberry Volcano a place of early interest for geothermal exploration.

Serious geothermal exploration in the vicinity of Newberry Volcano began in 1994 when a number of exploration wells were drilled. Although initial results were inconclusive, there was sufficient indication of geothermal potential to encourage continued exploration activities. By 2012, nearly two dozen exploratory and temperature gradient wells had been drilled, reaching depths between 396 m (1300 ft) and 3536 m (11,600 ft). Well 55-29, which reached a depth of 3066 m (10,060 ft) and a temperature in excess of 316°C (>600°F), showed promise as an EGS demonstration well.

The initial phase of the EGS demonstration effort involved collecting data relevant to the properties of the well and the potential reservoir. Using a combination of a borehole televiewer, elevation maps produced using LiDAR, and microseismic data, the local fracture, fault stress, and seismicity characteristics of the site were compiled (Cladouhos et al. 2011a, 2011b). Core samples and cuttings were also examined for mineralogy, alteration, fracture properties, and rock strength.

This research established that the minimum horizontal stress was oriented approximately 95° from truth North and was consistent with natural fractures and faults that indicated normal faulting was the dominant deformation mode, with fault planes running approximately north–south.

Seven surface seismometers and eight borehole seismometers were installed to monitor seismic activity during stimulation. By mapping the locations of the seismic events during stimulation, it would be possible to establish the geometry of the stimulated region. In addition, the failure mechanism (dilation, shearing, etc.) could also be determined from first motion studies. A strong motion sensor was installed to record any ground shaking. This system was able to detect three natural seismic events over two months prior to the initiation of stimulation activities. This two month data sample of the natural background seismicity is inadequate to establish the perennial level of natural seismic activity, but does establish that this region is subject to natural seismic activity.

Injection for the purposes of determining the characteristics of the system, as well as to undertake stimulation through hydro-shearing, was initiated on October 17, 2012, and terminated on December 12, 2012 (Cladouhos et al., 2013; Petty et al., 2013). Figure 13.6 shows a record of the changes in borehole temperature during two measurement periods when active work was being conducted at the site.

Stimulation was undertaken in three stages. The first stage (October 17 through November 25) was done using only injected water at elevated pressures. Stages two and three involved injecting TZIM at specific times, namely, November 25 and 28 for stage 2 and December 3 and 4 for stage 3.

Some of the data obtained during the stage 2 stimulation are shown in Figures 13.12 and 13.13. In the lower panel in Figure 13.13, the variation in wellhead pressure (in MPa) shows several distinct patterns that are observed during stimulation efforts. First, pressure is commonly held at constant values for relatively long periods of time (hours), during which rock behavior is monitored and other activities, such as TZIM introduction into the well, take place. These periods of constant pressure are separated by very short intervals over which the pressure is changed from one value to another. For example, a long period of low pressure (about 3.6 MPa during day 1) was followed by a period of higher pressure (about 9.5 MPa during day 2). Injection of TZIM occurred during both of these periods (upper panel).

Also shown in the lower panel is the rate at which water was injected into the well (in liters per second). This parameter is important because it provides a measure of permeability change. If, for example, the injection rate stays constant at a given pressure, it means that water is leaving the borehole at a constant rate, which is controlled by the permeability. If the rate of injection increases at constant pressure, it is an indication that water is escaping from the borehole at increasing rates and, therefore, permeability is increasing. The ratio of injection rate to pressure is called the injectivity and is shown in the upper panel in Figure 13.13. Note that the injectivity at the initiation of the stimulation effort was close to 0.

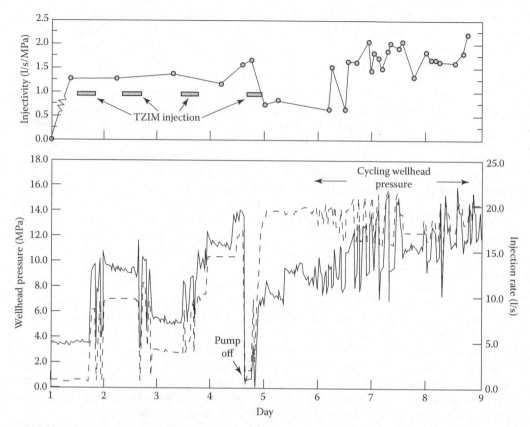

FIGURE 13.13 The stage 2 stimulation history at the Newberry EGS demonstration site. Lower panel shows the variation, by day, in the wellhead pressure (which is a measure of the energy put into the well, in MPa) and the injection rate (in l/s) which is a measure of the rate at which water exits the well through downhole permeability. Also shown is the period during which the pressure was cycled to stimulate the well. The upper panel shows the injectivity (in l/s/MPa) and the periods during which TZIM was injected. (Modified from Cladouhos, T.T. et al., *Geothermal Resources Council Transactions*, 37, 133–140, 2013.)

The injectivity at the beginning of the stage 2 stimulation was approximately 1.25 l/s/MPa (Figure 13.13). At the end of the stage 2 stimulation, injectivity had nearly doubled to 2.25 l/s/MPa. This increase took place primarily in response to two things. One is the injection of TZIM during the early period of the stimulation. This accomplished sealing off of all zones temporarily, but because of the thermal stability behavior of the material, only the shallower zones remained sealed during stimulation. As the TZIM in the deeper zones degraded and allowed the fractures to be accessed for hydroshearing, cycling of the pressure (Figure 13.13) between about 11 and 15 MPa resulted in an increase in injectivity, indicating that the deeper fracture permeability was steadily improving.

Over the seven-week stimulation period, the total water volume injected was 41,000 m³. The computed changes in permeability went from a very low value of 10^{-17} m² (0.01 mD) before stimulation to about 3.2×10^{-15} m² (3.27 mD) after the third stage of the stimulation (Cladouhos et al. 2013).

During the stimulation, 179 microseismic events were recorded. The seismicity (Figures 13.12 and 13.14) indicates that two areas were seismically active at low levels during the stimulation. One area is shallow, and is scattered throughout the region. This area may be related to leakage of fluid from the well casing and is unrelated to stimulation of the deeper geothermal resource (Cladouhos et al. 2013). The second area that exhibited seismic activity is mainly at depths of below 2500 m and lies within 200 m of the well bore. This region is the target EGS reservoir. The extent of the seismicity suggests that the stimulated region has a volume of approximately 1.5 km and is elongate in a NE–SW direction.

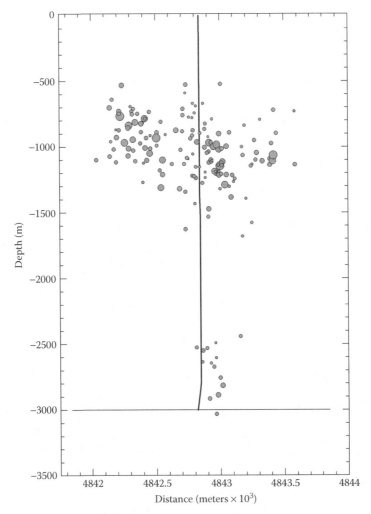

FIGURE 13.14 Cross section through the Newberry site, running approximately north–south. The circles indicate the location of microseismic events, the magnitude indicated by the circle size. Note the upper and lower zones of stimulated microseismic activity. See text for explanation.

The Newberry EGS effort is ongoing as of the date of publishing of this book. Total injectivity achieved was approximately 4.7 l/s/MPa, which is an adequate injectivity for development. Further testing and stimulation is planned for the future.

PROBLEMS

13.1. In what way is an EGS project different from one that will develop a conventional hydro-thermal geothermal system?

13.2. What are the benefits of an EGS system?

13.3. Describe the physical processes that distinguish hydro-shearing from hydro-fracturing. What is the benefit of using hydro-shearing over hydro-fracturing?

13.4. Referring to Figure 13.1, what is the region that has the largest area of relatively shallow high temperature resource? What is its geological context and characteristics?

13.5. In Figure 13.4, there is a broad range of temperature conditions that are possible at a depth of 10 km. What methods would you employ to determine, at any given location, the likely temperature at that depth?

13.6. According to Figure 9.5, if a sandstone experiences a confining pressure of 500 MPa, what fluid pressure would have to be applied to it to accomplish hydro-fracturing?

13.7 What are the factors that control seismic activity? What is the rupture area that is required to cause a magnitude 5 earthquake?

13.8. Referring to Figure 13.9, would compound A be a suitable temporary zonal isolation material for a stimulation effort where the temperature ranged between 150°C and 250°C? Why?

13.9. The potential EGS resource is very large. If you were an investor, what criteria would you use to determine where to first deploy this technology? Why?

13.10. Of the principle challenges faced by EGS, what, in your opinion, is the easiest to overcome? What is the hardest? Why?

13.11. If you were to develop an EGS site for a 60-year lifetime, what drilling schedule would you institute to assure that the facility was sustainable for that time period?

13.12. What reservoir management strategies can be followed to reduce the amount of time needed to replenish the heat removed from a reservoir. Use Equation 13.6 to quantify your discussion.

13.13. The time it takes for a temperature drop to propagate from an injection well to a production well can have an important influence on the sustainability of a resource. Using Equation 13.5, document how reservoir geometry and well spacing influence this temperature *breakthrough* behavior.

REFERENCES

Armstead, H.C.H. and Tester, J.W., 1987. *Heat Mining*. London: Chapman & Hall, 400 pp.

Bard, J.A., Joseph, A., Ramsy, D.W., MacLeod, N.S., Sherrod, D.R., Chitwood, L.A., and Jensen, R.A., 2013. Database for the geologic map of Newberry Volcano, Deschutes, Klamath, and Lake Counties Oregon. *U.S. Geological Survey Data Series* 771. http://pubs.er.usgs.gov/publication/ds771.

Baria, R., Jung, R., Tischner, T., Nicholls, J., Michelet, S., Sanjuan, B., Soma, N., Asanuma, H., Dyer, B., and Garnish, J., 2006. Creation of an HDR reservoir at 5000 m depth at the European HDR project. *Proceedings of the 31st Workshop on Geothermal Reservoir Engineering*, Stanford University, Stanford, CA, SGP-TR-179, January 30–February 1. p. 8.

Cladouhos, T.T., Clyne, M., Nichols, M., Petty, S., Osborn, W.L., and Nofziger, L., 2011b. Newberry Volcano EGS demonstration stimulation modeling. *Geothermal Resources Council Transactions*, 35, 317–322.

Cladouhos, T.T., Petty, S., Callahan, O., Osborn, W.L., Hickman, S., and Davatzes, N., 2011a. The role of stress modeling in stimulation planning at the Newberry Volcano EGS demonstration project. *Proceedings of the 36th Workshop on Geothermal Reservoir Engineering*, Stanford University, Stanford, CA, January 31–February 02, SGPTR-1191, 630–637.

Cladouhos, T.T., Petty, S., Larson, B., Iovenitti, J., Livesay, B., and Baria, R., 2009. Toward more efficient heat mining: A planned enhanced geothermal system demonstration project. *Geothermal Resources Council Transactions*, 33, 165–170.

Cladouhos, T.T., Petty, S., Nordin, Y., Moore, M., Grasso, K., Uddenberg, M., and Swyer, M.W., 2013. Stimulation results from the Newberry Volcano EGS demonstration. *Geothermal Resources Council Transactions*, 37, 133–140.

Clement, T.P., Sun, Y., Hooker, B.S., and Petersen, J.N., 1998. Modeling multispecies reactive transport in ground water. *Ground Water Monitoring and Remediation*, 18, 79–92.

Elsworth, D., 1989. Theory of thermal recovery from a spherically stimulated HDR reservoir. *Journal of Geophysical Research*, 94, 1927–1934.

Elsworth, D., 1990. A comparative evaluation of the parallel flow and spherical reservoir models of HDR geothermal systems. *Journal of Volcanology and Geothermal Research*, 44, 283–293.

Erbas, K., Dannowski, G., and Schrötter, J., 1999. Reproduzierbarkeit und Auflösungsvermögen faseroptischer Temperaturmessungen für Bohrlochanwendungen—Untersuchungen in der Klimakammer des GFZ. Scientific Technical Report STR 99/19, Geo Forschungs Zentrum, Potsdam, Germany 54 pp.

Glassley, W.E., Nitao, J.J., and Grant, C.W., 2003. Three-dimensional spatial variability of chemical properties around a monitored waste emplacement tunnel. *Journal of Contaminant Hydrology*, 62–63, 495–507.

Gringarten, A.C. and Sauty, J.P., 1975. A theoretical study of heat extraction from aquifers with uniform regional flow. *Journal of Geophysical Research*, 80, 4956–4962.

Henninges, J., Zimmermann, G., Buttner, G., Schrötter, J., Erbas, K., and Huenges, E., 2003. Fibre-optic temperature measurements in boreholes. *The 7th FKPE Workshop "Bohrlochgeophysik und Gesteinsphysik,"* GeoZentrum Hannover, Germany, October 23–24.

Parkhurst, D.L. and Appelo, C.A.J., 1999. Users' guide to PHREEQC (Version 2). US Geological Survey Water Resources Investigations Report 99-4259. U.S. Geological Survey, Denver, CO.

Petty, S., Nordin, Y., Glassley, W., Cladouhos, T.T., and Swyer, M., 2013. Improving geothermal project economics with multi-zone stimulation: Results from the Newberry Volcano EGS demonstration. *Proceedings of the 38th Workshop on Geothermal Reservoir Engineering*, Stanford University, Stanford, CA, SGP-TR-198, p. 8.

Phanikumar, M.S. and McGuire, J.T., 2004. A 3D partial-equilibrium model to simulate coupled hydrological, microbiological, and geochemical processes in subsurface systems. *Geophysical Research Letters*, 31, L11503, 1–4.

Pritchett, J.W., 1998. Modeling post-abandonment electrical capacity recovery for a two-phase geothermal reservoir. *Geothermal Resources Council Transactions*, 22, 521–528.

Pruess, K. and Azaroual, M., 2006. On the feasibility of using supercritical CO_2 as heat transmission fluid in an engineered hot dry rock geothermal system. *Proceedings of the 31st Workshop on Geothermal Reservoir Engineering*, Stanford University, Stanford, CA, January 30–February 1.

Rinaldi, A.P., Rutqvist, J., and Sonnenthal, E.L., 2012. TOUGH-FLAC coupled THM modeling of proposed stimulation at the Newberry Volcano EGS Demonstration. *Proceedings of the Tough Symposium*, Berkeley, CA, September 17–16.

Smith, M.C., 1983. A history of hot dry rock geothermal energy systems. *Journal of Volcanology and Geothermal Research*, 15, 1–20.

Steefel, C.I., DePaolo, D.J., and Lichtner, P.C., 2005. Reactive transport modeling: An essential tool and a new research approach for Earth sciences. *Earthy and Planetary Science Letters*, 240, 539–558.

Tester, J.W., Anderson, B.J., Batchelor, A.S., Blackwell, D.D., DiPippio, R., Drake, E.M., Garnish, J. et al., 2006. *The Future of Geothermal Energy*. Cambridge, MA: MIT Press. 372 pp.

Wyborn, D., 2011. Hydraulic stimulation of the Habanero Enhanced Geothermal System (EGS), South Australia. 5th British Columbia Unconventional Gas Technical Forum, Victoria, Canada, April 5–6.

Xu, T. and Pruess, K., 1998. Coupled modeling of non-isothermal multiphase flow, solute transport and reactive chemistry in porous and fractured media. 1. Model development and validation. Lawrence Berkeley National Laboratory Report LBNL-42050. Lawrence Berkeley National Laboratory, Berkeley, CA, 38 pp.

Xu, T. and Pruess, K., 2001. Modeling multi-phase non-isothermal fluid flow and reactive transport in variably saturated fractured rocks. 1. Methodology. *American Journal of Science*, 301, 16–33.

ADDITIONAL INFORMATION SOURCES

United States Department of Energy, Energy Efficiency and Renewable Energy, Geothermal Technologies Program (https://www1.eere.energy.gov/geothermal/enhanced_geothermal_systems.html).

This website provides access to the current program emphasis for geothermal activities within the US Department of Energy. Funding efforts and priorities are indicated, which can provide insight into the direction the geothermal research program is heading. Note that a significant effort is currently underway in the realm of EGS applications.

14 Use of Geothermal Resources
Economic Considerations

The economics of geothermal utilization depend on many factors, each of which is specific to the type of application. Ground source heat pump installations, for example, face many market challenges that differ significantly from those that affect direct-use applications or power generation. These differences reflect differences in the scale of the project, the characteristics and state of the technology, the customer base, the competitive milieu, and the policy and regulatory environment. In this chapter, the focus will be on the large-scale considerations that affect the economic viability of geothermal power generation. Particular attention will be paid to metrics and criteria used to evaluate competitive position in the market place, relative to other available technologies. The purpose of this discussion is to provide perspective regarding geothermal power generation costs relative to other options. Although the economics of other types of geothermal applications will not be specifically discussed because they have been treated directly in Chapters 11 and 12 that detail their technological and scientific dimensions, this chapter closes with a discussion of the necessary steps for completing an economically successful geothermal project of any type.

ECONOMICS OF GEOTHERMAL POWER

The production of electrical power requires that facilities be constructed and operated in a manner that is economically viable for the local market conditions. When developing a power producing facility, the key factors that affect the economic viability are as follows:

- Upfront capital costs
- Operating expenses
- The lifetime of the facility
- Fuel costs
- Average rate of power production

For a facility to be competitive in the power generation market, these factors, when taken together, must result in a cost for electricity that the market will accept. Several of these factors will be discussed in the remainder of this chapter.

Evaluating this cost for electricity has been done in numerous ways. The approach we will follow is that which results in the ability to compare, in an objective and consistent fashion, the real costs of electricity. This approach results in a measure that is expressed in cost/power delivered. The unit we will use is $/kWh and is called the *levelized cost of electricity* (LCOE). Other approaches have been proposed. Some of these consider, in addition to the items mentioned above, the environmental consequences of power production, or the energy returned on energy invested (EROEI), and life-cycle analyses. These will be discussed in Section "Alternative Economic Models". For the moment, we will consider those elements that allow LCOE to be calculated.

UPFRONT CAPITAL COSTS ASSOCIATED WITH GEOTHERMAL POWER

As discussed in Chapters 6, 7, and 13, geothermal power production confronts the formidable task of finding high-temperature reservoirs that are hundreds to thousands of meters underground.

In addition, these reservoirs must be capable of supporting flow of fluid through them at rates of tens to hundreds of liters per second, for sustained periods of time. Such efforts are arduous and time-consuming.

Once a potential site is identified, drilling must take place that allows assessment of the actual power production potential of the site. And, if viable, a power generation facility must be constructed and hooked into the local power grid. How the relative proportions of these various costs factor into the overall cost of putting in place a geothermal power plant has been considered by numerous researchers. A recent analysis by Cross and Freeman (2009) showed that the costs of developing a geothermal system were distributed as follows:

- Initial exploration (~1%)
- Permitting, exploratory drilling (~15%)
- Drilling production and injection wells (~35%)
- Plant construction and transmission (~49%)

These costs, however, vary significantly, depending upon the extent to which new development of a so-called *greenfield* is required and the quality of the resource. A greenfield is one that has not been previously explored or developed.

The total cost for putting in place a geothermal facility, from initial exploration to final power production, has ranged between $1500/kW and to well over $7000/kW, depending upon the temperature of the resource, local geological conditions, success rate for production well drilling, local infrastructure, and extent of previous development. Another variable for the overall project cost is the economic situation in the country in which development is occurring—the costs for a given size power plant can vary by a factor of 2, depending upon whether the plant is located in Central America, Asia, North America, or Europe (IRENA 2013).

The impact on cost of the power generating equipment is dependent upon the resource temperature, as shown in Figure 14.1. This reflects the fact that the engineering attributes of a power turbine become more demanding as the temperature of the resource drops, because efficiency is a direct function of the ΔT achieved during power generation.

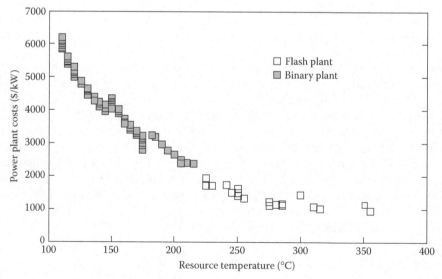

FIGURE 14.1 The relationship between the cost of power plants and the temperature of the geothermal reservoir. The cost of both binary plants (gray boxes) and flash plants (white boxes) is shown. (Modified from NREL, *Renewable Electricity Futures Study*, vol. 2, National Renewable Energy Laboratories, Golden, CO, 2012.)

Capacity Factors

In general, a power generating facility converts some form of energy (kinetic, thermal, and nuclear) into electricity. Each *conversion technology* has particular characteristics that determine efficiencies, size of the power generating facility, and so on. One of these characteristics has come to be known as the *capacity factor*.

The capacity factor of a given conversion technology is the ratio between the amount of power that technology produces over a given amount of time to the amount of power that would be produced if the technology was producing its maximum amount of power over that entire time period,

$$c_f = \frac{p_{realized}}{p_{ideal}} \tag{14.1}$$

where:

c_f is the capacity factor which has a value between 0 and 1

$p_{realized}$ is the power output that is realized over the given time period

p_{ideal} is the power output that would be achieved if the power plant had operated at 100% of its rated capacity for 100% of the time

Multiplying c_f by 100 converts the capacity factor to percent. The capacity factor is based on historical experience, aggregated over many different power plants. As such, it is subject to change and to local and regional physical and economic drivers that may make the performance at a given facility different from the aggregated value. The capacity factor is an indication of how closely an installed conversion technology comes to producing the amount of power an installed generating facility is rated to produce. A capacity factor of 0.9 (or 90%) for a specific conversion technology indicates that performance has shown that technology is likely to achieve 90% of its rated output over time. Table 14.1 lists the capacity factors (in percent values) for various conversion technologies, which are also plotted in Figure 14.2.

The capacity factor for geothermal power production is the highest of all energy conversion technologies, regardless of whether one is considering flash or binary geothermal facilities.

TABLE 14.1
Capacity Factors for Various Power Generation Technologies

Power Generation Technology	Capacity Factor (%)
Geothermal binary	95
Geothermal flash	93
Biomass combustion	85
Biomass—IGCC	85
Nuclear	85
Coal—IGCC	60
Hydro—small scale	52
Wind—onshore	34
Solar—parabolic trough	27
Solar—photovoltaic	22
Ocean wave	15

IGCC, integrated gasification combined cycle.

Source: Integrated Energy Policy Report, California Energy Commission, Sacramento, CA. CEC-100-2007-008-CMF, 234, 2007.

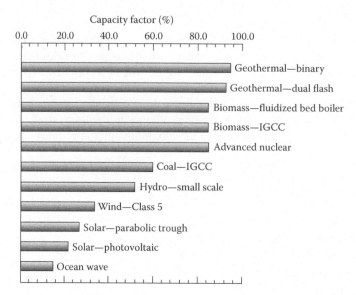

FIGURE 14.2 Capacity factors, as percent, for various conversion technologies. (Data are from the Integrated Energy Policy Report, California Energy Commission, Sacramento, CA. CEC-100-2007-008-CMF, 234, 2007.)

This exceptional performance is the consequence of several factors that make geothermal energy production attractive. One factor is that geothermal power production requires no fuel cycle, as discussed in Chapter 1. The absence of a fuel cycle insulates geothermal power production from the necessity of adjusting output to fuel availability and to the volatility of fuel prices. There is no need to reduce or eliminate power production during refueling (as is required for nuclear power) nor is there a need to adjust performance based on the feedstock or fuel quality.

Another important element that influences the capacity factor of geothermal power is the obvious fact that heat in the subsurface is constantly available. This distinguishes geothermal power from wind and solar renewable technologies, which are variable in output because of weather effects and the day–night cycle. This allows geothermal power facilities to operate constantly, resulting in their ability to approach their theoretical maximum power output.

This factor is the principal reason why geothermal power has traditionally been considered a *baseload* resource. Baseload power is power that can be consistently relied upon to maintain a constant level of electricity in a power grid. Because it is not affected by intermittency imposed by externalities such as weather, diurnal cycles, or other factors, it can function indefinitely without interruption at a predictable and reliable level. Recent technological advances have also made geothermal power production flexible, allowing it to follow the load demand that naturally occurs in an electrical grid. This will be discussed in more detail in Chapter 16.

Another factor that helps geothermal power to maintain a high capacity factor is the fact that geothermal power plants operate under relatively modest temperature (<350°C) and pressure conditions. Power plants that burn fossil fuels must function at temperatures in excess of 2000°C. Nuclear power plants must accommodate an environment with strong radiation fields. The relatively modest physical conditions that geothermal power plants experience result in lower stresses on the materials composing the facility, allowing long lifetimes and less disruptive maintenance efforts.

A factor that prevents geothermal plants from achieving a capacity factor of 1.0 (i.e., 100%) is the *parasitic load*. A parasitic load results from the need to utilize some of the generated power for demands required to keep the generating plant functioning. For example, large pumps are needed to move the production fluids to the turbines, as well as to pump the reinjected fluids into the subsurface. This load and others related to the operation and maintenance (O&M) of the facility reduce the amount of generated power that is fed into the electrical grid. As a result, the amount of power available for nonsite uses is diminished by 10%–20%, depending upon the characteristics of the resource. It is the combination of these factors that allow geothermal power generating facilities to achieve the high capacity factors of which they are capable.

LEVELIZED COSTS

The ability to compare the cost of generating power using different conversion technologies is crucial for developing investment strategies, incentive policies, and budgets. One standard method for doing this is through *levelized cost* analysis. The levelized cost for power generation is the minimum cost at which the power generated by the conversion technology must be sold in order for the facility to break even. Levelized cost takes into account the cost of constructing a facility, the cost of financing, operation costs, maintenance costs, the cost of fuel, and the lifetime for operations and power production. Such an analysis can also include incentives that have been put in place as a result of policy decisions, taxes, and other costs or benefits that accrue to a technology. But, for accurate and reasonable comparisons, the same factors must be used in an analysis when comparing different conversion technologies. Because publicly owned utilities, investor-owned utilities, and commercial enterprises have different tax structures, incentives, and pricing controls, it is also important that a levelized cost analysis compare enterprises that function within the same financial environment.

A number of formulations for computing levelized cost have been proposed. One example of a levelized cost calculation is as follows:

$$\text{LCOE} = \frac{\sum [(I_t + M_t + F_t)/(1+r)^t]}{\sum (E_t/(1+r)^t)} \tag{14.2}$$

where:

I_t is the investment expenditure in the year t
M_t is the operations and maintenance expenditures in the year t
F_t is the fuel expenditure in the year t
E_t is the amount of electricity generated in the year t
r is the discount rate

The summation is done from $t = 1$ to n, where n is the life of the generating system (IRENA 2013). Clearly, an important advantage for geothermal power, and one of the reasons it has such a low levelized cost is the fact that the fuel expenditure is 0 and the amount of electricity generated is very high (i.e., the capacity factor is greater than 80%). Figure 14.3 shows the levelized costs for various conversion technologies, as compiled by the International Renewable Energy Agency for the year 2012 (IRENA 2013).

The most significant point that emerges from Figure 14.3 is the relatively low levelized costs for biomass, geothermal, hydropower, and wind conversion technologies. For geothermal energy, this reflects, as noted above, a variety of aspects about the technology that allow it to produce energy at low cost. These include the absence of a fuel cycle, the extremely high capacity factor, and the low costs of maintenance and operation. Wind is comparable in levelized cost because it has low initial capital expenditures and no fuel costs despite its lower capacity factor. All of these renewable energy technologies are strongly competitive with conventional fossil-fueled energy systems. On this basis alone, there is no economic reason to favor fossil-fueled systems over these renewable energy technologies.

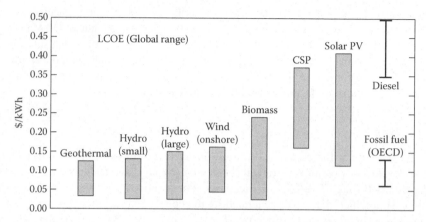

FIGURE 14.3 LCOE for various renewable energy conversion technologies throughout the world. The ranges reflect global variation, as well as in-country variation. For comparison, LCOE for diesel and other fossil-fueled technologies are also shown. (Data from IRENA, *Renewable Power Generation Costs in 2012: An Overview*, International Renewable Energy Agency, Abu Dhabi, United Arab Emirates, 88, 2013.)

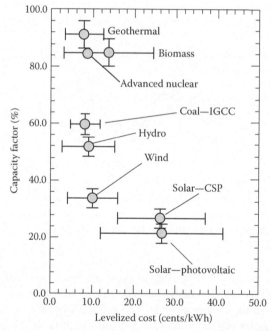

FIGURE 14.4 Comparison of capacity factor (plotted as percent) and levelized cost (in cents/kWh) for the same suite of conversion technologies as represented in Figures 14.2 and 14.3.

Figure 14.4 provides a composite view of the relationship between capacity factor and levelized cost. It is clear in the figure that there is a broad range in both capacity factor and levelized cost, but consistently, the conversion technologies with the highest capacity factors have among the lowest levelized costs. However, it is also clear that capacity factor alone is not a good indicator of levelized cost.

As noted above, levelized cost is subject to a variety of influences that reflect both market conditions and the performance of the technology. Research and development efforts that impact the efficiency of the conversion technology or that impact the cost of generation will change the levelized costs. Although predicting specific research and development outcomes is impossible, it has proven informative to use retrospective analyses of the evolution of various technologies to forecast future performance. One such approach considers the relationship between investment and performance to evaluate how a technology will perform in the marketplace.

ECONOMICS OF R&D INVESTMENT IN GEOTHERMAL ENERGY

TECHNOLOGY EVOLUTION AND S-CURVES

The relationship between the return on investment made in a technology and the maturity of that technology is complex. Despite the fact that a variety of factors influence that relationship, it has been repeatedly demonstrated (Ayres 1994) that a general pattern emerges that has become known as experience curves or, in a more recent and refined approach, technology S-curves (Schilling and Esmundo 2009). An idealized example of this pattern is shown in Figure 14.5.

The form of the curve describes the responsiveness of the technology to research and develop investments. In general, the early stages of a technology evolve slowly, requiring a relatively high investment for a given improvement in performance and return on investment. As experience is gained in the technology, lessons learned can guide investment toward those research and development challenges that will give the greatest improvement in performance. The result is that the slope of the curve becomes steeper, reflecting a development stage in which a relatively small investment results in a relatively large improvement in performance and return on investment. Once a technology reaches a high level of maturity, improvement in performance diminishes relative to the investment in research and development.

There are numerous ways in which this simple model of technology evolution may not be appropriate for a given technology. In some cases, a previously unrecognized process may be adapted for use in the technology, providing a new investment pathway for affecting improvements in performance. In some instances, a new market niche may be found that requires additional investment in research and development. As materials evolve, a new manufacturing process or a new material may enter the market that allows unexpected improvement in market penetration or performance. All of these occurrences can modify the form of the evolutionary curve. Nevertheless, as a general model for understanding the existing state of the art, and as a means to view the likely future of the technology, an S-curve analysis can be informative.

PROJECTED ENERGY COSTS

Plotted in Figure 14.6 is the cost, which is primarily O&M expenses, of generating power from geothermal sources between 1980 and 2005 (Schilling and Esmundo 2009). For each year,

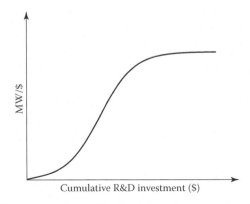

FIGURE 14.5 Schematic diagram showing an idealized form for a technology S-curve.

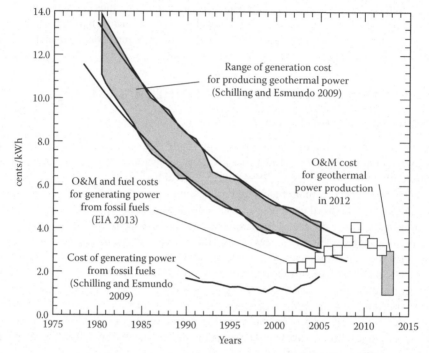

FIGURE 14.6 Cost of geothermally generated power for 1980–2005. Also, the O&M costs for geothermal power production are shown for 2012 (http://www1.eere.energy.gov/geothermal/faqs.html). For comparison, the cost of power generated using fossil fuels is also shown. The solid line for fossil fuels represents a weighted average of fuel costs that accounts for the proportion of each fossil fuel used and its energy content. The boxes represent the power generation costs (O&M and fuel costs) for steam-fired fossil fuel plants, as reported by the US Energy Information Agency (Table 8.4; US Energy Information Agency, Electric Power Annual Report, 2013, http://www.eia.gov/electricity/annual/). (Data from Schilling, M.A. and Esmundo, M., *Energy Policy*, 37, 1767–1781, 2009; from Schilling, M.A. and Esmundo, M., *Energy Policy*, 37, 1767–1781.)

the upper and lower values are used to define a range, which is shaded. Exponential curves were fitted to data for the upper and lower bounds and are drawn as the smooth curves. Also shown is the O&M costs reported by the Energy Efficiency and Renewable Energy Office of the US Department of Energy for 2013. For comparison, the cost of energy production using fossil fuels is also shown, using different metrics. Note that the latest cost figures for fossil-fueled power generation are directly comparable to the cost shown for geothermal energy in 2013, because these costs are based solely on O&M and fuel costs. These comparisons document that geothermally produced power is directly competitive with fossil-fueled technologies. Indeed, geothermal power generation costs are generally less than that for power generated using fossil fuels.

Figure 14.7 shows the resulting S-curve for geothermal energy. The data points are from Schilling and Esmundo (2009). The solid arrowed curve and the dashed curves are various fits to the data points. The best fit that was achieved to the data points resulted in an upper limit to the curve of 276 kWh/$ (Schilling and Esmundo 2009), which is significantly greater than the current fossil fuel value which is approximately 100 kWh/$. However, numerous factors can influence where the curve reaches its maximum value, as indicated by the hypothetical cases represented by the dashed lines.

The form of the curve defined by the range of the observed data points follows that of the early stages of an evolving technology. Currently, the trend is well within that segment of the curve that reflects rapid improvement in technology performance relative to the amount of investment

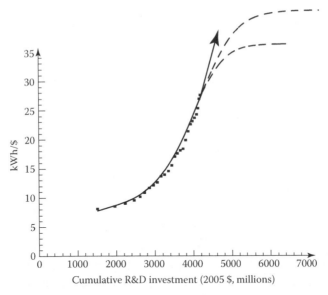

FIGURE 14.7 The amount of energy obtained per dollar compared with the cumulative amount of investment in R&D for geothermal energy since 1980. The graph is scaled to a beginning year of 1974, since R&D investment predates the available data. As a result, the S-curve does not begin at the origin. The arrowed line indicates the best fit to the data points. The dashed curves are hypothetical trends. (Data points from Schilling, M.A. and Esmundo, M., *Energy Policy*, 37, 1767–1781, 2009.)

in research and development. Whether the current trend will continue for a significant period of time cannot be predicted. However, along with the trends in Figure 14.6, there is an indication that significant improvement in both cost reduction and performance is likely to continue with relatively modest investment.

With these factors regarding the economic aspects of geothermal power generation, it is appropriate to consider the steps necessary to achieve a functioning and economically viable generation facility or direct-use application. The following discussion considers the basic steps that are usually required to complete a project, once the project concept has been defined.

DEVELOPING A GEOTHERMAL PROJECT

Geothermal projects that require drilling of wells, whether for power production or other applications, follow a development timeline that usually consists of six steps. The only exception to this are installations of ground source heat pump systems, which usually are developed in the same manner as water well projects and are discussed in detail in Chapter 12. The sequence of steps is generally as follows:

- Obtain rights to use the resource and the permits that allow exploration.
- Conduct an exploration and resource assessment program.
- Drill exploration wells to refine resource assessment.
- Drill production and injection wells and complete a feasibility study.
- Undertake construction of facility.
- Begin operation.

Each of these steps is discussed in the subsequent sections.

Rights to Develop a Resource and Permitting

Obtaining the rights to explore or develop a resource usually requires establishing a contract with the owner of the subsurface resource. Because of federal and local regulations and laws, it is not always the case that the owner-of-record of a parcel of land also owns the resources that are in the subsurface below that property. Careful research is often required to discover who the legal owner of a resource may be. Even if the owner-of-record is not the owner of the land under which the resource exists, it is prudent to discuss the exploration activities with the landowner and obtain their consent to explore. This is particularly true in areas where sites of cultural value exist. If a potential resource area has sites that have historical, traditional, or archaeological value, interested parties should discuss appropriate ways to mitigate exploration and development impacts.

Exploration programs can utilize a variety of techniques to develop a preliminary resource assessment. Many of these are noninvasive and do not require permits. Water sampling, remote sensing, and gravity surveys are examples of noninvasive techniques. If exploration activities will perturb the land surface or wildlife habitat, there will be permitting requirements. The nature and complexity of the permitting process will depend on landownership and the specific impacts and will vary from site to site. If federal or state lands are involved in the United States, it is likely that some form of environmental impact statement will be required. Recent efforts to streamline permitting processes have been undertaken by the US Bureau of Land Management and the US Forest Service (Peterson and Torres 2008) for federal lands. Other efforts to provide guidance regarding permitting requirements have also been pursued by a variety of agencies (e.g., Blaydes 2007) in an effort to reduce the time required to obtain permits.

Regardless of the site considered, the time to obtain the requisite permits is commonly in excess of a year. For large projects involving complex analysis of environmental impacts, it is likely that the time required will be significantly more than that.

Initial Resource Assessments

Conducting a resource assessment program during the early stages of a project satisfies two requirements. It is important to develop a preliminary scientific database that will support rigorous analysis of the size and characteristics of the resource being considered. Such information will provide a preliminary indication of whether the resource is sufficiently large to support the intended use. It will also provide a basis for identifying potential complications in the project that could increase cost and risk. Factors such as adverse chemical characteristics of the reservoir waters, high seismic risks, and insufficient cooling water can often be established at the early stage of the assessment. This initial assessment then becomes the basis for hypothesis testing when the drilling program begins for the refined resource assessment.

A rigorous, scientifically based resource assessment is also a prerequisite for obtaining project funding. Despite the attractiveness of superficial manifestations of heat, such as boiling springs and bubbling mud pots, prudent investors will seek information that demonstrates a depth of knowledge and scientific objectivity that establish the credibility of the project developer. A well-designed and conducted preliminary resource assessment can establish that credibility.

A preliminary resource assessment should establish, with defined uncertainty values, the following for a power project:

- The nature of the resource (hydrothermal, dry steam, etc.)
- The temperature of the resource
- The size of the resource
- Depth to the resource
- Estimated flow rates
- Reservoir lifetime, recharge strategy, and management challenges

- Drilling requirements (hard rock, lithological complexity, etc.)
- Environmental challenges (H_2S abatement, water supply for cooling, disposal of water or wastes, etc.)
- Access to transmission corridors
- Potential power purchase partners
- Local, state, and federal incentives

For direct-use projects, the same topics should be addressed, with the exception of the transmission and power purchase agreements (PPAs), which would be replaced by consideration of the access to markets for the product.

Several of the items on the above list are difficult or impossible to establish without subsurface information. Likely flow rates and depth to the resource are two such items. However, estimates can be obtained for these parameters through discussions with those who have worked in the area, particularly geologists and drillers. If the area being considered for development has already seen exploration and development in the past, historical data obtained during these efforts are often available. Most jurisdictions have some type of regulatory structure that requires providing a data repository with well-logging information and other subsurface data that may be available for public inspection.

Although the preliminary resource assessment will require collecting field data, the magnitude of that effort can be minimized by conducting a thorough review of the existing scientific and technical literature. Literature searches of scientific journals, state geologists reports, US Geological Survey reports, historical records regarding mineral claims, and so on can determine what additional data should be collected, what type of data it should be, and where it should be collected.

An important resource that has recently become available is the US National Geothermal Data System. This resource is intended to be a permanent repository of data from a variety of resources that cover geothermal systems.

The process of developing a preliminary resource assessment is equivalent to constructing a conceptual model of the geological system. This conceptual model will act as a hypothesis that will guide development of a testing program to evaluate the validity of the model and allow it to be refined. To do this requires data from the subsurface. The preliminary resource assessment thus becomes the basis for designing a drilling program to test and refine the initial conceptual model. Development of a preliminary resource assessment should require less than a year.

REFINING THE RESOURCE ASSESSMENT THROUGH EXPLORATION DRILLING

A drilling program designed on the basis of a good preliminary resource assessment will target locations that are likely to have high geothermal gradients. How highly resolved the target area will be depends upon the level of detailed information generated during the preliminary assessment. Regardless of the resolution, it is common that at least two exploration holes will be drilled. More holes will be necessary if the goal of an assessment is to define the extent of a geothermal resource.

Once target sites have been selected and prepared and target depths established, a drilling rig that has the capability to reach the necessary depths and can handle the expected conditions is deployed. Usually, these services are obtained through contract.

The holes that will be drilled vary in type and purpose. Slim holes, with diameters of 7.5–15 cm, can be drilled to depths of several thousand meters. The purpose of such holes is to inexpensively access depths where temperature gradients can be measured. These measurements allow significant refinement of the preliminary resource assessment by reducing the uncertainty associated with estimates of the temperature, extent, and depth of the reservoir. These *temperature gradient holes* are an integral part of an exploration program. If done with thorough cognizance of the geological framework, and if done in sufficient quantity, data they provide will become a crucial element in establishing a more extensive drilling program.

Cuttings from these wells or larger diameter wells provide samples of the geological rock units in the subsurface and their characteristics. Particularly important is the porosity and permeability of the reservoir rocks, and the extent to which permeability is fracture-controlled. This information allows refinement of estimates of likely flow rates.

These cuttings are also critical for providing insight into the geological and alteration history of the site. As discussed in Chapter 6, hydrothermal alteration is invariably associated with geothermal systems. The history of that alteration can be deduced from the mineralogical record preserved in the rocks and obtained through the cuttings. Whether a system has existed at the site can be established through such research. In addition, the state of evolution of the system, that is, whether it is in decline or has had multiple episodes of hydrothermal activity or is in an increasing heat-up phase, can be deduced through studies of these cuttings.

If it is possible to allow the wells to flow for a period of time, it may be possible to obtain relatively uncontaminated fluid samples. Chemical analysis of the collected fluids can reduce uncertainty about the chemical properties of the reservoir fluid and help determine if mitigation strategies will be required for using or disposing of the geothermal fluid. Depending on the number of wells and the complexity of the geology, it may take up to a year to complete a refinement of the resource assessment.

PRODUCTION WELLS AND FEASIBILITY STUDY

At this point, the resource assessment effort will have concluded whether there is an accessible resource suitable to meet the needs of the project. The next step is to drill a production well into the reservoir in order to establish that flow rates, temperatures, and other operational conditions are suitable. Drilling production wells often requires satisfying additional permitting and regulatory requirements. Any such requirements should be known and in the process of being satisfied before termination of the exploratory drilling program, if that drilling program has provided positive indications of a usable resource.

The characteristics of production wells were discussed in Chapter 9. Undertaking this part of the project is perhaps the most challenging, because it is at this point that the success of a project is determined. Heavy investment in equipment and materials is required and careful execution of the effort is needed to assure that the well will be robust enough to have a lifetime that will meet the needs of the project. Casing and cementing materials must be selected to match the chemical and environmental rigors of the emplacement environment. Temperature measurements must be made periodically to determine the thermal structure of the reservoir. Flow testing must be done to establish the hydrological properties of the reservoir.

Once these data have been collected, a feasibility study can be completed that provides the final refinement of the resource assessment. If flow rates, temperatures, and other conditions are adequate to meet the project goals, financing can be obtained to complete the last phases of project development, which are drilling of production and injection wells and construction of the facility.

Historically, the overall duration of project development has been between 5 and 7 years. It is anticipated, however, that the time needed to complete these steps may decrease significantly if geothermal applications become more common. As experience grows, industry and regulatory agencies gain knowledge regarding the best practices to employ to reduce development time while assuring compliance with necessary requirements for safe and sustainable project development.

ALTERNATIVE ECONOMIC MODELS

The economic considerations outlined in Section "Economics of Geothermal Power" are common approaches for analyzing the viability of geothermal systems and for comparing costs of energy in a "apples-to-apples" fashion. However, although these approaches are much better than simple methods that have been used in the past, there are newer approaches that allow other, more inclusive perspectives to be brought to the economic analysis arena. In the following discussion, we will consider two of these approaches. These are presented in a descriptive format, with only occasional reference to geothermal systems because there is still a significant amount of foundational research yet needed to allow comparative analysis of different renewable technologies using these approaches. Rather, they are presented with the intent to broaden perspectives concerning how to value renewable energy technologies, with the understanding that a foundational element to these approaches is the primary importance of sustainability.

LIFE-CYCLE ANALYSES

Life-cycle analysis (LCA) is an effort to identify the environmental impacts and resource use, social costs and benefits, and energy usage that develop as a result of the manufacture and use of a specific product. This is sometimes called a cradle-to-grave analysis and falls within the realm of what has become known as Industrial Ecology (Allenby 1995; Jensen et al. 1997; Zamagni 2012). The incentive for conducting such an analysis is to provide objective and thorough representation of the cost or environmental impacts of a product. For example, quantifying the greenhouse gas emissions associated with driving a car is often only done by calculating the mass of greenhouse gases emitted at the tailpipe of the vehicle. A thorough LCA, however, would also consider the emissions, on a per liter basis, related to exploration for the petroleum from which the gas was produced, transportation of the crude oil, refining of the crude oil to gasoline, and transport of gasoline to the distribution system, among other things.

LCA studies of renewable resources have recently been undertaken but often with considerable variability between results (NREL 2013). The difficulty such assessments face is defining the boundaries of what will be included in the study and how to quantify it. This is particularly true of efforts to evaluate the costs of introducing renewable resources into energy generation and transmission systems. This reflects the fact that the cost of generating power using a particular technology does not reflect the costs of incorporating that generation facility into an existing transmission system, also known as *integration costs*.

For example, integrating variable energy resources into a grid requires that the transmission infrastructure be capable of accommodating the variability of the energy output. This requires that back-up generation be available to quickly come online when weather conditions, for example, result in a rapid reduction in power output. When that back-up power is not needed, the power plants that are not variable must either be shut off or ramped down. In addition, transmission lines must be capable of accommodating the maximum amount of power a variable resource can generate. But, when that variable resource is not generating at full capacity, the line will not be operated at full capacity because most back-up systems are not located in the same sites as variable resources, thus leaving the transmission lines operating significantly below capacity. Because the costs of a transmission line can be on the order of $1 million to $4 million per mile (Western Electricity Coordinating Council 2012), building capacity for transmission that is only fractionally used can be an additional cost.

Many of the additional costs are not directly reflected in the price paid for power by the user, because these costs are often accounted for through other means. PPAs are the governing document for establishing the price a consumer will pay for power. These PPAs are often regulated by the state or regional agencies that are guided by policies that reflect a broad range of political, economic, and social motivations. Hence, the price paid for power by a consumer is commonly a compromise between the actual costs of generating and distributing that power and the market price, which is subsidized through various incentives, tax credits, transferred costs, and allowances. A full LCA of each energy generation technology, harmonized to take into account a consistent suite of factors, could establish the actual comparable costs paid by consumers for power generation using any specific technology. A recent attempt to elucidate the attributes that are of value for geothermal energy production demonstrated, for example, that geothermal energy has positive attributes that are not valued when price comparisons are usually conducted for generating technologies (Matek and Schmidt 2013). This analysis provided first enumeration of these benefits, including lowest integration costs, displacement of fossil fuels, reduced emissions, lower footprint, and greatest job creation. In the future, it is likely that LCA assessments will be consistently and thoroughly developed for all energy technologies, providing the ability to inform decisions regarding the most sustainable, cost-effective approaches to power generation.

ENERGY RETURNED ON ENERGY INVESTED

To obtain energy for any use, whether it be power generation or heating a greenhouse, requires energy. In other words, energy must be invested in order to obtain energy for other uses. This has led to the concept of energy returned on (energy) investment (Hall 1972; Odum 1973; Hall et al. 1986), variously shortened to EROI or EROIE. This value is computed as a ratio of energy produced from a resource or commodity to that used to extract or develop that energy:

$$\frac{\text{Energy returned to society (joules)}}{\text{Energy consumed in obtaining that energy (joules)}} \tag{14.3}$$

Although this equation is simple, its application is complex (Hall et al. 2009). It was originally intended to be applied at the point at which an energy resource was obtained, such as at the wellhead for an oil well or the mine mouth of a coal mine. This EROI is usually given the subscript mm, for mine mouth, that is, $EROI_{mm}$. However, although this $EROI_{mm}$ has a clear boundary, it is also evident that a more comprehensive perspective may want to take into account the energy consumed to deliver the energy to the user, that is, a point of use EROI, also called $EROI_{pou}$, or that the denominator should include, as well the energy required to use the energy, a so-called $EROI_{ext}$, where ext stands for "extended" EROI.

Finally, there is the issue of energy "quality" and how to account for it. For example, petroleum from the ground is highly variable in its characteristics—some, such as "light" crude oil, has a low density and has low concentrations of unwanted compounds, whereas "heavy crude" is higher in unwanted compounds that must be treated and removed during the refining process. In addition, some of the products of the refinement process, such as gasoline, have a very high energy density and are easily transported, whereas other potential fuels, such as cooking oils, have a much lower energy density and are more difficult to handle. These and many other issues have led to the concept of a "quality correction" to the EROI. However, how best to accomplish this correction remains a matter of debate.

Nevertheless, the concept and use of EROI is gaining increasing validity and interest because of its implications. It is intuitively obvious that any society in which the overall $EROI_{mm}$ is less than 1.0 is not sustainable—such a society will incessantly be consuming more energy than it is producing, eventually resulting in an inadequate supply of joules to power its needs. Hall et al. (2009) have concluded that the minimum $EROI_{mm}$ that is required for maintaining transportation and related systems for a sustainable society is approximately 3. It is likely that a higher $EROI_{mm}$ is needed to assure adequate support for important other components of a functioning society.

Figure 14.8 compares the latest EROI values compiled by Gupta and Hall (2011) for a range of technologies. The large range for several of the resources indicates the effect of different assumptions on the calculated EROI as well as the range of values that can be expected due to differences in individual site characteristics. For example, the broad ranges for "Geothermal" and "Hydro" reflect the fact that some locations are highly favorable for these technologies and require minimal energy to develop, whereas other sites require more energy for development and/or provide less energy output.

An important point of this type of analysis that is made clear in Figure 14.8 is that geothermal, hydropower, and wind can provide EROI values that are comparable to or better than oil and coal, thus making them logical replacements for those fossil fuel resources in the appropriate applications.

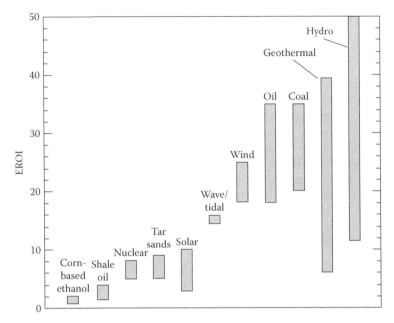

FIGURE 14.8 EROI for various energy resources. Note that "Hydro" extends beyond the maximum value for the graph. (Data are compiled from Gupta, A.K. and Hall, C.A.S., *Sustainability*, 3, 1796–1809, 2011.)

SYNOPSIS

The factors that affect the economics of geothermal power production make it a competitive energy conversion technology. The capacity factor, which reflects the proportion of time a conversion technology is available to generate power, is the highest of all other technologies, often exceeding 90%. The levelized cost of power, which accounts for the balance of expenses that a generation technology must absorb in order to generate power, is among the lowest of competing power generation technologies. Combined, these factors indicate that geothermal power production is currently an economically competitive energy conversion technology. When considered in terms of its maturity with regard to R&D investment and improvement in performance, geothermal power falls at the early stages of an S-curve technology analysis. This indicates that significant improvement may be possible in performance for a relatively small investment, compared to historical trends. The trend in cost reduction for geothermal power generation has continued for some years, resulting in geothermal energy production generally being less expensive than production from fossil fuels. To be economically viable, a geothermal project must address the necessary steps to successfully be completed in a timely fashion. A number of sequenced actions are required which, under normal circumstances, take about 5–7 years to complete. These include acquiring the rights to explore and use geothermal heat from the subsurface, an adequate assessment of the resource, and a feasibility study to demonstrate the resource is suitable for the targeted project goals. When put in place, geothermal power is a highly competitive resource compared to all other energy resources, even when considered through the perspective of newer economic analytical approaches including LCA and EROI methodologies.

PROBLEMS

14.1. What are capacity factors, what determines their magnitude, and how do they vary among power generating technologies?

14.2. How is levelized cost computed? Is it constant through time? What can influence it?

14.3. Describe the economic and output evolution of a technology as it goes from its initial state of innovation to a mature industry. What factors influence this process? What can change the evolutionary path?

14.4. What is a technology S-curve? Where does geothermal energy sit in the technology S-curve analytical approach?

14.5. What are the projected costs of geothermal power, compared to power generated from fossil fuels? What can influence these projections?

14.6. What are the key steps that are usually required to successfully develop a geothermal project?

14.7. How might the amount of time required to accomplish the steps in problem 14.6 be reduced without jeopardizing sound project development?

14.8. If you were to compare the life-cycle costs of a coal-fired power plant to those of a geothermal power plant, which elements of that analysis would be unique to each technology? Which ones would be similar?

14.9. In Figure 1.6, the EROI of oil has decreased between ~1925 and ~2005. What factors are causing this decrease?

REFERENCES

Allenby, B., 1995. *Industrial Ecology*. Englewood Cliffs, NJ: Prentice Hall.

Ayres, R.U., 1994. Toward a non-linear dynamics of technological progress. *Journal of Economic Behavior and Organization*, 24(1): 35–69.

Blaydes, P., 2007. California geothermal energy collaborative: Expanding California's geothermal resource base—Geothermal permitting guide. California Energy Commission Report CEC-500-2007-027. 54 pp.

Cross, J. and Freeman, J., 2009. 2008 Geothermal Technologies Market Report. US Department of Energy. http://www1.eere.energy.gov/geothermal/pdfs/2008_market_report.pdf.

Gupta, A.K. and Hall, C.A.S., 2011. A review of the past and current state of EROI data. *Sustainability*, 3, 1796–1809.

Hall, C.A.S., 1972. Migration and metabolism in a temperate stream ecosystem. *Ecology*, 53, 585–604.

Hall, C.A.S., Balogh, S., and Murphy, D.J.R., 2009. What is the minimum EROI that a sustainable society must have? *Energies*, 2, 25–47.

Hall, C.A.S., Cleveland, C., and Kaufmann, R., 1986. *Energy and Resource Quality: The Ecology of the Economic Process*. New York: Wiley Interscience.

Integrated Energy Policy Report, 2007. California Energy Commission, Sacramento, CA. CEC-100-2007-008-CMF. 234 pp. http://www.energy.ca.gov/2007publications/CEC-100-2007-008/CEC-100-2007-008-CMF.PDF.

IRENA, 2013. *Renewable Power Generation Costs in 2012: An Overview*. Abu Dhabi, United Arab Emirates: International Renewable Energy Agency. 88 pp.

Jensen, A.A., Hoffman, L., Moller, B.T., and Schmidt, A., 1997. Life cycle assessment. European Environmental Agency (London), *Environmental Issues Series*, No. 6. 116 p.

Matek, B. and Schmidt, B., 2013. *The Values of Geothermal Energy: A Discussion of the Benefits Geothermal Power Provides to the Future US Power Sector*. Washington, DC: Geothermal Energy Association, 19 pp.

NREL, 2012. *Renewable Electricity Futures Study*, vol. 2. Golden, CO: National Renewable Energy Laboratories.

NREL, 2013. Life cycle assessment harmonization. National Renewable Energy Laboratories, Golden, CO: National Renewable Energy Laboratories. http://www.nrel.gov/analysis/sustain_lcah.html.

Odum, H.T., 1973. *Environment, Power and Society*. New York: Wiley Interscience.

Peterson, J.G. and Torres, I., 2008. Final programmatic environmental impact statement (PEIS) for geothermal leasing in the Western United States. Vols. 1, 2, and 3. US Department of the Interior FES 08-44. http://www.blm.gov/wo/st/en/prog/energy/geothermal/geothermal_nationwide/Documents/Final_PEIS.html.

Schilling, M.A. and Esmundo, M., 2009. Technology S-curves in renewable energy alternatives: Analysis and implications for industry and government. *Energy Policy*, 37, 1767–1781.

US Energy Information Agency, 2013. Electric power annual report. http://www.eia.gov/electricity/annual/.

Western Electric Coordinating Council, 2012. Capital Costs for Transmission and Substations. Black & Veatch Project No. 176322. http://www.wecc.biz/committees/BOD/TEPPC/External/BV_WECC_TransCostReport_Final.pdf.

Zamagni, A., 2012. Life cycle sustainability assessment. *The International Journal of Life Cycle Assessment*, 17(4), 373–376.

FURTHER INFORMATION

Additional sites that provide access to a broad range of data related to geothermal systems are as follows:

Geothermal Resources Council (https://www.geothermal-library.org/).

Great Basin Center for Geothermal Energy (http://www.gbcge.org/).

Oregon Institute of Technology (http://geoheat.oit.edu/database.htm).

Southern Methodist University Geothermal Laboratory (http://smu.edu/geothermal/database/smugeodatabase.html).

International Geothermal Association (http://www.geothermal-energy.org/publications_and_services/news.html; https://pangea.stanford.edu/ERE/db/IGAstandard/default.htm).

The IGA maintains a database published conference papers that can be searched for data relating a geothermal technology and site characteristics. Access to the database and papers are free.

US National Geothermal Data System (http://geothermaldata.org/).

The NGDS provides free access to an extensive array of data that relate to geothermal resources throughout the United States. Data are provided by a range of contributors, including academic researchers, state and federal agencies, and private sector participants.

15 Use of Geothermal Resources
Environmental Considerations

Throughout this book we have discussed many of the environmental benefits and challenges associated with using geothermal energy. In this chapter, we will focus on several issues that are of particular interest because they are of general concern to the public and regulatory agencies. The intent of this discussion is to provide a knowledge base from which further discussions can evolve. The areas of concern that we will consider are emissions to the atmosphere of unwanted gases; introduction to the environment of unwanted chemicals and compounds; and the potential for resource recovery, seismicity and stimulation, ground subsidence, water use, and land use.

EMISSIONS

As detailed in Chapter 5, geothermal systems are complex geochemical environments in which a range of chemical processes occur. Conceptually, these processes represent interactions in which the atmosphere, the geological framework, and subsurface fluids evolve toward their lowest Gibbs energy state. These processes can be schematically represented as an interacting three-part system:

$$\text{Atmosphere} \quad \Leftrightarrow \quad \text{minerals} \quad \Leftrightarrow \quad \text{water} \tag{15.1}$$

The $\Leftrightarrow$ symbol indicates interactions that reflect exchange and chemical reactions between parts of the system. These chemical changes can influence the physical attributes of the system by changing the rock porosity, fracture apertures, and temperature, among other things. Any perturbation in one part of the system will be reflected in some measurable adjustment in other parts of the system. If, for example, the temperature of the water increases, minerals with which the water is in contact will dissolve or precipitate, reflecting the temperature dependence of the respective mineral solubilities, which are site specific. Likewise, the temperature dependence of dissolved gas concentrations will change in response to temperature perturbations, and this will, in turn, result in gases being liberated to or absorbed from the coexisting atmosphere. For considerations regarding gaseous emissions from geothermal power plants, the behavior of three types of gaseous emissions is particularly important—greenhouse gas emissions (especially CO_2), hydrogen sulfide (H_2S), and toxic metals such as mercury (Hg) are discussed elaborately in three subsections. In the discussion, attention will be focused primarily on flash geothermal generating plants because binary generating plants have no emissions as they do not expose the geothermal fluid to the atmosphere. However, their operation can influence subsurface processes. These effects will be discussed at appropriate points.

CARBON DIOXIDE

The solubility of CO_2 in geothermal fluids is controlled by the temperature of the fluid and the minerals along the flow path. If there are no carbonate minerals (such as calcite or dolomite) along the flow path, the only sources for CO_2 will be atmospheric gases that are at depth or release of gases from cooling magmas. In instances where carbonate minerals are present along the flow path,

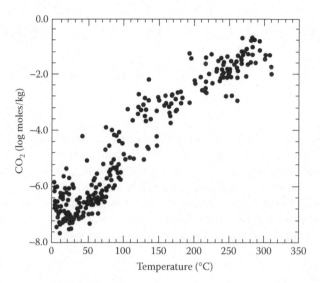

FIGURE 15.1 The concentration of CO_2 in geothermal waters and springs, as a function of the reservoir temperature. (Data from Arnórsson, S., Environmental impact of geothermal energy utilization. In *Energy, Waste and the Environment: A Geochemical Perspective*, eds. R. Giere and P. Stille, Geological Society, London, 297–336, 2004.)

CO_2 concentrations will be significantly higher. Figure 15.1 shows measurements of CO_2 concentration in geothermal fluids as a function of temperature (Arnórsson 2004). These concentrations show that the variability is strongly influenced by the temperature of the resource, with high temperature reservoirs having the highest CO_2 concentrations in the geothermal fluids. This variability likely reflects the impact of temperature on the mass action expression for various mineral reactions. For example, one possible reaction that can influence CO_2 concentration at temperatures of 200°C–300°C (Glassley 1974; Arnórsson 2004) is likely to be

$$3Ca_2Al_2Si_3O_{10}(OH)_2 + 2CO_2 \Leftrightarrow 2Ca_2Al_3Si_3O_{12}(OH) + 2CaCO_3 + 3SiO_2 + 2H_2O$$

Prehnite   gas   clinozoisite   calcite   quartz   water

where the mineral names and liquid and gas phases are indicated. Numerous reactions involving different mineral species and involving a carbonate and CO_2 can be written that result in the same behavior of the geochemical system, namely, a change in the dissolved CO_2 concentration as the temperature changes.

Fluid circulation in natural geothermal systems results in solutions that change temperature along their flow paths. In the above reaction, the equilibrium concentration of CO_2 in the fluid will increase as the temperature of the fluid goes up if the four solid mineral phases coexist (Glassley 1974). Hence, in a naturally convecting geothermal system, one would expect to see the circulating fluid increase its dissolved gas load of CO_2 as it was heated along the flow path and then release the CO_2 as it circulated to cooler regions.

Shown in Figure 15.2 are the measured CO_2 emissions from the ground and soils in natural settings in which convecting subsurface fluids are likely to exist. The emissions are based on field measurements and scaled to the total emission expected from that setting per year, in kg of CO_2. All of these settings are associated with active or recently active volcanic systems. Natural emissions vary by 3 orders of magnitude, which happens to correspond to the natural variability in CO_2 concentrations at a given temperature, as shown in Figure 15.1.

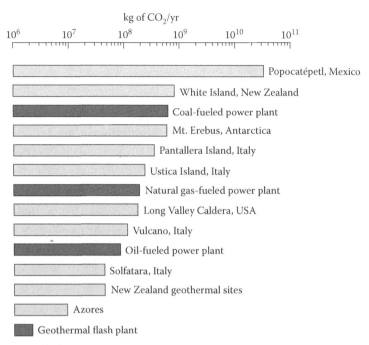

FIGURE 15.2 CO_2 emissions (kg/yr) for a suite of natural sites at which emissions have been measured. All of the locations are volcanic centers from a variety of tectonic settings. Also shown for comparison (indicated in dark color blocks) are the emissions from power plants that utilize coal, natural gas, oil, and geothermal heat as energy sources. For the power plants, the values shown represent kg of CO_2 emitted per MW throughout a year. These values are computed from the Environmental Protection Agency eGRID website for emissions by fuel source, and using the Energy Information Agency values for 2007 capacity factors. (Baubron, J.-C., et al., Measurement of gas flows from soils in volcanic areas: The accumulation method (abstract). *Proceedings of the International Conference on Active Volcanoes and Risk Mitigation*, Napoli, Italy, September 14–17, 1991; Gerlach, T.M., *Nature*, 315, 352–353, 1991; Seaward, T.M. and Kerrick, D.M., *Earth and Planetary Science Letters*, 139, 105–113, 1996; Chiodini, G., et al. *Applied Geochemistry*, 13, 543–552, 1998; Delgado, H. et al., *EOS Transactions of the American Geophysical Union* 79, Fall Meeting Supplement, 926, 1998; Sorey, M.L. et al., *Journal of Geophysical Research*, 103(15), 303–315, 323, 1998; Wardell, L.J. and Kyle, P.R., *EOS Transactions of the American Geophysical Union* 79, Fall Meeting Supplement, 927, 1998; Cruz, J.V. et al., *Journal of Volcanology and Geothermal Research*, 92, 151–167, 1999; Etiope, G. et al., *Journal of Volcanology and Geothermal Research*, 88, 291–304, 1999; Favara, R. et al., *Applied Geochemistry*, 16, 883–894, 2001; Gerlach, T.M. et al., *Chemical Geology*, 177, 101–116, 2001; Evans, W.C. et al., *Journal of Volcanology and Geothermal Research*, 114, 291–312, 2002; Ármannsson, H., CO_2 emission from geothermal plants, *International Geothermal Conference*, September Session #12, Reykjavík, Iceland, 56–62, 2003.)

In the ideal case, geothermal power plants would reinject 100% of the geothermal fluid from a production well. To do so would conserve water, maintain fluid mass in the reservoir, and minimize energy losses. However, in the cooling process and throughout the steam cycle, a significant percentage of fluid loss occurs, particularly in the condensing and cooling cycles. In addition, noncondensable gases in geothermal steam, such as CO_2, diminish the efficiency of energy conversion in the steam-turbine system. Hence, as previously noted, noncondensable gases are usually removed from the steam before it enters the turbine. These processes inevitably result in some amount of CO_2 release from flash geothermal power plants. Shown in Figure 15.2 is the CO_2 emission from a typical flash plant, in kg of CO_2 per year per MW of electricity produced. Also shown are the emissions from coal-, oil-, and natural gas-fueled power plants, per MW of electricity produced per year.

As is evident from Figure 15.2, geothermal power plants emit 15 to 150 times less CO_2 per MW produced than fossil-fueled power generating systems. This fact, along with the absence of any CO_2 emissions from binary power plants, makes electricity production from geothermal power plants an attractive alternative to fossil fuel-based generating systems. In 2012, 37% of the US electrical generation came from burning coal and 30% from burning natural gas (Energy Information Administration 2013). For every MW of power generated by a geothermal power plant that displaces the equivalent power production by this fossil fuel mix, the CO_2 emissions are reduced by more than 99%. In addition, because a geothermal power plant uses the local water, the actual CO_2 emissions from a geothermal power plant will be controlled by the chemistry of the reservoir fluid. As a result, the emissions will reflect the geochemical controls at the site. This suggests that geothermal power plants will compatibly blend in to the natural emissions background, rather than be an exaggerated point source for CO_2 releases, unlike fossil-fueled power plants.

It is important to note that reinjection of cooler fluids with lower CO_2 contents into a geothermal reservoir can perturb whatever steady-state or equilibrium conditions may have existed in the reservoir prior to exploitation. As a result, it is possible that mineral equilibria similar to the reaction listed above involving calcite, prehnite, clinozoisite, and quartz will shift toward replenishment of CO_2 in the aqueous phase. How this may affect the long-term geochemistry and mineralogy of the site must be considered on a site-by-site basis.

Reflecting interest in capturing CO_2 and disposing of it in the subsurface, recent research has suggested that CO_2 may be a fluid used in the reinjection process in geothermal plants (Preuss 2006; Xu and Preuss 2010). This research, which has concentrated on applications at EGS sites, has shown through computer modeling that using CO_2 as a heat transfer agent in geothermal systems can result in sequestering carbon in the subsurface through chemical reactions in which the CO_2 reacts with minerals to form new mineral phases that effectively remove CO_2 from the circulating fluid. This carbon capture and sequestration process has obvious benefits for reducing greenhouse gases. As well, this approach has the benefit of reducing reliance on water at geothermal plants, which is becoming an increasingly fragile resource.

The use of CO_2 in geothermal systems faces two challenges. One challenge is assuring that the mineral reactions that sequester CO_2 into solid phases do not also result in reduction of permeability by minerals growing and blocking fluid flow paths. This challenge will require further research to establish how best to manage and monitor the behavior of CO_2 along the flow path. The second challenge is that such a scheme effectively adds a requirement that CO_2 be captured and transported to a geothermal generating facility. Currently, geothermal systems do not rely on such external supply chains, allowing them to operate in an independent fashion, insulated from other market and transportation pressures. Adding CO_2 to the system would introduce a complexity that would need to be managed and accounted for when planning and developing a geothermal site. Although these two issues are important, they are not at all insurmountable. The advantages of using CO_2 are potentially very significant and justify research into this strategy for power generation.

HYDROGEN SULFIDE

H_2S occurs naturally in the air and in the subsurface. Under normal conditions, it is present in the air at ground level at <1 part per billion (1 ppb) to several hundred ppb, depending upon the local environmental conditions. H_2S is a highly reduced sulfur compound that forms in conditions where the oxygen partial pressure is very low. Hence, environments such as swamps, where anaerobic decay is underway, can have relatively high H_2S levels and is the reason such environments often have a rotten egg smell. In the subsurface, where the abundance of free oxygen can be very low, sulfur occurs as compounds that are usually quite reduced as well. Minerals such as pyrite (FeS_2) and pyrrhotite (FeS) are common reduced sulfur compounds, whereas H_2S is the common gaseous form of sulfur. It is a reflection of these conditions that geothermal fluids have some H_2S present in the solution.

H$_2$S, at concentrations of 500 ppb or higher, can cause unconsciousness and death. At levels of 10–50 ppb, most people find it to have an offensive "rotten egg" odor. At levels between 50 ppb and a few hundred ppb, H$_2$S can quickly deaden olfactory sensors so that the smell is not perceived. For these reasons, strict emissions regulations are imposed on it and abatement is required if concentrations are above regulatory limits. This can be an issue during drilling of a well (as discussed in the Case Study for Chapter 9), as well as throughout the power generation complex.

Strategies for abating H$_2$S in geothermal emissions must take into account several factors. One is that, as a noncondensable gas, it is preferable that it be removed from geothermal steam before the steam is introduced to the turbines, for the same efficiency reasons that other noncondensable gases such as CO$_2$ are removed. However, because of its high volatility, H$_2$S will partition itself between the separated steam and the noncondensable phase. Thus, H$_2$S should be expected to be present both in the noncondensable separated phase and in the steam at the cooling and condensing stage, if it is present at the production wellhead. As a result, it is important to determine H$_2$S concentrations throughout the steam path during the generation process to establish where and to what extent H$_2$S must be removed and neutralized to meet regulatory requirements.

In addition, H$_2$S concentrations can be highly variable from one well to another within a geothermal system, owing to the heterogeneous distribution of minerals and gases that control H$_2$S concentrations. For example, as with the carbonate mineral controls on CO$_2$, there are a number of reactions involving sulfur-bearing minerals that control the amount of H$_2$S in the fluid phase. A schematic reaction such as

$$FeS + H_2O \Leftrightarrow FeO + H_2S$$
$$\text{Pyrrhotite} \quad \text{Fe-bearing}$$
$$\text{mineral}$$

demonstrates one of many possible reaction paths that can influence H$_2$S abundance. The multiplicity of paths results from the fact that sulfur has several oxidation states that can form in natural settings, each of which is a function of the oxidation state of the system. Because the local oxidation state can vary by several orders of magnitude over distances of a few meters, and the abundance of Fe-bearing minerals can also vary significantly, it is not surprising that there is great variability in H$_2$S content in fluids from wells in the same geothermal field.

Abatement of H$_2$S takes advantage of a variety of possible reactions that result in its oxidation. The most commonly used pathways involve various means to produce elemental sulfur or SO$_2$. The overall reaction for reduction to elemental sulfur is

$$2H_2S + O_2(\text{air}) \Leftrightarrow 2S + 2H_2O$$

A similar oxidation reaction can be written for producing SO$_2$. Either path results in a more environmentally acceptable output from the geothermal plant. The reaction path that is used will depend upon the form in which the H$_2$S exists (e.g., within the liquid condensate stream, as part of a separated noncondensable gas phase, etc.). The primary challenge in engineering the abatement scheme is to employ reactors at appropriate locations in the facility such that the H$_2$S is oxidized as economically as possible. The concentration of H$_2$S and its form, whether gaseous or liquid, will dictate the approach that needs to be used. The operating cost for various schemes for removing H$_2$S ranges from about 20 cents/kg of H$_2$S to more than \$30/kg of H$_2$S (Nagl 2008). Removal of more than 90% of the H$_2$S can be readily accomplished.

MERCURY

Emissions of gaseous Hg from geothermal power plants are usually well below regulatory standards. However, instances do exist in which geothermal resources occur in close proximity to geological environments where Hg occurs at elevated levels. Such sites are usually associated with geological environments in which cinnabar (an ore for mercury) or other mineral deposits occur. Although mercury is a metal, it boils at a low temperature and is thus preferentially partitioned into the gas phase, if it is detectably present. Sites where this has been observed include The Geysers and Piancastagnaio, Italy (Baldacci and Sabatelli 1998).

Mitigation of mercury emissions is usually accomplished using either a condensation and cooling method that allows mercury separation or sorption of mercury on to a mineral substrate, such as a carbon material or a zeolite, that is impregnated with sulfur. Such systems are easily employed in conjunction with an H_2S mitigation scheme, thus minimizing engineering and design costs. Recovery of mercury using such systems is usually well beyond ninety percent.

SOLUTE LOAD AND RESOURCE RECOVERY

As discussed in Chapter 5, geothermal fluids can range in composition from dilute solutions to concentrated brines (Table 5.1). Reinjection is usually the preferred means for disposal of geothermal fluids after they have passed through a generating facility or other application. However, in instances where economic or other considerations do not support immediate reinjection, strategies can be developed for handling the spent geothermal water. In such cases, it is necessary to establish whether the water is sufficiently dilute to meet regulatory standards for surface disposal without treatment. Such standards usually consider, among other things, the total dissolved load, the alkalinity, and concentrations of certain metals that can be of environmental concern which, in the case of geothermal waters, are mainly arsenic (As) and boron (B). For waters that meet regulatory standards, the primary challenge is finding a disposal location that can handle the volumetric output, once the water has been sufficiently cooled to prevent an environmental impact.

For mitigation of As, numerous approaches are possible, most of which rely on oxidizing the reduced form of As (As^{3+}) that occurs in geothermal fluids to As^{5+} by reacting it with some appropriate aqueous species that will result in its precipitation of an As-bearing mineral from solution. An example of such a reaction is

$$Fe_2(SO_4)_3 + As^{3+} + 2H_2O \Leftrightarrow FeAsO_4 \bullet 2H_2O(s) + Fe^{3+} + 2SO_2 + O_2(g)$$

where:
   (s) indicates a solid precipitate
   (g) indicates a gas

The necessity to remove B from geothermal fluids is less frequent, but nevertheless can be an issue in some instances. Successful treatment has been accomplished using ion exchange resins (Kabay et al. 2004) and nanofiltration with reverse osmosis (RO) (Dydo et al. 2005).

Geothermal waters with high solute loads are of growing interest because they often contain high concentrations of valuable resources. Although concentrated geothermal fluids have long been perceived as analogs for solutions that formed precious and base metal ore deposits (Helgeson 1964; Skinner et al. 1967; Muffler and White 1969; McDowell and Elders 1980; Bird and Helgeson 1981), it has only been recently that attention has turned toward extracting these metals directly from geothermal fluids (Gallup 1998, 2007; Gallup et al. 2003; Entingh and Vimmerstedt 2005; Bourcier et al. 2006). This change in interest reflects two rapidly evolving developments. One of these developments is the growing ability in the materials science field to design nanomaterials and ion-selective membranes suitable for application with geothermal brines. These materials incorporated into RO technologies and other applications have improved the ability to extract useful resources (e.g., Bourcier et al. 2006). The other factor is the rapidly evolving market for resources contained in geothermal brines. The electronics and energy storage industries are particularly important markets for the resources in geothermal brines (Entingh and Vimmerstedt 2005).

TABLE 15.1

Brine Concentrations of Selected Elements (from Indicated References)

Resource	Brine Concentration (mg/kg)
Silica (Si)[a]	>950
Lithium (Li)[d]	327
Gold (Au)[c]	0.08
Silver (Ag)[a]	1.4
Manganese (Mn)[e]	1560
Zinc (Zn)[e]	790

Sources: [a] Bloomquist, R.G., Economic benefits of mineral extraction from geothermal brines. *Washington State University Extension Energy Program*, 6, 2006

[b] McKibben, M.A. and Hardie, L.A., Ore-forming brines in active continental rifts. In *Geochemistry of Hydrothermal Ore Deposits*, ed. H.L. Barnes, Wiley, New York, 877–935, 1997

[c] Gallup, D., *Ore Geology Reviews*, 12, 225–236, 1998

[d] Ellis, A.J. and Mahon, W.A.J., *Chemistry and Geothermal Systems*, Academic Press, New York, 392, 1977

[e] Skinner, B.J. et al., *Economic Geology*, 62, 316–330, 1967.

Listed in Table 15.1 are concentrations of precious and base metals in some geothermal brines. The table is selective in that the concentrations shown are among the highest reported and therefore are not necessarily from the same well nor indicative of what can be expected in all geothermal systems. They are shown as an indication of the concentrations that have thus far been observed. The commercial value for each of these resources varies with time and market conditions. An early estimate, conducted by Entingh and Vimmerstedt (2005), demonstrated that, for the concentrations listed in Table 15.1, market value for these resources would be in the tens of millions of dollars, if the resources could be recovered and the flow volumes were those typical of a 50 MW geothermal energy generating facility.

Recovery of resources is an attractive complement to power production, because it would provide an additional revenue stream to a producing facility and would also make use of a resource that is readily available, thus reducing the need for surface mining, which can have serious environmental impacts. However, the technology for resource recovery remains a challenge. Most importantly, separation of a resource into its pure state is difficult because of the chemical complexity of geothermal brines. A substantial research and development effort is required to obtain a useful technological approach at each site because the solutions at a particular location tend to have unique characteristics. As well, separation of resources carried out on site will be a parasitic load on the power generation because recovery of these materials requires energy input. Nevertheless, successful approaches have been developed for separation of lithium, silica, manganese, and zinc. Separation technologies for other resources are likely to be available in the near future (Bloomquist 2006).

An additional commodity that can be obtained from geothermal brines is water in arid settings (Gallup 2007). With relatively modest investment, water that has been used in a geothermal application can be cleaned to meet a broad range of standards that can satisfy industrial as well as drinking water needs. In this sense, the water that is the solvent for the metals discussed above also becomes a resource that can be recovered. However, as discussed later in Section "Water Use", availability of water can also be a constraint for geothermal development. As a result, whether or not water can be usefully extracted for other uses will depend on local environmental factors.

SEISMICITY

Seismicity associated with resource use and extraction has attracted growing public interest because of the expanded use of hydraulic fracturing, or "fracking," for obtaining gas and oil from rocks with low permeability. As noted in Chapter 13, permeability enhancement for improving fluid flow in the subsurface can be accomplished through hydro-fracturing, but also through hydro-shearing. Both of these techniques are invariably accompanied by seismicity. Understanding the nature of the risk for property damage and environmental degradation in the form of aquifer contamination and release of unwanted contaminants requires knowledge of the nature and mechanics of this *induced* seismicity.

Seismic activity associated with geothermal applications results from several effects: injection of cool water into hot geothermal reservoirs, extraction of fluid from reservoirs, and high-pressure injection of fluid to enhance reservoir permeability. The seismic events associated with these processes are generally very small in magnitude. From a seismological perspective, the magnitude of the vast majority of events is less than 2.0 and they usually are not felt. However, larger magnitude events have been recorded. The largest event that was associated with geothermal power production was a magnitude 4.6 at The Geysers field in California in 1982 (Peterson et al. 2004). The immediately following sections addresses the fundamentals of the relevant rock mechanics issues. For detailed discussions, see the source material recommended at the end of this chapter.

MECHANICS OF SEISMIC EVENTS

Shear Stress, Normal Stress, and Frictional Strength

As discussed in the Sidebars 4.1, 13.1, and 15.1, rock failure occurs when the internal strength of a rock is exceeded by the stress to which it is subject. For real rocks in a geothermal setting, evaluating rock strength and its relationship to local stresses can be complex. Most rocks in such a setting will possess several sets of fractures, each with specific characteristics. Among the important characteristics will be the length of the fractures, their roughness and planarity, the extent to which they have been cemented by secondary minerals deposited by fluids migrating along the fractures, and their orientation. How the rock responds to an imposed stress field will depend upon the interplay among these variables, as well as the magnitude of the stress, its orientation, and the rate at which stress is applied.

The criterion for failure is based on the ratio of the shear stress (τ) to the normal stress σ_n. For our purposes, we will define the frictional strength of a material as

$$\mu_f = \frac{|\tau|}{\sigma_n} \tag{15.2}$$

where μ_f is the fracture or fault frictional strength.

Because the internal friction of a rock is determined by the weakest part of the rock, μ_f essentially is a definition of the lower limit to the internal friction of a fractured or faulted rock. The value of μ_f is determined by the properties of a rock mass and will vary from rock type to rock type as well as from place to place within a given rock mass. If this ratio of shear stress to normal stress is exceeded for a fracture or fault, the rock will fail by slipping along the fracture or fault.

In the simplest case where there is no stress other than that due to gravity, σ_1 is vertical and equal to σ_n and the minimum principal stress direction, σ_3 is horizontal. Figure 15.3 shows the required stresses for failure to occur for three different situations, assuming rock properties are those typical of granite. The curve labeled σ_1 defines the stress required for the rock to fail, assuming that the initial rock is unfractured. The increasing stress required for failure to occur

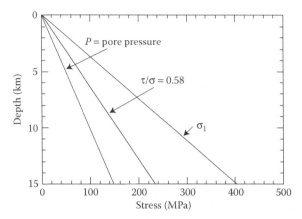

FIGURE 15.3   The stress required to overcome the frictional strength of a granite, as a function of depth. Three cases are considered. In each case, failure will occur at stress values higher than defined by the line. The line labeled σ_1 defines failure stresses for unfractured granite. The line labeled $\tau/\sigma = 0.58$ defines failure stresses for a fractured granite with fractures optimally oriented to fail. The line labeled P = pore pressure defines the stress conditions for failure for unfractured granite in which the pore pressure is hydrostatic. (Modified from Lockner, D.A., Rock failure. In *Rock Physics and Phase Relations*, ed. T.J. Ahrens. American Geophysical Union, Washington, DC, 127–147, 1995.)

as one goes to deeper depths in the earth reflects the effects of increasing confining pressure. As the confining pressure increases, greater force is required for cracks to propagate once they form. If the rock is fractured, and if the fractures or faults are favorably oriented with respect to the normal stress direction, failure will occur at values of the stress defined by the line labeled $\tau/\sigma = 0.58$ (real rock fractures will fail at conditions approximating this value). This line indicates the stresses beyond which the fracture or fault frictional strength (Equation 15.2) is exceeded. This failure is conditional, requiring the existence of a set of fractures or a fault with the proper orientation. For many rocks, a favorable orientation for fracture failure is approximately 30° between the maximum principle stress direction and the fracture plane. The exact angle depends on the fracture properties. Stresses between this curve and the curve labeled σ_1 are the stress conditions for fractured or faulted rock in which the fracture or fault is at a higher or lower angle to σ_1 than the optimal angle.

Pore Water

Pore water diminishes internal and fracture or fault frictional strength through two mechanisms. One mechanism is through its effect on the effective stress. In the situation we are considering, with a vertical maximum principle stress, the normal stress is reduced by the hydrostatic stress (which results from the pressure exerted on the rock by the pore water), resulting in an effective normal stress (recall discussion in Chapter 13) determined by

$$\sigma_n^{\text{effective}} = \sigma_n - P_p \tag{15.3}$$

where P_p is the pore pressure.

The line labeled as P = pore pressure defines the stress condition required for failure of unfractured rock in which the pore pressure is equivalent to the hydrostatic pressure.

Pore water affects frictional strength also through chemical mechanisms. At the molecular level, ionic interactions between polar water molecules and the mineral framework, as well as interactions between certain solute species (such as acids) and the crystal lattice, result in relatively rapid rearrangement of bonds. This can weaken the rock at fracture tips, allowing crack growth at stress values below those that would otherwise be necessary for a fracture to propagate.

The curves in Figure 15.3 are specific to the rock that was studied (granite) and the stress conditions (vertical maximum principal stress). Other rock types or granites in different settings with different geological histories will have curves that are different, in terms of absolute values, from those depicted. Nevertheless, the relationship to stress and the relative locations of the curves will be similar for all rocks.

The general nature of these relationships provides insight into the factors that determine the conditions under which failure of an intact rock or of a fault or fracture may occur. Specifically, it is obvious that the large number of variables that influence failure (fracture and fault orientation; fracture and fault properties such as frictional strength, pore pressure, stress) and the difficulty of obtaining detailed descriptions of these variables several kilometers in the subsurface precludes the ability to precisely predict the conditions under which failure and hence seismic activity will occur.

Predicting the magnitude and manner of the energy release, and hence whether or not a felt seismic event will occur, is just as problematic. The energy release and the nature of the movement associated with rock failure is a direct function of the size of the area over which a fracture or fault slips. This is a reasonably intuitive relationship—if the unfractured rock in Figure 15.3 is subjected to a stress of 300 MPa while it is under pressure equivalent to a depth of 2 km, it will fail unequivocally. If it fails by forming a set of 1000 fractures, each with 0.1 cm^2 in surface area with equal movement on each one, the collective movement could amount to a slip of 100 cm and yet the sample would exhibit very little movement. If instruments were monitoring the sample, a swarm of microseismic events would be detected. If, however, only 1 fracture formed, but the same total movement had to be accommodated, that is, a slip of 100 cm, a much more obvious seismic event would be recorded.

A conclusion that implicitly develops from this discussion of the mechanics of rock failure is that it will be difficult to precisely predict where and when seismic activity will occur and what its magnitude will be. However, several factors specific to geothermal projects allow forecasts to be made with some confidence regarding potential seismic activity and risk.

SEISMIC ACTIVITY ASSOCIATED WITH GEOTHERMAL PROJECTS

As noted in Section "Mechanics of Seismic Events", seismic activity has been associated with injection of cool water into hot geothermal reservoirs, extraction of fluid from reservoirs, and high-pressure injection of fluid to enhance reservoir permeability. With the basic mechanics of rock failure as background, each of these is considered and discussed elaborately in the below subsections. Further discussion of these topics, as they relate to geothermal projects, can be found in a well-documented study published by the National Research Council (2013).

Seismicity Associated with Injection of Cool Water

Most minerals respond to heating by expanding. The extent of expansion depends on the mineral structure. Laboratory measurements of the change in molar volume as a function of temperature allow determination of the volumetric coefficient of thermal expansion (α_V) which is defined as

$$\alpha_V = \frac{(\Delta V/V_0)}{\Delta T} \tag{15.4}$$

where ΔV is the change in volume from a reference state, V_0, over the temperature interval ΔT

Table 15.2 lists the volumetric coefficient of thermal expansion for the potassium feldspars sanidine and microcline and the sodium feldspars low and high albite. These minerals are among the most common rock-forming minerals and their behavior can provide a guide to the response of the rocks that make up the bulk of a geothermal reservoir.

TABLE 15.2
Coefficients of Thermal Expansion for Feldspar Minerals

Minerals	α (T^{-1})	Reference Volume (Å^3)
Microcline[a]	1.86×10^{-5}	722.02
Sanidine[b]	1.92×10^{-5}	723.66
Low albite[a]	3.07×10^{-5}	664.79
High albite[c]	3.15×10^{-5}	666.98

Sources: [a] Hovis and Graeme-Barber 1997
[b] Hovis et al. 1999
[c] Stewart and von Limbach 1967.

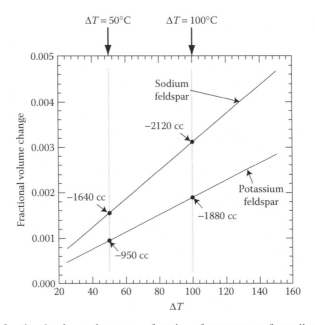

FIGURE 15.4 The fractional volume change as a function of temperature for sodium feldspar and potassium feldspar. The sodium feldspar and potassium feldspar curves are the respective average values for low and high albite, and sanidine and microcline from Table 12.2. The negative values in the plot are the change in volume (in cc) that would occur for a cubic meter of the pure mineral phases, upon cooling through a ΔT of 50° and 100°.

Shown in Figure 15.4 are the fractional changes in volume that occur for sodium and potassium feldspars as a function of temperature change. Because the changes for sanidine and microcline are very similar, and those for low and high albite are very similar, the respective averages for these minerals are shown. Also shown in the figure are the number of cubic centimeters by which one cubic meter of the pure mineral phases would shrink if it were cooled by 50°C and 100°C. Such volume changes, although small, are not insignificant. It is this effect that has been hypothesized to be the underlying cause of very small earthquakes in regions where cool water is injected into hot geothermal reservoirs.

For example, Stark (2003) has documented that during injection of recharge waters into The Geysers geothermal reservoir, the temperature difference between the reservoir and the injected fluid will be in excess of 100°C. Stark (2003) noted that microseismic events are temporally and spatial associated with fluid injection (see Figure 10.17). He concluded that the thermoelastic effects associated with injecting cool water into this geothermal system are the likely cause of this relationship. The volume changes associated with this thermoelastic process are most likely accommodated by the growth of existing fractures and the formation of new microcracks.

Normally, such processes cannot lead to earthquakes beyond a magnitude of about 2.5. The reason for this can be understood by considering the size of area affected by the cooling process associated with water injection. The volumetric changes indicated in Figure 15.4 would amount to slip of 10–15 cm if uniform contraction of a rock volume took place. If the contraction were constrained to be solely in one plane, the maximum amount of contraction would be 1–2 m. Because the area affected is restricted to the immediate vicinity of the injection well, and because the amount of slip is small, the amplitude of motion propagated to the ground surface would be very small. Such seismic signals can be detected using sensitive seismometers and can be useful for mapping where fracturing is occurring, but would have little consequence otherwise.

The main exception to this is when there exists a significant component of natural stress. As noted previously in Chapter 2, the earth is a heat engine that drives the movement of tectonic plates. These movements stress the plates to varying degrees and in different ways. High stress conditions can occur where two plates slide against each other, as is the case with the well-known San Andreas Fault in California. High stress can also be generated where plates converge, such as at subduction zones. Finally, stress may accumulate in the interior of plates due to changes in plate motion and other effects that slightly warp the plates. Hence, in many settings, there will exist some preexisting state of stress, also called in situ *stress*. The magnitude of the *in situ* stress is normally well below the value of μ_f for the rocks, fractures, and faults that compose the local geology. Under those conditions, there will be no seismic activity. However, if the *in situ* stress field results in the ratio of shear stress to normal stress being close to values of μ_f for local fractures and/or faults, it is possible that thermoelastic effects associated with injection of cool water could trigger larger than anticipated slip. This consequence could result either from an increase in the total stress due to the thermoelastic effects or the thermoelastic effects could perturb the orientation of the *in situ* stress field. If either situation results in a stress condition that results in failure, the magnitude of the resulting ground motion will depend on, among other things, the dimensions of the area over which rupture or slip occurs.

Rupture Area and Magnitude

An empirical relationship between magnitude and the area over which rupture occurs was developed by Wells and Coppersmith (1994)

$$M = 4.07 + 0.98 \log(A) \qquad (15.5)$$

where:
   M is the magnitude of the event
   A is the rupture area (km^2)

Shown in Figure 15.5 is the relationship between magnitude and rupture area. Most seismic events associated with injection of cool water have magnitudes less than 2.5 (shaded box in Figure 15.5), suggesting rupture areas of less than 0.2 km^2. If such ruptures occur on surfaces that are more or less equant in form, the slip areas would have dimensions of ~100–150 m on a side or less.

To avoid the possibility of inducing larger, damaging seismic events through geothermal energy development, these relationships and observations suggest several important attributes to evaluate for a site. Most areas of interest for geothermal applications have been studied by geologists. The research results often include evaluation of the seismic history and some consideration of the local stress field. Data from such studies should be used to evaluate the magnitude and orientation of the local stress field, how it changes with depth, and the uncertainties associated with those evaluations.

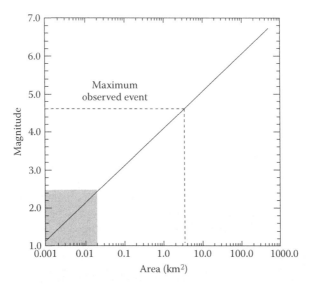

FIGURE 15.5 The relationship between earthquake moment magnitude and rupture area (in km²), based on the empirical relationship of Wells and Coppersmith (1994). The shaded area encloses the region that encompasses the majority of seismic and microseismic events observed in geothermal developments. The dashed line indicates the largest observed earthquake at a geothermal site and its inferred rupture area.

On the basis of that research, it can be established what additional research should be done to minimize uncertainties and data deficiencies.

In conjunction with analysis of the stress field, the mechanical properties of the rock types in the area should be evaluated. Such studies include an analysis of the orientation, mechanical properties, and mineralogy of fracture sets. Establishing the approximate values for μ_f and $\sigma_n^{effective}$ *in situ* provides a basis for estimating the likelihood of inducing rupture.

The third important effort for characterizing a site is a thorough analysis of the largest fault or fracture features that have the potential to rupture. Establishing the location and size of these features can help delineate the most advantageous location to minimize risk associated with a geothermal installation.

The combination of quantitatively evaluating the stress field, rock properties, and potential structures that could fail will provide an important and credible method for reducing seismic risk. Concomitantly, it is important to consistently monitor microseismic activity before, during, and after construction and operation activities have commenced in order to identify unusual activity that may indicate unexpected effects. Often such effects can be mitigated by modification of the rate, timing, and operational parameters for injection.

Seismicity Associated with Fluid Extraction

Seismicity associated with extraction of geothermal fluids can be understood by considering the influence pore pressure has on rock behavior. In areas of limited recharge, removal of a geothermal fluid will change the *in situ* stress field by lowering the pore pressure in a restricted region. That effect will locally reorient the principal stress directions. Rupture may occur if a fracture set exists in the area that is oriented appropriately and μ_f is exceeded when the stress field changes orientation and the effects of changing pore pressure propagate through the local geological framework. Mitigation of any potential negative impacts can be accomplished by following the same data collection, analysis, and monitoring activities as described for injection of cool water.

In addition, it is important to balance fluid extraction and injection so that pore pressure variability is minimized. By compiling sufficient information regarding the local rock properties and state of stress, the difference in pore pressure due to imbalance in the volume of fluid extracted versus that which is reinjected can be maintained within limits that are likely to prevent or minimize significant slip and seismic activity.

Seismicity Associated with High-Pressure Injection of Fluid to Enhance Reservoir Permeability

In order to improve recovery of oil and gas deposits, the oil and gas industry developed a technique of fracturing rock formations at depth using high-pressure fluid. The technique is called hydro-fracturing and can result in a significant increase in formation permeability, as discussed in Chapter 13. This process is also called *reservoir stimulation* or *reservoir enhancement*. The geothermal industry adopted this approach early on as a means of improving permeability near production wells and as a means for developing a permeability pathway between an injection well and a producing well. For a period of time hydro-fracturing was a crucial element in the development of enhanced geothermal systems (EGS), as discussed in some detail in Chapter 13. However, hydro-shearing has become a more useful and applicable technology for improving permeability in EGS developments.

To review, hydro-fracturing is accomplished by using a pump capable of injecting fluids at pressures between 2 and 20 MPa (300–3000 pounds per square inch) into a well that is lined with high strength steel pipe. In the part of the subsurface where increased permeability is desired, the steel pipe is perforated, allowing the high-pressure fluid to enter the enclosing rock. Depending upon the rock properties and fracture sets, the rock will either form new fractures or preexisting fractures will open. The pressure will be maintained for hours to days, depending on the distance over which the new permeability is sought.

Research has shown that the primary effect of this process is to cause preexisting fractures to fail by shearing (Tester et al. 2006; Kraft et al. 2009). Fluid injection reduces the normal stress which effectively increases the ratio of shear stress (τ) to normal stress (σ_n). At sufficiently high pressures, the resulting τ to σ_n ratio exceeds the fracture frictional strength μ_f and the fracture slips or ruptures. The size of the microseismic events suggest that the rupture area is commonly on the order of a hundred m^2 to a few thousands of m^2.

Induced fracturing produces swarms of many small seismic events. From an operational perspective, these events are useful for monitoring the location and progress of the hydro-fracturing. The use of high sensitivity seismometers on the surface and installed in boreholes provides the ability to map the location, orientation, and extent of the fracturing.

Figure 15.6 shows the concentration and distribution of seismic events for a hydro-fracturing effort undertaken at the European EGS site at Soultz-sous-Forêts in eastern France from 2000 to 2004 (Baria et al. 2006). The colored region encompasses tens of thousands of seismic event. The strongest events had magnitudes of 2.6 and 2.9. Thousands of events were recorded but were not sensed as ground motion by the local inhabitants. The vast majority of the seismicity had magnitudes less than 2.0. Although a broad range of injection rates, injection volumes, and pressures were used for this effort, the seismic moment magnitudes were small, suggesting that the overall risk of large events was small.

However, 150 km south of this region, near the city of Basel, a similar EGS effort resulted in a magnitude 3.4 seismic event. Although no structural damage was reported, the unexpected shock resulted in concern regarding seismic risk. This event occurred several hours after injection had been stopped. Similar *post shut-in* seismic events have been observed elsewhere including at the site in Soultz-sous-Forêts.

Such seismic events associated with fluid injection are not a new phenomenon. Nicholson and Wesson (1990) documented that injection of fluids can induce earthquakes in a variety of settings. The largest induced earthquake was a magnitude 5.5 event in 1967 associated with waste disposal at the Rocky Mountain Arsenal in Colorado. Bachmann et al. (2009) have shown that the seismicity associated with fluid injection, specifically that at the Basel site, exhibits characteristics similar to

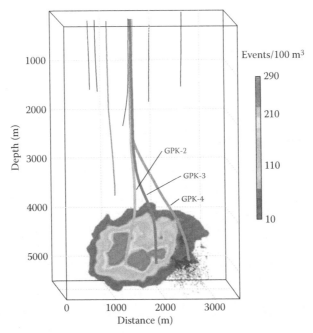

FIGURE 15.6   **(See color insert.)** Density contour map of the number of seismic events associated with stimulation of the crystalline rock reservoir at the European Hot Dry Rock (EGS) project in Soultz-sous-Forêts, France. The contour map reflects the results from two periods of hydro-fracturing carried out on the wells labeled GPK-2 and GPK-3. Also shown are individual events (dots) associated with hydro-fracturing from GPK-4. The other wells shown in the figure were used for seismic monitoring. (Modified from material in Baria, R. et al. Creation of an HDR reservoir at 5000 m depth at the European HDR project, *Proceedings of the 31st Workshop on Geothermal Reservoir Engineering*, Stanford University, Stanford, CA, SGP-TR-179, January 31–February 1, 8, 2006.)

aftershock sequences observed for natural earthquakes, which suggests that the physical processes are similar.

Mitigation of seismic risk associated with fluid injection for hydro-fracturing purposes can be accomplished by coordinating site studies, as described above in Section "Seismicity Associated with Injection of Cool Water," with several operational approaches. These aspects are briefly described by Baria et al. (2006). They include detailed monitoring of the microseismic response to changes in injection rate, volume, and pressures. Analysis of these responses, along with thorough monitoring of the seismic behavior after shut-in can provide guidelines for injection rates, volumes, and pressures that will minimize seismic risk.

In addition, a thorough analysis of seismicity associated with all aspects of subsurface activity within the energy technology industry (National Research Council 2013) documented the relationship between seismicity and resource extraction efforts. That report developed specific recommendations that relate to managing risk and mitigating effects. Key recommendations that were developed included the following:

- Classify wells by use and type, and strictly assure construction requirements for each well type are met. This includes consideration of casing type, cementing, and well finishing.
- Assure that the type and composition of the injected fluid is recorded and that the injection well satisfies criteria specific to the fluid type.
- Assure that drinking water aquifers are protected by the wells.
- Assure microseismic monitoring and data processing are adequate.

- Assure a systematic approach is in place to investigate instances of induced seismicity that meet certain criteria.
- Evaluate seismic risk in a region and have in place guidelines that are relevant for the population density, structure types, and likely seismic magnitudes.
- Put in place a "best practices" protocol so that project development can proceed as efficiently as possible while minimizing and addressing risk.

GROUND SUBSIDENCE

Withdrawal of fluid from the subsurface without balanced reinjection can impact the local hydrology and subsurface stress regime. In locations where the rock framework has high strength, water in interconnected pores and fractures experiences little or no load from the rock. In such a case, the only force acting on the water at depth is from the overlying water mass and the pressure is hydrostatic. Such sites are usually composed of granite, gneiss, or other crystalline rock. Water removed from such a location will have no discernible effect on the elevation of the land surface.

Rocks with low strength, and which therefore are compressible, will exert some force on water in pores and fractures. Because water is incompressible, it will exert a corresponding force on the impinging rock that can be as high as the lithostatic pressure. Under such circumstances, the water becomes an intrinsic element in the subsurface structural framework. Rocks that behave this way are usually unconsolidated sediments, porous volcanic rocks, or rocks with a high clay content. If water is removed from such a setting, the overlying rock mass will settle to some extent, depending upon the amount of water removed and the compressibility of the rock. On the ground surface, this effect will be recognized as subsidence.

The rigidity of a rock material is known as the bulk modulus (K) and has units of GPa. The compressibility of a material is the inverse of the bulk modulus. Table 15.3 lists experimentally determined values of the bulk modulus as a function of porosity for an artificial glass. Although the bulk modulus for rocks varies considerably, the values in the table provide a reasonable indication of the magnitude and variability of K. Plotted in Figure 15.7 are the values in Table 15.3 along with values from other rock types.

TABLE 15.3
Bulk Modulus for Porous Glass, as a Function of Porosity

Porosity	K (GPa)	Compressibility (GPa^{-1})
0.00	45.9	2.18E–02
0.05	41.3	2.42E–02
0.11	36.2	2.76E–02
0.13	37.0	2.70E–02
0.25	23.8	4.20E–02
0.33	21.0	4.76E–02
0.36	18.6	5.38E–02
0.39	17.9	5.59E–02
0.44	15.2	6.58E–02
0.46	13.5	7.41E–02
0.50	12.0	8.33E–02
0.70	6.7	1.49E–01

Source: Walsh, J.B., et al., *Journal of the American Ceramic Society*, 48, 605–608, 1965.

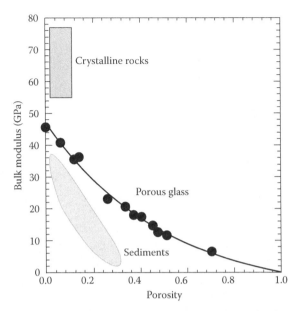

FIGURE 15.7 The variation in bulk modulus (GPa) as a function of porosity, for porous glass (solid dots), sedimentary sandstones (light gray field), and various crystalline rocks (dark gray box). The line drawn through the porous glass data points is an exponential decay curve. (From Walsh, J.B. et al., *Journal of the American Ceramic Society*, 48, 605–608, 1965; Han, D.-H. et al., *Geophysics*, 51, 2093–2107, 1986.)

Crystalline rocks such as granite, gneiss, and other similar rocks have values of K that are sufficient to prevent significant subsidence if fluid is removed from fractures and pores. Hence, such rocks are good geothermal reservoirs, when considered from the point of view of land surface stability. Porous rocks with less rigidity are likely to exhibit some compression if they compose part or all of a setting in which fluid pressure is diminished as a result of fluid extraction. The extent to which subsidence can occur will vary depending upon the change in pressure and the rock bulk modulus.

Perhaps the best documented record of subsidence in a geothermal field is that for the Wairakei, New Zealand, region where measurements and operational approaches over several decades allow inferences to be drawn regarding the response of the geothermal system to development (Allis et al. 2009). Data that have been collected are from surveys of land surface elevations using a network of installed benchmarks. Careful surveying using this method can establish subsidence rates as small as an mm/year. It is important to recognize that these surveys identify where changes in land elevation have occurred but cannot establish the cause of the changes. Additional data that provide a description of the mechanism that causes subsidence must be obtained in order to determine the reason for the subsidence.

The results of the repeated surveys have identified a region of about 50 km^2 over which subsidence has occurred. In most of this area, subsidence has been less than 1 m. However, localized "bowls" of subsidence have been identified where subsidence as great as 15 m has been documented. This is the largest reported subsidence that is potentially associated with geothermal activities anywhere in the world. The repeated releveling has also shown that the subsidence rate has changed over time. Between the late-1950s and the mid-1970s, the subsidence rate in some areas accelerated, reaching a maximum rate of over 400 mm/yr. Subsequently, those rates have declined to about 10–25 mm/yr. Figure 15.8 shows the distribution of subsidence rates in the area, as of 2009. Two zones of subsidence rates in excess of 60 mm/yr are indicated, each occupying an area of about 1 km^2.

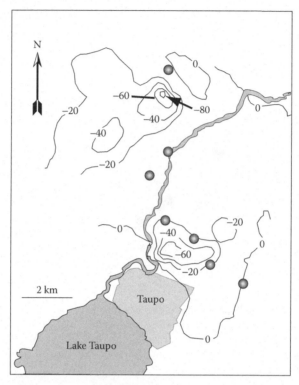

FIGURE 15.8 Contours of the subsidence rates (mm/yr) in the vicinity of Taupo, New Zealand. The locations of wells for geothermal power generation are shown by the partially filled circles. At each of these sites, there usually are multiple wells. (Data from Allis, R. et al., *Geothermics*, 38, 169–180, 2009.)

The locations of geothermal wells do not exactly match the positions of the most intense subsidence. It is nevertheless inferred that a drop in pressure in the deeper steam zone due to fluid extraction contributed to subsidence. The pressure drop over the period of production has been about 1.5 MPa. Reinjection of fluid was not initiated until late in the production history, at which point subsidence slowed. As of 2006, reinjection amounted to about 15% of the produced fluid mass. The pattern of subsidence is, however, somewhat enigmatic. The subsurface geology in the region consists of relatively flat-lying sequences of various volcanic rocks. Some of these are quite porous and easily altered to clays and other soft, weak minerals. The location of the subsidence bowls may indicate locations in the subsurface where such alteration has been localized and is pervasive. It is also possible that there exist in the subsurface locations where the rock sequence has substantial tilt to it, allowing altered sequences to collapse downslope. Currently, distinguishing between these and other possible mechanisms for the bowl formation is not possible. Furthermore, the amount of subsidence suggests an unusually compressive rock unit. There currently are no obvious candidates for this material. Further work is underway to establish the exact causes of this unusually high degree of subsidence.

The releveling efforts that established the subsidence history around Wairakei are labor intensive and time-consuming. Recent advances in remote sensing methods are making it possible to conduct surface elevation measurements at high precision using satellite-generated data. Eneva et al. (2009) report the results of a survey of the geothermal area of the Salton Sea that holds great promise. The technique uses radar signals from orbiting satellites to conduct interferometric synthetic aperture radar (InSAR) studies. InSAR studies allow ground deformation to be measured by constructing difference maps based on repeated surveys. Such differential InSAR (DInSAR) studies have been well documented (Massonnet and Feigl 1998; Bürgmann et al. 2000). In areas where vegetation is present and rapidly changes because of seasonality or agricultural activity, DInSAR cannot be readily employed. However, DInSAR methods can be adapted and modified in such instances if permanent radar scatterers can be identified that are independent of the vegetative effects. This method is described by Eneva et al. (2009) and Falorni et al. (2011), as applied to mapping ground motion in a portion of the Salton Sea area.

Ground subsidence generally has little impact in geothermal regions, particularly if reinjection of the produced mass is employed. However, remote sensing techniques such as satellite interferometry may provide the means to rapidly assess if subsidence problems are developing and how best to mitigate them.

WATER USE

As noted in Chapter 10, a significant quantity of water is required to achieve the energy production rates necessary to efficiently extract energy from a geothermal fluid. The flow path followed by a geothermal fluid at a generating site generally involves the following sequence of stages:

- High-pressure mixed steam and liquid flow from the wellhead to the separator.
- High-pressure steam is separated from condensed liquid and piped to the turbine.
- Steam expands into the turbine, dropping in pressure and temperature with some condensation.
- Condensate is separated from steam as steam flows through multiple turbine stages.
- Steam exits the turbine through the turbine exhaust and is condensed in the condenser.
- Condensed water from the condenser is cooled in the cooling tower.

Cooling towers spray water into flowing air to result in evaporative cooling. Throughout this process, about 60%–80% of the original steam that enters the turbine ends up evaporating to the atmosphere. The remaining fluid is collected and reinjected.

For long-term sustainability of the geothermal resource, it is important to replenish the extracted fluid. Although this will occur naturally, the time frame over which that will be accomplished can vary from a few hours or days to many years, depending upon the natural recharge rate. If the extraction rate from a geothermal resource is very high and the rate at which water is naturally replenished is also high, it is possible that draw down of water from other resource reservoirs can occur. Such an effect might be expressed as a drop in the local water table, diminution of geothermal surface manifestations such as hot springs and geysers, and other effects. If the recharge rate is very slow, it is possible that subsidence of the land surface could occur, as well as reduction in the geothermal resource productivity. For these reasons, it is important to balance fluid extraction by reinjection.

The loss of a large fraction of the geothermal fluid to evaporative cooling is a significant challenge to maintaining mass balance. The Geysers site has developed an agreement with local water districts for disposal of their wastewater by injecting it into the geothermal reservoir. This undertaking has moved The Geysers system much closer to achieving mass balance and thus sustainability. This solution, however, is not likely to be useful for many sites located in areas removed from significant population centers that need to dispose of wastewater in an environmentally appropriate way. In areas where there is an abundance of surface water, collection and injection of that water into the reservoir can provide the mass needed to replenish the reservoir and maintain sustainability. For sites located near coastlines or other nonpotable water, such as brackish groundwater, those water resources can be used to make up the loss to evaporation.

Consideration is also currently being given to a hybrid approach that uses air cooling for the coolest part of the year. Such a system would be designed for maximum efficiency at the expected ambient air temperature range during the coolest season. As the diurnal temperature range changes during the warmer months, air cooling would be supplemented by incremental addition of water cooling. Such an approach allows all of the geothermal fluid to be reinjected for some of the year and minimizes losses to the atmosphere during the remainder of the year. Mishra et al. (2011) analyzed in detail the impacts on water budgets and implications for cooling technologies for geothermal energy production. They showed that water-stressed regions need to employ advanced technologies for cooling to assure that water management is done in a way that assures sustainable energy production and water availability for nonpower generation purposes.

Regardless of the approach used, it is important that careful attention be given to maintaining the long-term stability of the geothermal resource by careful management of the water cycle throughout the power generation facility and the surrounding ecosystem. Although geothermal resources are inherently renewable on the geological timescale, their use on human timescales requires that they be managed in a way that astutely balances all parts of the system to assure sustainability.

LAND USE

Construction and operation of a geothermal facility requires land, thus impacting the local landscape. The construction phase, as with many development efforts, produces the largest footprint a project may cause but, in most cases, is significantly reduced once construction and testing have been completed and operations begun. The construction phase for a geothermal facility usually consists of two basic components, drilling a well or wells, and construction of the geothermal facility. These construction components are the same for all applications, regardless of whether one is dealing with ground source heat pumps, direct-use, or power generation.

Drilling a geothermal well requires land access for a drilling rig of sufficient size to achieve the needed depth. This may require road building to get to the site selected for the well. At the well site, a drilling pad must be excavated that is large enough to allow maneuvering the drilling rig or construction of the well-drilling platform. The site must also have adequate storage area for the drill pipe and casing that will be used. Excavation will also be required for a sump or multiple sumps, which are pits into which the drilling mud is pumped as it exits the well and is recycled. For very deep wells in geothermal systems for power generation, it may also be necessary to have a facility for cooling the drilling mud. The total area required to accommodate these facilities will vary from 1,000 to 10,000 m^2, depending upon the depth to be drilled and the equipment to be used.

Site restoration must follow drilling. This effort includes removing the mud from the sumps and backfilling them, removal of all drilling equipment and supplies, and re-forming the terrain, if appropriate. At completion, the wellhead and apron around it will occupy an area significantly less than 100 m^2. For direct-use and ground source heat pump installations, the wellhead and apron will occupy an area less than 5 m^2.

Construction of a generating facility must include all aspects of the conversion technology that are required to get the power to a distribution system. In addition, feed stocks for the fuel cycle and habitat fragmentation are also an integral part of the overall footprint a conversion technology imposes on the land. McDonald et al. (2009) undertook an analysis of land area impacted by various energy conversion technologies. Their analysis included consideration of various policies that might be emplaced to reduce carbon emissions and projected the results to 2030, which is one of the years for which the United States Energy Information Administration has developed forecasts for energy production. Figure 15.9 summarizes the results of that study. For the energy conversion technologies they evaluated, geothermal power production had the second lowest environmental footprint. Other analyses taking different approaches have consistently reached the same conclusion.

From a land-use perspective, geothermal energy use applications, whether they be for direct-use, ground source heat pumps, or power generation, have among the lowest footprint of any available technology. An important part of this reduced footprint is the fact that there is no fuel cycle that is linked to the operation of the geothermal plant. This reduces the footprint significantly, because fuel storage is unnecessary. If a full life-cycle analysis were conducted, the relative footprint size of a geothermal system, compared to that of fossil-fueled plants, would be reduced even more.

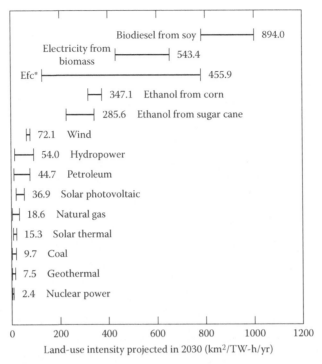

FIGURE 15.9 Land use of energy technologies based on an analysis of the impact of the technology on habitat disruption and changes in use. The bars represent the maximum and minimum impact, by the year 2030, determined on the basis of potential policies that would influence development of the technology. The numbers by each bar are the midpoint values. Efc, Ethanol from cellulose. (Data are from McDonald, R.I. et al., *PLoS ONE*, 4, 1–11. http://www.plosone.org/article/info:doi/10.1371/journal.pone.0006802, 2009.)

SYNOPSIS

The environmental impact of converting energy to electricity or some other useful form inevitably disturbs the environment. For this reason, it is imperative that aggressive, scientifically based monitoring, analysis, and mitigation efforts be considered an integral part of any energy development. The importance of renewable energy resources such as geothermal is that their environmental impacts can be minimal, if properly managed. The impacts from geothermal energy are primarily from emissions, water disposal, seismicity, ground subsidence, water use, and land use. Emissions from flash geothermal power plants (there are no emissions from binary plants) are due to gases dissolved in the geothermal fluid. The volumetrically most significant of these is CO_2, with occasional impacts from H_2S and Hg. CO_2 emissions are quite small, amounting to less than 1% of those from fossil-fueled power conversion technologies. Because the amount of dissolved CO_2 is determined by the local geology, which emits measurable CO_2 as a background component to the atmosphere, it is likely that, in most cases, a geothermal power plant will represent a relatively small increment to the natural emissions at a site. The other gaseous components are routinely removed from the output of a geothermal plant, if concentrations might exceed regulatory limits. Disposal of geothermal water is generally by reinjection, but some components in brines can represent a potential resource of significant economic value. Resource recovery is currently an active area of research and development. Currently, silica, zinc, and manganese are capable of being economically recovered. Seismicity induced by development and operations activities associated with geothermal power development occurs in response to fluid extraction, reinjection, and reservoir enhancement. The magnitude of ground shaking from such induced events is very low under most circumstances. Larger amplitude events can happen only if the local stress regime and geological structures have sufficient capability for rupture over a large area. Thorough analysis of a site can identify such conditions and guide activities to avoid significant rupture events. Subsidence of the land surface can result if there is inadequate replenishment of geothermal fluids in areas where the subsurface rock has low compressive strength. As with the case involving seismicity, thorough site analysis and monitoring can guide operational activities, resulting in greatly reduced risk. Water use in areas where water availability is limited can impose important challenges on geothermal operations if water cooling is involved. Planning, design, and operation of a geothermal facility must thoroughly consider the water mass balance through the entire operations cycle in order to assure sustainable use. Use of seawater, wastewater, brackish groundwater, and surface water, where appropriate, can significantly mitigate the impact on water budgets. Development of hybrid cooling technologies and operational approaches are also under development to reduce water use. Finally, the land-use footprint of a technology can be an important environmental issue. Habitat disruption and displacement of other uses are two important impacts that can result from developing an energy-use facility. Geothermal power, as with any application, has a land-use impact. That impact, however, is the smallest of all energy conversion technologies, per MW generated, with the exception of nuclear power.

PROBLEMS

15.1. What studies and measurements can be done before a geothermal project is initiated to determine the likely contaminants that would need to be dealt with?

15.2. What are the environmental emissions that must be considered for geothermal power projects? How do they compare, quantitatively, with those from fossil-fueled power generation systems?

15.3. What are ways in which water is used for power production? What methods could be employed to reduce water consumption?

15.4. What is resource recovery? What role can it play in geothermal power generation systems?

15.5. What causes subsidence? How can it be mitigated?

15.6. What are the key factors that determine the sustainability of a geothermal project?

REFERENCES

Allis, R., Bromley, C., and Currie, S., 2009. Update on subsidence at the Wairakei-Tauhara geothermal system, New Zealand. *Geothermics*, 38, 169–180.

Ármannsson, H., 2003. CO_2 emission from geothermal plants. *International Geothermal Conference*, September Session #12, Reykjavík, Iceland, September 11–13, pp. 56–62.

Arnórsson, S., 2004. Environmental impact of geothermal energy utilization. In *Energy, Waste and the Environment: A geochemical Perspective*, eds. R. Giere and P. Stille. London: Geological Society, pp. 297–336.

Bachmann, C., Wössner, J., and Wiemer, S., 2009. A new probability-based monitoring system for induced seismicity: Insights from the 2006–2007 basel earthquake sequence. *Annual Meeting of the Seismological Society of America*, Seismological Research Letters, vol. 80, April 8–10, Monterey, CA, p. 327.

Baldacci, A. and Sabatelli, F., 1998. Perspectives of geothermal development in Italy and the challenge of environmental conservation. *Energy Sources, Part A: Recovery, Utilization, and Environmental Effects*, 20, 709–721.

Baria, R., Jung, R., Tischner, T., Nicholls, J., Michelet, S., Sanjuan, B., Soma, N., Asanuma, H., Dyer, B., and Garnish, J., 2006. Creation of an HDR reservoir at 5000 m depth at the European HDR project. *Proceedings of the 31st Workshop on Geothermal Reservoir Engineering*, Stanford University, Stanford, CA, SGP-TR-179, January 31–February 1, 8 pp.

Baubron, J.-C., Mathieu, R., and Miele, G., 1991. Measurement of gas flows from soils in volcanic areas: The accumulation method (abstract). *Proceedings of the International Conference on Active Volcanoes and Risk Mitigation*, Napoli, Italy, September 14–17.

Bird, D.K. and Helgeson, H.C., 1981. Chemical interaction of aqueous solutions with epidote-feldspar mineral assemblages in geologic systems; II, Equilibrium constraints in metamorphic/geothermal processes. *American Journal of Science*, 281, 576–614.

Bloomquist, R.G., 2006. Economic benefits of mineral extraction from geothermal brines. *Washington State University Extension Energy Program.* 6 pp.

Bourcier, W., Ralph, W., Johnson, M., Bruton, C., and Gutierrez, P., 2006. *Silica extraction at Mammoth Lakes, California.* Lawrence Livermore National Laboratory Report UCRL-PROC-224426. Livermore, CA, 6 pp.

Bürgmann, R., Rosen, P.A., and Fielding, E.J., 2000. Synthetic aperture radar interferometry to measure Earth's surface topography and its deformation. *Annual Reviews of Earth and Planetary Sciences*, 28, 169–209.

Chiodini, G., Cioni, R., Guidi, M., Raco, B., and Marini, L., 1998. Soil CO_2 flux measurements in volcanic and geothermal areas. *Applied Geochemistry*, 13, 543–552.

Cruz, J.V., Couthinho, R.M., Carvalho, M.R., Óskarsson, N., and Gíslason, S.R., 1999. Chemistry of waters from Furnas volcano, São Miguel, Azores: Fluxes of volcanic carbon dioxide and leached material. *Journal of Volcanology and Geothermal Research*, 92, 151–167.

Delgado, H., Piedad-Sànchez, N., Galvian, L., Julio, P., Alvarez, J.M., and Càrdenas, L., 1998. CO_2 flux measurements at Popocatépetl volcano: II. Magnitude of emissions and significance (abstract). *EOS Transactions of the American Geophysical Union* 79 (Fall Meeting Supplement), 926.

Dydo, P., Turek, M., Ciba, J., Trojanowska, J., and Kluczka, J., 2005. Boron removal from landfill leachate by means of nanofiltration and reverse osmosis. *Desalination*, 185, 131–137.

Ellis, A.J. and Mahon, W.A.J., 1977. *Chemistry and Geothermal Systems.* New York: Academic Press, 392 pp.

Energy Information Administration, 2013. Monthly Energy Review 2013, Table 7.2a. US Department of Energy. http://www.eia.gov/totalenergy/data/monthly/pdf/sec7_5.pdf.

Eneva, M., Falorni, G., Adams, D., Allievi, J., and Novali, F., 2009. Application of satelite interferometry to the detection of surface deformation in the Salton Sea geothermal field, California. *Geothermal Resources Council Annual Meeting*, October 4–7, Reno, NV.

Entingh, D. and Vimmerstadt, L. 2005. Geothermal chemical byproducts recovery: Markets and potential revenues. Princeton Energy Resources International, LLC. Technical Report 9846-011-4G, National Renewable Energy Laboratory, Golden, CO.

Etiope, G., Beneduce, P., Calcara, M., Favali, P., Frugoni, F., Schiatterella, M., and Smriglio, G., 1999. Structural pattern and CO_2-CH_4 degassing of Ustica Island, Southern Tyrrhenian basin. *Journal of Volcanology and Geothermal Research*, 88, 291–304.

Evans, W.C., Sorey, M.L., Cook, A.C., Kennedy, B.M., Shuster, D.L., Colvard, E.M., White, L.D., and Huebner, M.A., 2002. Tracing and quantifying magmatic carbon discharge in cold groundwaters: Lessons learned from Mammoth Mountain, USA. *Journal of Volcanology and Geothermal Research*, 114, 291–312.

Falorni, G., Morgan, J., and Eneva, M., 2011. Advanced InSAR techniques for geothermal exploration and production. *Geothermal Resources Council Transactions*, 35, 1661–1666.

Favara, R., Giammanco, S., Inguaggiato, S., and Pecoraino, G., 2001. Preliminary estimate of CO_2 output from Pantelleria Island volcano (Sicily, Italy): Evidence of active mantle degassing. *Applied Geochemistry*, 16, 883–894.

Gallup, D., 1998. Geochemistry of geothermal fluids and well scales, and potential for mineral recovery. *Ore Geology Reviews*, 12, 225–236.

Gallup, D., 2007. Treatment of geothermal waters for production of industrial, agricultural or drinking water. *Geothermics*, 36, 473–483.

Gallup, D., Sugiaman, F., Capuno, V., and Manceau, A., 2003. Laboratory investigation of silica removal from geothermal waters to control silica scaling and produce usable silicates. *Applied Geochemistry*, 18, 1597–1612.

Gerlach, T.M., 1991. Etna's greenhouse pump. *Nature*, 315, 352–353.

Gerlach, T.M., Doukas, M.P., McGee, K.A., and Kessler, R., 2001. Soil efflux and total emission rates of magmatic CO_2 at the Horseshoe Lake tree kill, Mammoth Mountain, California, 1995–1999. *Chemical Geology*, 177, 101–116.

Glassley, W.E., 1974. A model for phase equilibria in the prehnite-pumpellyite facies. *Contributions to Mineralogy and Petrology*, 43, 317–332.

Han, D.-H., Nur, A., and Morgan, D., 1986. Effects of porosity and clay content on wave velocities in sandstones. *Geophysics*, 51, 2093–2107.

Helgeson, H.V., 1964. *Complexing and Hydrothermal Ore Deposition.* New York: MacMillan Company, 128 pp.

Hovis, G.L., Brennan, S., Keohane, M., and Crelling, J., 1999. High-temperature X-ray investigation of snaidine-analbite crystalline solutions: Thermal expansion, phase transitions and volumes of mixing. *Canadian Mineralogist*, 37, 701–709.

Hovis, G.L. and Graeme-Barber, A., 1997. Volumes of K-Na mixing for low albite-microcline crystalline solutions at elevated temperature: A test of regular solution thermodynamic models. *American Mineralogist*, 82, 158–164.

Kabay, N., Yilmaz, I., Yamac, S., Samatya, S., Yuksel, M., Yuksel, U., Arda, M., Sağlam, M., Iwanaga, T., and Hirowatari, K., 2004. Removal and recovery of boron from geothermal wastewater by selective ion exchange resins. I. Laboratory tests. *Reactive and Functional Polymers*, 60, 163–170.

Kraft, T., Mai, P.M., Wiemer, S., Deichmann, N., Ripperger, J., Kästli, P., Bachmann, C., Fäh, D., Wössner, J., and Giardini, D., 2009. Enhanced geothermal systems: Mitigating risk in urban areas. *EOS Transactions of the American Geophysical Union*, 90, 273–274.

Lockner, D.A., 1995. Rock failure. In *Rock Physics and Phase Relations*, ed. T.J. Ahrens. Washington, DC: American Geophysical Union, pp. 127–147.

Massonnet, D. and Feigl, K.L., 1998. Radar interferometry and its applications to changes in the Earth's surface. *Reviews of Geophysics*, 36, 441–500.

McDonald, R.I., Fargione, J., Kiesecker, J., Miller, W.M., and Powell, J., 2009. Energy sprawl or energy efficiency: Climate policy impacts on natural habitat for the United States of America. *PLoS ONE*, 4, 1–11. http://www.plosone.org/article/info:doi/10.1371/journal.pone.0006802.

McDowell, S.D. and Elders, W.A., 1980. Authigenic layer silicate minerals in borehole Elmore 1, Salton Sea Geothermal Field, California, USA. *Contributions to Mineralogy and Petrology*, 74, 293–310.

McKibben, M.A. and Hardie, L.A., 1997. Ore-forming brines in active continental rifts. In *Geochemistry of Hydrothermal Ore Deposits*, ed. H.L. Barnes. New York: Wiley, pp. 877–935.

Mishra, G.S., Glassley, W.E., and Yeh, S., 2011. Realizing the geothermal electricity potential—Water use and consequences. *Environmental Research Letters*, 6, 1–8.

Muffler, L.J.P. and White, D.E., 1969. Active Metamorphism of Upper Cenozoic Sediments in the Salton Sea Geothermal Field and the Salton Trough, Southeastern California. *Geological Society of America Bulletin*, 80, 157–182.

Nagl, G.J., 2008. Controlling H_2S Emissions in Geothermal Power Plants. Merichem Company. http://www.gtp-merichem.com/support/technical_papers/geothermal_power_plants.php.

National Research Council, 2013. *Induced Seismicity Potential in Energy Technologies*. Washington, DC: National Academies Press, 300 pp.

Nicholson, C. and Wesson, R.L., 1990. Earthquake hazard associated with deep well injection. USGS Bulletin 1951, Washington, DC: US Geological Survey, 74 pp.

Peterson, J., Rutqvist, J., Kennedy, M., and Majer, E., 2004. Integrated High Resolution Microearthquake Analysis and Monitoring for Optimizing Steam Production at The Geysers Geothermal Field, California. California Energy Commission, Geothermal Resources Development Account Final Report for Grant GEO-00-003, 41 pp.

Preuss, K., 2006. Enhanced geothermal systems (EGS) using CO2 as working fluid—A novel approach for generating renewable energy with simultaneous sequestration of carbon. *Geothermics*, 35, 351–367.

Seaward, T.M. and Kerrick, D.M., 1996. Hydrothermal CO_2 emission from the Taupo Volcanic Zone, New Zealand. *Earth and Planetary Science Letters*, 139, 105–113.

Skinner, B.J., White, D.E., Rose, H.J., and Mays, R.E., 1967. Sulfides associated with the Salton Sea geothermal brine. *Economic Geology*, 62, 316–330.

Sorey, M.L., Evans, W.C., Kennedy, B.M., Farrar, C.D., Hainsworth, L.J., and Hausback, B., 1998. Carbon dioxide and helium emissions from a reservoir of magmatic gas beneath Mammoth Mountain, California. *Journal of Geophysical Research*, 103(15), 303–315, 323.

Stark, M., 2003. Seismic evidence for a long-lived Enhanced Geothermal System (EGS) in the northern geysers reservoir. *Geothermal Resources Council Transactions*, 27, 727–731.

Stewart, D.B. and von Limbach, D., 1967. Thermal expansion of low and high albite. *American Journal of Science*, 267A, 44–62.

Tester, J.W., Anderson, B.J., Batchelor, A.S., Blackwell, D.D., DiPippio, R., Drake, E.M., Garnish, J. et al., 2006. *The Future of Geothermal Energy*. Cambridge, MA: MIT Press, 372 pp.

Walsh, J.B., Brace, W.F., and England, A.W., 1965. Effect of porosity on compressibility of glass. *Journal of the American Ceramic Society*, 48, 605–608.

Wardell, L.J. and Kyle, P.R., 1998. Volcanic carbon dioxide emission rates: White Island, New Zealand and Mt. Erebus, Antarctica (abstract). *EOS Transactions of the American Geophysical Union* 79 (Fall Meeting Supplement), 927.

Wells, D.L. and Coppersmith, K.J., 1994. New empirical relationships among magnitude, rupture length, rupture width, rupture area, and surface displacement. *Bulletin of the Seismological Society of America*, 84, 974–1002

Xu, T. and Preuss, K., 2010. Reactive transport modeling to study fluid-rock interaction in enhanced geothermal systems (EGS) with CO_2 as working fluid. *Proceedings of the World Geothermal Congress*, Bali, Indonesia, April 25–30, pp. 25–29.

FURTHER INFORMATION

Energy Information Administration (EIA: http://www.eia.doe.gov/).

> The EIA is an office in the United States Department of Energy. It provides an updated analysis of energy use in the United States. It maintains a website that provides sector analysis, efficiency information, and other useful data that are needed for informed discussion regarding environmental issues.

International Energy Agency (IEA: http://www.iea.org/).

> The IEA is an intergovernmental organization that was setup during the oil crisis in 1973 and 1974 to advise regarding energy policy. Since then it has evolved to focus on "climate change policies, market reform, energy technology collaboration and outreach to the rest of the world, especially major consumers and producers of energy like China, India, Russia and the OPEC countries." Their website is a significant source for data on international and national energy issues.

Kristmannsdottir, H. and Ármannsson, H., 2003. Environmental aspects of geothermal energy utilization. *Geothermics*, 32, 451, 461.

> This article provides an excellent summary of the environmental issues regarding geothermal use in the country of Iceland. It is a good model for how other countrywide analyses could be formulated.

Wolfson, R., 2008. *Energy, Environment, and Climate*. New York: W.W. Norton&Co., 532 pp.

> This book provides an excellent overview of the environmental issues associated with energy use, particularly as they relate to climate. The overview is broad, encompassing many facets of the energy challenge, and is a good starting point for deeper study.

SIDEBAR 15.1   Measuring Seismic Events

When rock failure occurs, energy is released. The characteristics that are observed from such an event are the amount of energy that is released and the intensity of shaking that occurs.

Richter (1935) established that the relative amount of energy released during an earthquake could be deduced from the amplitude of the trace of the event recorded on standardized seismometers and the distance to the earthquake. His relationship was

$$M_L = \log A + B \tag{15S.1}$$

where A is the amplitude of the trace (mm).

This relationship was then used by Gutenberg and Richter (1956) to compute the actual energy release:

$$\log E_Q = 2.9 + 1.9 M_L - 0.024 M_L^2 \tag{15S.2}$$

where E_Q is the energy release (J).

Figure 15S.1 shows how energy release depends on magnitude. Note that for each increase in magnitude, there is an increase in energy released that is much more than a factor of 10. Modifications of this relationship continue to be refined as new models of faulting are developed (see Abe [1995] for a summary).

For ground shaking, there is no single measurement associated with a specific earthquake or amount of energy released because the amount of shaking depends upon the local geology. A person standing on unconsolidated sediments, such as landfill, for example, would experience much more movement of the ground surface for a given earthquake than someone nearby standing on a bedrock outcrop. Experience over the years has resulted in the development of a qualitative scale that describes the amount of damage that would be done and the sensation that a person would experience for a given amount of shaking. This scale is called the Modified Mercalli Intensity (MMI) scale. The MMI scale goes from I to XII. Table 15S.1 gives a brief summary of what some of the intensity values mean.

For the seismic events associated with geothermal development, the vast majority fall at magnitudes of less than 2.0 and MMI values of less than II. Whether larger events occur remains a matter of debate.

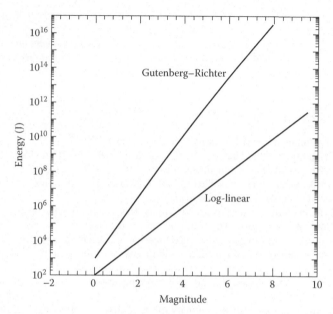

FIGURE 15S.1   The relationship between energy released (E) during an earthquake, in Joules, and the magnitude of the earthquake. The curve labeled "Gutenberg–Richter" is derived from Equation 12S.2. The line labeled "Log-linear" shows the E versus magnitude relationship for a different model that assumes a log-linear relationship between energy release and magnitude. The difference between the curves provides an indication of the uncertainty in establishing the absolute amount of energy released during an earthquake.

TABLE 15S.1
Description of Modified Mercalli Intensity Scale

MMI Intensity	Description	Ground Motion Velocity (cm/s)
I	Not felt except by a very few, especially in favorable conditions	–
III	During the day, felt by many who are indoors but only noticed by a few outdoors. If at night, some people awakened. Vibration like that of a passing light truck	1–2
VI	Felt by everyone. Damage slight. Some heavy furniture moved	5–8
IX	Considerable damage, even in well-built structures. Some buildings collapse. Conspicuous ground cracking	45–55
XII	Total destruction	>60

MMI, modified Mercalli intensity
Source: Bolt, B., *Earthquakes*, W.H. Freeman, New York, 1988.

Thus far, of those events purported to be associated with geothermal activities, the largest magnitude appears to have been around 4.5.

REFERENCES

Abe, K., 1995. Magnitudes and moments of earthquakes. In *Global Earth Physics*, ed. T.J. Ahrens. Washington, DC: American Geophysical Union Handbook of Physical Constants, pp. 206–213.
Bolt, B., 1988. *Earthquakes*. New York: W.H. Freeman.
Gutenberg, B. and Richter, C.F., 1956. Magnitude and energy of earthquakes. *Annals of Geofisks (Rome)*, 9, 1–15.
Richter, C.F., 1935. An instrumental earthquake magnitude scale. *Geological Society of America Bulletin*, 25, 1–32.

16 Geothermal Energy Future
Possibilities and Issues

Throughout this book there has been an effort to provide an historical context for the various elements that contribute to the development and use of geothermal energy. Knowledge of the events that have contributed to the growth and success of geothermal use provides perspective on underlying issues that influence whether an application or concept will succeed. For geothermal energy that historical context is brief, barely spanning 100 years. Even so, significant advances in understanding and technological capability have motivated rapid growth, particularly in the last 50 years. And yet, as suggested by much of the material in this book, that growth may well be little more than the initial stage of geothermal energy applications playing a much larger role in the global energy landscape. This chapter considers some of the important new resources and technologies that are likely to influence growth in the near future.

But, prior to considering specific technologies that could contribute to growth, it is important to understand how growth in an industry occurs, and how it is manifest. In Section "History of Geothermal Emergence in the Market Place", we will examine the history of growth for geothermal power generation. That perspective will provide context for the remainder of the chapter, which will examine several promising areas for potential geothermal growth.

HISTORY OF GEOTHERMAL EMERGENCE IN THE MARKET PLACE

As discussed in Chapter 10, employing geothermal energy to produce electricity did not emerge as a significant effort until the early 1950s when the first commercial plants in New Zealand and California came online. Since that time, geothermal power production has increasingly enhanced its presence in the power generation economy. However, that growth has not been steady. Figure 16.1 shows the growth of geothermal power in the United States between 1975 and 2012. The key features of this graph are that the growth has been marked by periods of stagnation and rapid expansion and that those periods correlate with important policy and technological changes. For example, the ramp-up of geothermal power production between about 1980 and 1990 occurred in response to policies put in place after the energy crisis of the 1970s. The incentives put in place as a result of those policies, as well as changes in the energy market, lead to rapid expansion of geothermal power production using existing dry steam power generation technology. As the geothermal power market expanded, new technology along with the incentives that were in place made development of flash plant applications economically appealing. The growth in dry steam applications ended around 1985, at which point most of the readily developable resources at The Geysers in California had been put into production. At about that time, however, the use of hydrothermal resources, employing flash plant technologies, began to grow. The use of flash technology enabled continued growth through about 1990, at which point power production by both dry steam and flash technologies leveled out.

Because of the temporary nature of many of the incentives that were put in place in response to the energy crisis of the 1970s, the growth of power production that was stimulated by the incentives began to wane as the incentives expired.

However, the development and significant deployment of binary technology beginning around 1990 resulted in continued growth of geothermal power production. Continued growth in geothermal power production to date has come primarily through expanding deployment of binary power generation technologies.

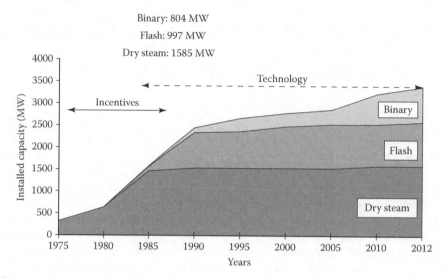

FIGURE 16.1 History of geothermal power generation in the United States between 1975 and 2012. Indicated in the figure are the magnitudes of power generation for dry steam, flash, and binary generation technologies in 2012, as well as how they have changed over time. The arrowed time intervals delineate the periods over which growth was most significantly affected by policy decisions, labeled "Incentives" and technological developments, labeled "Technology."

This history demonstrates the importance of new technologies for expanding geothermal power generation. In the remainder of this chapter, an overview is provided of certain research and development efforts that are currently underway, and which have the potential to significantly positively affect deployment of geothermal power generation. The research and development efforts that are described are not all activities that are underway. Rather, they represent activities that have been ongoing and of interest to the geothermal community for some years and have reached a level of development that suggest they may be commercialized in the not-too-distant future.

GEOPRESSURED RESOURCES

MAGNITUDE OF THE RESOURCE

Drilling for oil and gas in the southern United States demonstrated the presence of horizons in the subsurface where the fluid pressure exceeded that expected for a simple hydrostatic gradient. Since first recognized, these *geopressured* systems have been found in other regions that are usually associated with oil and gas fields (Chacko et al. 1998; Sanyal and Butler 2010). They are attractive as a potential geothermal resource because they may have temperatures suitable for power generation using binary generators. In addition, these zones are distributed in regions where industrial demand for power is high, as in the oil and gas industry. In many instances, such areas are also near major electric load centers, thus making them potential resources for municipal power (Figure 16.2).

The power generation potential for these systems is significant. Some fluids in these settings occur in the temperature range of 110°C–150°C. It has been reported (Garg 2007) that the recoverable thermal energy in the northern Gulf of Mexico Basin, a region of geopressured resources, is between 270×10^{18} and 2800×10^{18} J. Other regions, as indicated in Figure 16.2, are likely to possess significant additional potential. The total capacity for electrical power generation is estimated to be greater than 100,000 MW (Green and Nix 2006).

Many of these locations have high methane concentrations associated with them. This energy-rich hydrocarbon gas is an additional resource with an estimated recoverable energy content between 1×10^{18} and 1640×10^{18} J (Westhusing 1981; Garg 2007).

FIGURE 16.2 The distribution of potential geopressured regions in the United States (black) that have the potential to be used for geothermal power generation. The regions with potential geopressured zones are shown overlain on a map of the temperature in the subsurface at 6 km (see Figure 16.6 for the unshaded version). The black regions enclose those areas where the temperature in oil and gas wells exceeds 150°C. (Modified from U.S. Department of Energy, Energy Efficiency and Renewable Energy Office, http://www1. eere.energy.gov/geothermal/geomap.html.)

Why Geopressured Reservoirs Form

Geopressured zones form in sedimentary basins where subsurface fluid migration is impeded by greatly reduced permeability. Generally, these low-permeability, or sealed, zones block what would otherwise be a natural fluid migration pathway. They, therefore, form natural traps for migrating oil, gas, and other fluids. As fluids continue to migrate into those regions, driven by burial, compaction, and hydrological gradients, the pressure below these sealed zones increases beyond the normal hydrostatic gradient and approaches the local lithostatic pressure over geological time (see Sidebar 3.1). The elevated pressures in these regions are sufficient to drive rapid transport of fluid to the surface, thus eliminating the need for pumping and providing additional energy, as kinetic energy, to the fluid for extraction. In instances where the temperatures of the fluids exceed ~100°C, depending upon the actual ΔT that might be achieved locally, binary power generation is potentially economically feasible.

The primary mechanism that leads to the development of these sealed zones is recrystallization and growth of new minerals (authigenesis) in rock pores (Giorgetti et al. 2000; Nadeau et al. 2002). In sedimentary basins in which porous sandstones are interlayered with meters thick mud layers, burial of the sedimentary sequence over geological time will result in an increase in the formation temperature. Clay particles that make up the natural mudstones are sensitive to elevated temperatures. At about 60°C–80°C, certain clay minerals begin to go through a complex recrystallization, dissolution, and precipitation process that evolves up to about 120°C–130°C. During this process, unstable clays recrystallize and dissolve, and new clay minerals precipitate in pore spaces. Figure 16.3 shows how the overpressure, which is the pressure in excess of the expected hydrostatic pressure, increases in the interval over which a complex clay recrystallization, dissolution, and deposition process occurs in a specific example studied in detail by Nadeau et al. (2002).

Often associated with clay recrystallization is deposition of carbonate minerals, such as calcite and dolomite, and silica minerals such as quartz, chalcedony, and cristobalite. These secondary minerals form because the waters in these settings often have high concentrations of dissolved solids, with salinities occasionally exceeding 200,000 mg/l (Garg 2007). Such high salinities make it likely that the fluids are close to saturation in one or more mineral phases. Small temperature changes associated with tectonic activity or fluid migration can result in deposition of mineral phases along grain margins and in pore spaces.

Together, the deposition of clay, carbonate, and silica minerals results in a reduction in porosity and permeability of several orders of magnitude. The resulting seal can extend over distances sufficient to result in a reservoir that can be up to 4 km^3 (Garg 2007).

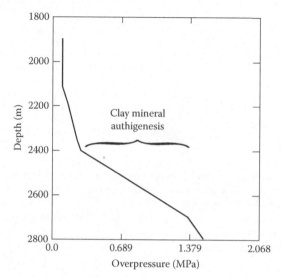

FIGURE 16.3 Overpressure (measured pressure—hydrostatic pressure, in MPa), as a function of depth in a geopressured well on the Norwegian continental shelf. The increase in overpressure in the depth interval from 2400 to 2700 m corresponds to the interval over which clay mineralogy is evolving. (From Nadeau, P.H. et al., *American Mineralogist*, 87, 1580–1589, 2002.)

EXAMPLE OF GEOPRESSURED SYSTEMS: LOS ANGELES BASIN

An example of the characteristics geopressured systems possess can be demonstrated by considering the features of oil fields in the Los Angeles Basin, California. Figure 16.4 shows the location of oil fields within the Los Angeles Basin. For each field shown, the areal footprint indicates the projected shape on the ground surface of the oil reservoir. Each reservoir is several thousand feet underground. Figure 16.5 shows the measured pressure for each field. The dashed line indicates nominal hydrostatic pressure. The solid line indicates pressures that are 10% above hydrostatic pressure. Any point that falls on or above the solid line indicates a field that is significantly overpressured. Also indicated in Figure 16.5 are those fields that have temperatures in excess of 91°C, which is the minimum required for current binary geothermal power generation technology. The large gray circles identify specific fields that are both geopressured fields and fields that have temperatures high enough to be considered possible geothermal resources.

These figures provide important information regarding these geopressured geothermal resources. The most important point is that oil and gas fields have the potential to provide geothermal power, but not all fields within a basin have conditions that will be suitable for power generation. Thorough analysis of the characteristics of each field must be undertaken to establish whether a particular field has sufficient thermal energy for such an application. A second important point is that the deeper the oil field, the more likely it is to be geopressured. This is consistent with the discussion presented in the Sidebar in Chapter 3, where it was noted that, as depth increases, pore pressure approaches that of the lithostatic load, due to the fact that the compressive stress eventually overcomes the inherent ability of the rock to resist deformation.

CHALLENGES TO DEVELOPMENT

Fluid Chemistry

There are, however, significant challenges that must be overcome before this resource can be economically utilized. One significant factor is that these solutions are often highly saline, with dissolved loads as high as 200,000 mg/l, as noted in Chapter 5. They often contain, in addition,

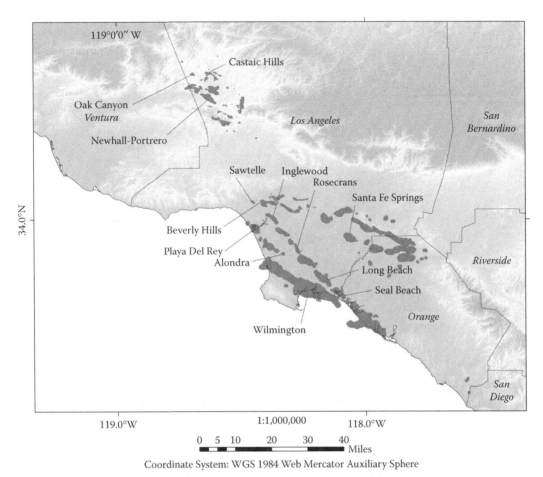

FIGURE 16.4 Map of the Los Angeles Basin, California, showing the location of developed oil fields (dark gray). Counties (in italics) and their boundaries are indicated. The names of some of the oil fields are also shown.

significant concentrations of CO_2. These solutes must be removed from the feed stream before the fluid enters the turbine and undergoes a reduction in temperature and pressure. This is required because the dissolved load will readily precipitate on the turbine blades, dramatically reducing turbine efficiency (see discussion of this process in Chapter 10 Sidebar).

Additionally, the complex chemical properties of these solutions affect the amount of heat in the solution that is available to do work. This results from the fact that dissolution of many of the compounds that make up the dissolved load requires energy (so-called heats of solution). As an example, if one mole of NaCl, a common salt, is added to a kilogram of water at 25°C, approximately 3836 joules of heat must be added to the water to keep the temperature at 25°C. However, the heat of solution varies with the amount of salt added. Figure 16.6 shows how the heat of solution changes as a function of number of moles of NaCl added to the water. Complicating this analysis is the fact that natural geothermal solutions contain many different solutes, each with their own heats of solution, which also are influenced by the concentration of other species. Hence, analysis of the heat budget at a particular site requires thorough analysis of the solution composition.

Removal of heat from such solutions can result in the solution becoming saturated in one or more of the dissolved salts, which can lead to their precipitation. If that happens, it will have the effect of reducing flow in the piping system of the geothermal plant. If the salt is removed from the solution

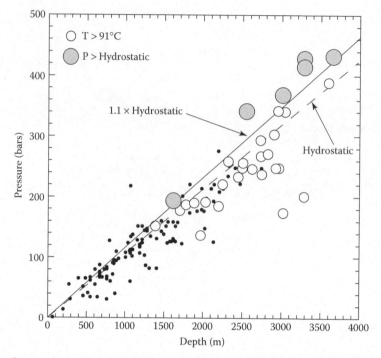

FIGURE 16.5 Pressure versus depth for fields in the Los Angeles Basin. The nominal hydrostatic pressure gradient is indicated by the dashed line. The solid line delineates conditions that are 10% above the nominal hydrostatic case. Fields that have temperatures above the minimum for binary power generation (in this case, taken to be 91°C) are shown by the open circles. Large gray circles are fields that are both geopressured and possess temperatures exceeding 91°C.

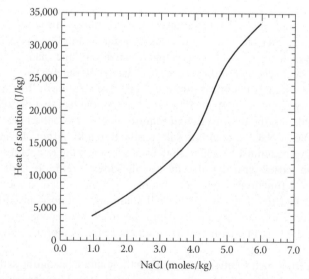

FIGURE 16.6 The heat of solution (J/kg of solution) as a function of the amount (moles/kg) of NaCl in solution.

prior to passing through the heat exchanger of a binary plant, the heat management of the fluid becomes a significant challenge, because removal of the salt also will result in the release of the heat of solution. Hence, when considering the temperature of the geothermal fluid, it must be recognized that some of the heat in that fluid is actually maintaining the salt in solution and is not necessarily available for power generation.

There is the potential of compensating for the lost energy resulting from resource extraction by recovery metals and other commodities that have economic value. As discussed in Chapter 15, the ability to economically recover essential metals and other compounds is a rapidly evolving field. Brines associated with geopressured reservoirs may provide recovery opportunities as the technology improves.

Reinjection

Regardless of the strategy developed for management of these fluids during power generation, they are generally too saline for surface disposal. As a result, reinjection of the fluids will be required. Although this may add additional costs, the resulting environmental protection justifies this strategy.

In addition to the environmental benefits obtained by reinjection, improving the sustainability of the resource is likely to be an additional benefit. The presence of a sealed horizon in the subsurface makes it likely that many geopressured zones are receiving little natural recharge. This is emphasized by the fact that flow tests on some wells in the Gulf Coast showed a significant drop in pressure when high rates of flow were maintained for relatively short periods (Garg 2007). Reinjection can make up some of the fluid volume extracted, potentially mitigating the reduction in pressure. It may be that careful design and implementation of waste brine reinjection could mitigate these effects.

In summary, the primary challenges that are faced when trying to develop an economically viable power generation system for a geopressured resource are as follows:

- Separate the dissolved solute load from the aqueous phase while minimizing the loss of thermal energy
- Separate and capture the dissolved methane gas phase from the aqueous phase
- Efficiently extract the thermal energy and kinetic energy from the fluid while maintaining sufficient pressure and flow rates

These scientific and engineering challenges are significant, but are not insurmountable. Current support for research and development in this area from a variety of agencies suggests that geopressured resources may become a significant contributor to geothermally produced electric power.

SUPERCRITICAL GEOTHERMAL FLUIDS

As was discussed in Chapters 3 and 10, generating power from geothermal fluids is a matter of extracting the maximum enthalpy from the fluid to accomplish work. Below the critical point of water, which is at 374°C and 22.1 MPa, the most attractive fluid for geothermal power generation is superheated steam, because it lies on the high enthalpy side of the two-phase region in a pressure–enthalpy diagram for H_2O (recall Figure 10.2). For such dry steam systems, nearly all the enthalpy in the steam is potentially available for power generation. Although such superheated steam fields exist, they are rare—The Geysers in California is one such rare field. Most geothermal fluids that are used for power generation lie on the lower enthalpy side of the two-phase region. Such fluids inevitably go through a phase separation episode as they are accessed and brought to the surface where they are introduced to a turbine-generator complex. This phase separation results in loss of enthalpy as partitioning of energy occurs between the liquid and steam phases.

However, it is obvious from Figure 10.2 that fluids that are above the critical point have the potential to provide high enthalpy steam without significant phase separation if they can be accessed and brought to the surface without much temperature drop. In addition, the physical properties of

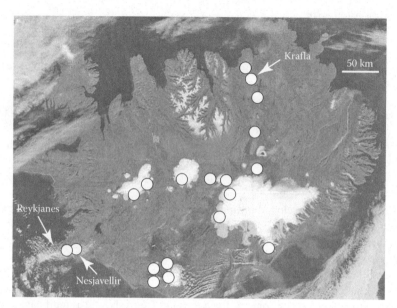

FIGURE 16.7 Satellite image of Iceland. The Krafla site is indicated, as is the Reykjanes Peninsula. The circles are the approximate locations of major eruptive, volcanic centers. (Modified from Friðleifsson, G.Ó. et al., *Geothermics*, 49, 119–126, 2014. Superimposed on Google Earth map of Iceland.)

water change significantly above the critical point. One of these changes is that buoyancy/density characteristics, as well as fluid viscosity, change dramatically, such that extremely high rates of energy and mass transfer can occur if supercritical fluids are accessed.

These supercritical geothermal fluids have been of interest for years. Such fluids have the potential to increase by many times the power output potential from a geothermal well, resulting in a tremendous increase in generation without an increase in the environmental footprint of the facility. However, it has only been recently that technological capabilities have developed sufficiently to allow serious consideration of accessing them.

The most significant effort in pursuit of supercritical geothermal fluids is the Iceland Deep Drilling Project (IDDP). The history and experience of the IDDP through 2012 is detailed in fifteen papers published in a special issue of *Geothermics* (2014, volume 49). The following summary is developed from papers contained in that special issue.

The IDDP was initiated in 2000. This project began after it was recognized that supercritical fluids in the Iceland Geothermal system might be accessible at depths of about 4 km or less (Steingrímsson et al. 1990). The first indication of this resource came from a well drilled in Nesjavellir geothermal field (Figure 16.7) in 1985, when very-high-pressure fluids were encountered at a depth of about 2200 m and were interpreted to be from temperatures greater than 380°C.

In 2008–2009, drilling of the first well in the IDDP project was undertaken. IDDP-1 was drilled in the Krafla geothermal field (Figure 16.7), where there was evidence of high-temperature resources at depth. This was based, in part, on the fact that volcanic eruptions in the region in the mid-1970s through 1984 had affected the geothermal fluids in that region, implying the presence of a magma source at depth. The target drilling depth was 4000–5000 m, where it was expected that supercritical fluids would be encountered.

Drilling began in November 2008. A variety of drilling problems developed below a depth of 500 m, and progress to depth was slow. In June 2009, after a series of challenging events, it was discovered that, at a depth of about 2100 m, quenched glass was brought up with the cuttings, indicating that drilling had encountered magma. The magma was rhyolitic in composition and was at a temperature of more than 900°C. This was the first time a geothermal well had ever encountered molten rock. Over the next two months, various strategies were attempted to continue drilling to depth, but the effort was terminated after magma was encountered two more times.

Although the target depths were not achieved, the drilling was successful in that it demonstrated the ability to successfully drill at temperatures similar to that encountered in Kakkonda (Chapter 9) and to access supercritical geothermal fluids. The well was completed so that it produced geothermal fluids from the 500°C zone just outside the magma body. Flow testing of the well proceeded for 2 years, producing fluids of about 450°C at the wellhead, at a pressure of 40–140 bars. This was the world's hottest geothermal well during that time, producing supercritical geothermal fluids consistently. The well was eventually shut down because of damage to the wellhead equipment.

A number of technological challenges were encountered and addressed, which provide the basis for further technology development. Among these are lessons demonstrating the need to improve cementing techniques and strategies, new data on the limits of instrument and equipment thermal stability, and new data regarding the performance of drill strings, cement slurry, and muds. The next phase of the IDDP is currently underway. Key issues that will be addressed in the next phase of drilling include the development of new tools that can survive in the very aggressive environment.

THERMOELECTRIC GENERATION

In 1821, Thomas Seebeck discovered that an electric current would be generated if two different metals were in contact and there was a temperature difference between them. This so-called Seebeck effect is the underlying principle that relates to thermoelectric generation. In its currently most common embodiment, a thermoelectric generator consists of semiconductor materials sandwiched between two materials, one of which is in contact with a heat source and the other in contact with a heat sink. Figure 16.8 is a schematic of such a device.

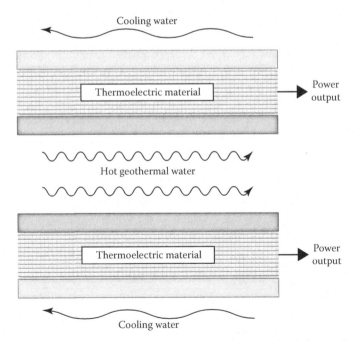

FIGURE 16.8 A schematic cross section through a thermoelectric generator. A thermoelectric semiconductor material is sandwiched between thermally conductive plates. Hot geothermal fluid flows along a set of plates, maintaining a high temperature side to the assembly. Cold water, in a counter flow, maintains a low temperature along the other plate. The ΔT across the semiconductor results in current flow and power output. In a functioning thermoelectric generator, multiple pairs of assemblies would be linked to increase power output.

The power output of the generator will depend on two key factors. One factor, which we have already discussed in numerous situations, is the ΔT between the hot and cold sides of the device. As with previous situations, the greater the ΔT, the greater the energy conversion efficiency. The other factor that is critical for a thermoelectric generator is the efficiency of the semiconductor material to produce current in response to the available heat.

Although most applications of thermoelectric generators are for small-scale power demands, the conceptualization of using these devices for geothermal applications has advanced substantially over the last few years. Bocher and Weidenkaff (2008) described a system for specifically such an application. Later work by Li et al. (2013) examined the applications of these systems in more detail. These and other studies point to the possibility of using such devices in geothermal power generation, either directly or as a means to extract extra energy for downstream warm fluids that have been exhausted from power generation/turbine complexes. In either case, although still in the research and development stage, such devices have the potential to provide power at the kW to MW range at modest cost and with very low operations and maintenance expense (Suter et al. 2012; Anthony Evans, pers. comm. 2013).

FLEXIBLE GENERATION

Geothermal power production has, since its inception, been viewed and operated as a baseload resource. In that capacity, it has become a reliable power generating technology. However, there is no technical reason why it cannot be operated in a flexible mode, allowing it to follow demand cycles on a power grid. This capability becomes increasingly important as other resources that are inherently variable in nature, such as solar and wind power, are added to a power grid. Although flexible generation in response to varying demand has previously been accomplished by fossil-fueled plants, that capacity can increasingly be taken on by geothermal power, as new flexible technology is added to the mix of geothermal capabilities.

Recent research has demonstrated the capabilities to accomplish flexible geothermal power generation (Linville et al. 2013, in Appendix 1 to Matek and Schmidt 2013). A schematic diagram of one realization of how this could be accomplished using a binary generation system is shown in Figure 16.9. In this system, flexible generation is accomplished by changing the amount of geothermal fluid flowing to the turbine by use of a bypass valve—synchronously opening the bypass valve and turning down the injection valve changes the flow rate through the turbine, which allows power output to be responsive to demand. Other means for accomplishing this could potentially include controlling the flow rate from the wellhead or temporarily venting steam to other applications.

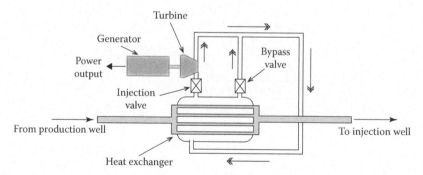

FIGURE 16.9 Schematic diagram of a method to accomplish flexible power generation. In this realization, a binary generating system is depicted. Geothermal fluid (gray piping) flows from a production well through a heat exchanger to the injection well. The binary working fluid (clear piping) flows as depicted by the double-headed arrows. The amount of flow through the turbine is controlled by balancing flow using the injection and bypass valves. (Modified from Linville, C. et al., The Value of Geothermal Energy Attributes: Aspen Report to Ormat Technologies, Aspen Environmental Group, 44, 2013.)

The response time of geothermal generating technology for ramping power generation up or down is equivalent to, or better than, that of existing fossil-fueled generating systems (Linville et al. 2013). That being the case, use of such systems as flexible generation in locations where geothermal resources are available makes them the most beneficial power systems for maintaining grid reliability at low cost. It is expected that, as these systems gain wider implementation, they will become an important component in power generation applications.

HYBRID GEOTHERMAL SYSTEMS

Combining geothermal systems with other renewable resources is becoming an area of interest in a variety of areas. Although this is an emerging technological field, creative engineering concepts are being proposed that make combined projects of growing interest. These applications are most effective when geothermal resources can be colocated with the other renewable technologies that may be locally available. Some examples of possible applications are described in the subsequent paragraphs. The list is not exhaustive—this is a research area that is rapidly evolving as new, creative applications are imagined.

Combining geothermal heat with biodigesters for producing either liquid (e.g., ethanol) or gas (e.g., methane) fuels is an area of growing interest. The concepts that have been put forward to consider using either waste heat from geothermal power applications in the downstream side of a turbine or geothermal fluids that have temperatures too low for power generation, to supply thermal energy to the biodigester system. These systems could also be combined with flexible geothermal power generators, as described above. Such applications can greatly reduce costs to the digester facility by providing reliable, inexpensive heat without relying on a fuel cycle.

Combining flexible geothermal power production with wind and solar systems can result in consistent power output in situations that otherwise would be highly variable. In addition, excess output from wind or solar systems, at times of low demand, could be used to enhance geothermal reservoirs by supplying heat (solar) or power for pumping (wind) to the fluid injection system that replenishes the reservoir. The resulting improvements in overall efficiency and reservoir management could greatly benefit the use of each of these resources.

Many other realizations of these hybrid applications can be imagined. The above brief list is intended to provide examples of ways in which energy use from renewable resources could be combined to improve reliability, efficiency, and reduce costs. Because these are in their formative stages of conceptualization, it is unclear what the overall magnitude would be of power these systems could provide. This must wait for future, detailed life-cycle analyses. However, preliminary indications are that such applications could significantly reduce reliance on all fossil fuels and improve overall system efficiencies, if thoughtful applications are carefully and thoroughly analyzed and implemented.

An important consideration in these applications is the ability to monitor and balance power use and generation so that, at any particular moment, the most efficient power producers are employed to meet the instantaneous demand. Such a capability requires a flexible and robust power distribution system (i.e., transmission grid). Although there is currently energetic discussion about *smart grids* (Borlase 2012), which would be a manifestation of this concept, there is considerable debate about how best to put such systems in place. Given the fact that generation technologies are rapidly evolving, as demonstrated above, this likely will be an area of discussion and research for the foreseeable future. At the same time, however, it is also clear that rapid adaptation of the existing grid to something more flexible and responsive is of great importance in order to realize the benefits of renewable power generation technologies.

These points also have implicitly suggested that centralized power generation, which is the backbone structure of our existing grid system, will likely evolve to embrace more localized or *distributed* generation. Many geothermal applications represent situations in which relatively local renewable resources can combine to provide power when they are colocated. In many instances, these applications could directly supply power to communities and regions through *microgrids*

(Lasseter 2007), while also providing excess power to large-scale grids. Such a distributed system of power generation can minimize transmission needs and costs, provide greater local autonomy over resource use, and better match power production to local demand. However, accomplishing this mixed centralized/distributed power generation scheme will require much more thorough analysis of how demand and generation can be monitored and matched to achieve the most efficient, reliable, and lowest-cost power generation and distribution infrastructure.

SYNOPSIS

The development of technology suitable for generating power from geothermal systems has improved significantly since the 1950s, when electrical power generation using geothermal energy resources grew to an international enterprise. The growth of the power generation industry has been steady since that time. Even so, geothermal resources of even greater magnitude are available but have yet to be developed. One such resource is moderate temperature fluids associated with oil and gas fields. These reservoirs, some of which are geopressured, could contribute significantly to meeting the energy needs in regions where these resources exist. Another resource that has the potential to expand by several times the geothermal generation potential is supercritical fluid systems. These very high temperature resources can provide exceptional efficiencies because the physical properties of water in the super-critical region result in delivering power to turbines at much greater efficiencies than conventional systems. Technological developments in the materials sciences, particularly in thermoelectric materials, suggest the possibility of generating electricity using geothermal fluids and mechanically simple thermoelectric generators. Although the efficiencies of these systems currently need further research and development, they hold significant promise for new generation technologies. New technology and engineering designs have also led to the development of geothermal generating systems that can be very responsive to demand, resulting in flexible generation. These systems are coming online and are likely to play a growing role in geothermal power generation installations. Finally, combining geothermal generators with other renewable resources is of growing interest in the research community. Such systems may result in widespread deployment at a variety of scales, from small communities to large regions. Current research in this area suggests such systems will become more common as smart grid technology expands and more robust local and regional transmission infrastructure grows

PROBLEMS

16.1. What is a geopressured geothermal resource and how do they form?

16.2. Where are geopressured resources located? How might this distribution affect the development of geopressured resources?

16.3. Why are geopressured resources difficult to develop? Consider in your discussion the effects of local climate.

16.4. What generation technologies are suitable for generating power at geopressured sites? What difficulties might this technology face in these settings?

16.5. What are the advantages of supercritical geothermal systems? What might be the challenges in developing such systems, from an operational and maintenance perspective?

16.6. If a community wanted to integrate a solar PV system and a 1 MW geothermal binary system, in what ways might the systems be combined? What would be the benefits of such a hybrid system? What would be the challenges that would have to be dealt with in order to make the system most reliable?

REFERENCES

Bocher, L. and Weidenkaff, A., 2008. Development of Thermoelectric Materials for Geothermal Energy Conversion Systems. GEO-TEP, Solid State Chemistry and Catalysis, Empa-Swiss Federal Laboratories for Materials Testing and Research, Dübendorf, Switzerland, 21 pp.

Borlase, S., ed., 2012. *Smart Grids: Infrastructure, Technology, and Solutions*. Boca Raton, FL: CRC Press, 607 pp.

Chacko, J.J., Maciasz, G., and Harder, B.J., 1998. Gulf Coast geopressured-geothermal program summary report compilation. Report, US Department of Energy Contract No. DE-FG07-95ID13366, U.S. Department of Energy, June.

Friðleifsson, G.Ó., Sigurdsson, Ó., Þorbjörnsson, D., Karlsdóttir, R., Gíslason, P., Albertsson, A., and Elders, W.A., 2014. Preparation for drilling well IDDP-2 at Reykjanes. *Geothermics*, 49, 119–126.

Garg, S., 2007. Geopressured geothermal well tests: A review. *Presentation to Geothermal Energy Utilization Associated with Oil & Gas Development Conference*, Southern Methodist University, Dallas, TX. http://smu.edu/geothermal/Oil&Gas/2007/SpeakerPresentations.htm.

Giorgetti, G., Mata, P., and Peacor, D.R., 2000. TEM study of the mechanism of transformation of detrital kaolinite and muscovite to illite/smectite in sediments of the Salton Sea Geothermal Field. *European Journal of Mineralogy*, 12, 923–934.

Green, B.D. and Nix, R.G., 2006. Geothermal—The energy under our feet. National Renewable Energy Laboratory Technical Report NREL/TP-840-40665, National Renewable Energy Laboratory, Golden, CO, 21 pp.

Lasseter, R.H., 2007. Microgrids and distributed generation. *Journal of Energy Engineering*, American Society of Civil Engineers, 133, 144–149.

Li, K., Liu, C., and Chen, P., 2013. Direct power generation from heat without mechanical work. *Proceedings of the 38th Workshop on Geothermal Reservoir Engineering*, Stanford University, Stanford, CA, SGP-TR-198, 11–13 pp.

Linville, C., Candelaria, J., and Elder, C., 2013. The Value of Geothermal Energy Attributes: Aspen Report to Ormat Technologies. Aspen Environmental Group, 44 pp. http://www.geothermal.org/PDFs/Values_of_Geothermal_Energy.pdf. Appendix to Matek, B. and Schmidt, B., 2013.

Matek, B. and Schmidt, B., 2013. The Values of Geothermal Energy. Geothermal Resources Council, Sacramento, CA and Geothermal Energy Association, Washington, DC, 19 pp.

Nadeau, P.H., Peacor, D.R., Yan, J., and Hillier, S., 2002. I-S precipitation in pore space as the cause of geopressuring in Mesozoic mudstones, Egersund Basin, Norwegian continental shelf. *American Mineralogist*, 87, 1580–1589.

Sanyal, S.K. and Butler, S.J., 2010. Geothermal power capacity from petroleum wells—Some case histories of assessment. *Proceedings of the World Geothermal Congress*. Bali, Indonesia, 6 pp.

Steingrímsson, B., Guðmundsson, A., Franzson, H., and Gunnlaugsson, E., 1990. Evidence of a supercritical fluid at depth in the Nesjavellir field. *Proceedings of the 15th Workshop on Geothermal Reservoir Engineering*, Stanford University, Stanford, CA, January 23–25, 81–88.

Suter, C., Jovanovic, Z.R., and Steinfeld, A., 2012. A 1 kW$_e$ thermoelectric stack for geothermal power generation—Modeling and geometrical optimization. *Applied Energy*, 99, 379–385.

U.S. Department of Energy, Energy Efficiency and Renewable Energy Office, http://www1.eere.energy.gov/geothermal/geomap.html.

Westhusing, K., 1981. Department of Energy geopressured geothermal program. *Fifth Conference on Geopressured-Geothermal Energy Proceedings*, eds. D.G. Bebout and A.L. Bachman, CONF-811026-1, October 13–15, Baton Rouge, LA, 3–6 pp.

FURTHER INFORMATION

In previous chapters, a number of information resources have been described that relate to the future direction of geothermal energy efforts. The reader is encouraged to review those, especially Chapter 13 on EGS resources. The following resources, in addition, are particularly relevant for other resources:

California Energy Commission, Sacramento, CA (http://www.energy.ca.gov/).

    The State of California has committed to achieving 12,000 MW of local, distributed generation using renewable resources by 2020. The California Energy Commission provides updates and information useful for following the efforts to achieve this goal.

Southern Methodist University Geothermal Laboratory, Dallas, TX (http://smu.edu/geothermal/).

    This institution's website contains a body of data and datasets available to the public at no charge. It is an excellent resource for information on geothermal energy and subsurface conditions within the United States.

Subject Index

Index of Locations